Trends in Biomathematics: Modeling, Optimization
and Computational Problems

Rubem P. Mondaini
Editor

Trends in Biomathematics: Modeling, Optimization and Computational Problems

Selected works from the BIOMAT
Consortium Lectures, Moscow 2017

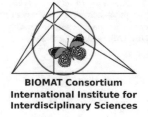

BIOMAT Consortium
International Institute for
Interdisciplinary Sciences

Editor
Rubem P. Mondaini
President, BIOMAT Consortium – International
Institute for Interdisciplinary Sciences
Rio de Janeiro, Brazil

Federal University of Rio de Janeiro
Rio de Janeiro, Brazil

ISBN 978-3-030-08175-1   ISBN 978-3-319-91092-5   (eBook)
https://doi.org/10.1007/978-3-319-91092-5

Mathematics Subject Classification (2010): 92Bxx, 92B05, 92-08, 62P10

Printed on acid-free paper

This Springer imprint is published by the registered company Springer Nature Switzerland AG
The registered company address is: Gewerbestrasse 11, 6330 Cham, Switzerland

Preface

The present book is a collection of papers which have been accepted for publication after a peer review evaluation by the Editorial Board of the BIOMAT Consortium (http://www.biomat.org) and ad-hoc international referees. These papers have been presented at the technical sessions of the BIOMAT 2017 International Symposium, the 17th Symposium of the BIOMAT Series which was held at the Institute of Numerical Mathematics of the Russian Academy of Sciences–Russia, from 30th October to 03rd November 2017. On behalf of the BIOMAT Consortium, we thank the Director of the Institute, Prof. Eugene Tyrtyshnikov, and the Co-Chairs of the BIOMAT 2017 Local Organizing Committee, Prof. Yuri Vassilevski and Prof. Vitaly Volpert, for their technical expertise in following the guidelines and fine tradition of the BIOMAT Consortium for preserving the excellency of the BIOMAT Symposium Series on this first BIOMAT Conference in Russia. Research collaborators, Ph.D. Students and Secretaries of the Institute like Alexander Danilov, Tatiana Dobroserdova, Konstantin Novikov, Roman Pryamonosov, Nina Gorodnova and Anna Zagumennykh have done their best to provide all local technical facilities to help the speakers during the scientific sessions as well as to follow the Scientific Programme of the BIOMAT 2017. We are so much indebted for their invaluable help since the Opening Session on Monday morning to the Closing Session on Friday evening.

Financial support in terms of accommodation, lunches and coffee-breaks has been provided by the Institute of Numerical Mathematics, the International Union of Biological Sciences, the Interdisciplinary Scientific Center Jean-Victor Poncelet, Russian Foundation for Basic Research, Parseco Foundation, and the Federal Agency for Scientific Organizations. We are also indebted for the special offer of accommodation of the Keynote Speakers and the staff of the BIOMAT Consortium in the hotel of the Steklov Mathematical Institute, Russian Academy of Sciences.

The BIOMAT Consortium has succeeded once more in its fundamental mission of enhancing the interdisciplinary scientific activities of Mathematical and Biological Sciences in Developing Countries with the organization of the BIOMAT 2017 International Symposium. Participants from Western and Eastern Europe, Asia,

Africa, North and South America had the usual opportunity of exchanging scientific feedback of their research fields with their colleagues from Russian Federation and delegates coming from other 13 countries: Serbia, France, Brazil, Cuba, China P.R., Morocco, India, UK, USA, Pakistan, Portugal, Hungary, Italy.

The Editor of the book and President of the BIOMAT Consortium is very glad for the collaboration and critical support of his wife Carmem Lucia on the editorial work, from the reception of submitted papers for the peer review procedure of BIOMAT Consortium Editorial Board to the ultimate publication of the Scientific Programme. He also thanks his research student Simão C. de Albuquerque Neto from the Federal University of Rio de Janeiro for his computational skills and technical expertise with LaTeX versions.

Moscow, Russian Federation   Rubem P. Mondaini
November 2017

Editorial Board of the BIOMAT Consortium

Contents

Evolution of Spatial Patterns in Host-Parasitoid Metapopulation

Brajendra K. Singh and Somdatta Sinha

1 Introduction

Inclusion of spatial dimension in mathematical models of physical and biological systems leads to evolution of interesting dynamical features [1, 2]. Particularly, the important role of space, in the survival and coexistence of species populations, has been fully established in both field and modeling studies in ecology [3, 4]. Space assumes more significance in disease spread, and therefore the variation in spatial structure is an important determinant of the distribution of any infectious disease. In addition, unequal distributions of resources in space can allow for the explicit or implicit introduction of spatial dimension in a population model. In many studies of spatially extended single and interacting species metapopulation models, natural (geographical) resources were heterogeneously distributed [5–14], even though the subpopulation patches were homogeneous. The primary goal of these studies was to show the stabilizing effect of population dispersal on the persistence of extinct-prone single-patch population dynamics, in the metapopulation scenario [7, 8]. In these studies, all patches were considered to be identical, which counters the scenario of continuous destruction and fragmentation of natural landscapes [15].

The classical discrete Nicholson-Bailey host-parasitoid (HP) population model [16] is unstable leading to death of both the host and parasitoid. Some studies [17–19] have considered the effect of environmental variability in spatial HP models to resolve the stability/persistence problem. In a study [20] to enumerate the effects of environmental variabilities in lattice metapopulations, it was shown that the HP

B. K. Singh
The University of Notre Dame, Department of Biological Sciences, Notre Dame, IN, USA

S. Sinha (✉)
Department of Biological Sciences, Indian Institute of Science Education and Research Mohali, Mohali, India
e-mail: ssinha@iisermohali.ac.in

© Springer International Publishing AG, part of Springer Nature 2018
R. P. Mondaini (ed.), *Trends in Biomathematics: Modeling, Optimization and Computational Problems*, https://doi.org/10.1007/978-3-319-91092-5_1

metapopulation dynamics showed complete reversal in spatial patterns, if a small amount of heterogeneity was introduced randomly in space. The spatial pattern, which was primarily synchronous in homogeneous landscapes, showed asynchrony in spatial dynamics with heterogeneity.

Variability in natural landscapes can be of several types. The most common one is habitat structure, where the physical components of any habitat patch can be different from their neighbouring patches due to presence of natural or man-made dispersal barriers [15]. Such landscape heterogeneity can occur due to increased socioeconomic activities by humans that induce varied levels of destruction and fragmentation of natural habitats [21]. Another naturally occurring heterogeneity can arise if the host and parasitoid populations living in these patches consist of different genotypes having different phenotypic traits [19, 22], such as different levels of 'risk of parasitism' for the host species distributed in different patches [12, 23]. These two common types of variabilities observed in the metapopulation scenario are known as *'landscape heterogeneity'* and *'demographic heterogeneity'*.

In this paper we model the effect of various patterns of landscape and demographic heterogeneities on the spatial dynamics of host-parasitoid metapopulations. These different patterns of heterogeneity coupled to different connectivity patterns of the habitat patches are shown to lead to the evolution of different spatial patterns in population distributions. These results are discussed in the light of possible role of different types of dispersal barriers in animal migration and disease spread.

2 Single and Spatial Host-Parasite Model

The discrete Host-Parasitoid model used in this study is a simple modified Nicholson-Bailey [20], where the host exhibits density-dependent logistic growth in absence of the parasitoid. In presence of the parasitoid, which induces infection due to which that fraction of the host population cannot reproduce, and the parasitoids grow in infected hosts only. In a lattice metapopulation scenario, where each node is a habitat that supports a HP subpopulation, with migration taking place amongst the nearest neighbouring patches, the discrete HP model takes the form:

$$H_{t+1}(s) = F(H', P'),$$
$$P_{t+1}(s) = G(H', P'),$$
(1)

where $F(\cdot) = H'\mu(1 - H')e^{-\beta P'}$ and $G(\cdot) = H'(1 - e^{-\beta P'})$.

The parameters, μ and β, are the host growth rate and the parasitoid attack rate, respectively. H' and P' are the post-dispersal host and parasitoid population densities at any site $s \equiv (x, y) \in L \equiv (l \times l)$, where L is a square lattice of l^2 number of habitat patches. Variants of this model have been studied by several authors [13, 14] while investigating different questions.

The dispersal functions to the nearest 8 neighbours in a square lattice is given by

$$H' = (1 - d_1)H_t(s) + \frac{d_1}{8} \sum_{j=1}^{8} H_t(j)$$

$$P' = (1 - d_2)P_t(s) + d_2 \sum_{j=1}^{8} P_t(j)\delta_t^j(s). \tag{2}$$

where d_1 and d_2 are the host and parasitoid dispersal coefficients, respectively. The dispersal of the host populations from any patch is independent of the population level of the destination patches, it simply depends on its own population size. The parasitoid dispersal is assumed to be dependent on both the host and the parasitoid densities of the neighbouring patches [13, 14]. The term δ_t^j denotes the proportion of dispersing parasitoid populations from the neighbouring sites (j's) to site s. The functional form of δ_t^j is given by:

$$\delta_t^j(s) = C_N \left(\frac{H_t^j(s)}{\sum_{i=1}^{8} H_t^j(i)} \right)^\eta,$$

where η is known as the 'aggregation index', and C_N is a normalising constant such that $\sum_{i=1}^{8} \delta_t^j(i) = 1$.

The other model parameters were kept at: $\mu = 4$, $\eta = 1$, $d_1 = 0.2$ and $d_2 = 0.1$. Spatial patterns (mostly, spirals) were highly probable for these values of d_1 and d_2 [24]. For this study we have used two specific values of β: $\beta = 4$ for which the single HP system shows quasiperiodic behaviour, and $\beta = 5$ for which it shows chaotic dynamics. The results are shown for square lattices of size of $l = 50$, and they were verified for $l = 128$ and $l = 256$. Zero-flux boundary conditions were used for all metapopulations. Host initial population sizes on all habitable sites were randomly selected from the range [0.2, 0.3]. Parasitoid populations were initially assigned to only 2% (or, 50) randomly selected habitable sites, sampled from the same range as used for the host. All results shown in the figures are the snapshots taken at the end of 10,000 generations. The simulation experiments were repeated 10 times with different sets of initial population values.

3 Heterogeneous Landscapes

Non-homogeneous distribution of life supporting resources are common in natural landscapes. Apart from the natural biotic and abiotic processes being responsible for this heterogeneity, they can also happen due to ever increasing socio-economic activities of anthropogenic origin, which is known to be one of the major causal factors for worldwide habitat modifications, fragmentations and destructions [21]. Thus, the heterogeneity in model landscapes could be of highly diverse nature.

Below we consider a few simple, yet plausible, cases of environmental heterogeneity to study their effects on the spatial patterns of the population distributions of the HP system as represented by (1) and (2). The spatiotemporal dynamics in these landscapes are compared with those where such variability is not present (i.e., the landscape is *homogeneous*).

In the first case (**Type I**), we consider the model landscape to have a small number of vacant patches (lattice sites) that cannot support the growth of HP populations and are not accessible for migration. All other patches are identically habitable. The distribution of such vacant patches (5% of the total) in the landscape can occur in several ways:

Type Ia: Individual vacant patches distributed randomly.

Type Ib: Clusters of (3 × 3) vacant patches distributed randomly.

Type Ic: Vacant patches forming a certain pattern in the landscape. Two forms of this patterned vacant regions are considered in this study: (1) an impermeable barrier of vacant sites across the width (or length) of the landscape dividing metapopulation into two parts; and (2) a barrier with one or more passages through which population dispersal can occur between the parts.

In the second case (**Type II**), the metapopulation occupies a landscape that has all habitable patches, but a few sites are demographically heterogeneous, i.e., species have different life history traits. These sites are inhabited by parasitoid populations that have a different attack rate (β) from those that inhabit the remaining sites. This kind of demographic heterogeneity is common in natural world, where species having different genotypes/phenotypes coexist [25, 26]. In this study we have used two values of β. With host growth rate $\mu = 4$, the HP model shows quasiperiodic population dynamics for $\beta = 4$, and chaotic dynamics for $\beta = 5$ [20].

The distribution of these demographically different patches (5% of the total) in the landscape can occur in different ways:

Type IIa: 5% randomly selected sites have the parasitoid with $\beta = 5$, while the rest have $\beta = 4$.

Type IIb: The opposite of **Type IIa**, i.e., 5% sites have the parasitoid with $\beta = 4$, while the rest have $\beta = 5$.

Type IIc: 4% of the total sites form a single sub-lattice of (10 × 10) sites, where the parasitoid populations have $\beta = 5$, while the rest of the sites have $\beta = 4$.

Type IId: The opposite of **Type IIc**, i.e., the single sub-lattice have parasitoid populations with $\beta = 4$, while the rest of the sites have $\beta = 5$.

4   Results

Here the spatial dynamics of the HP metapopulation in heterogeneous landscapes are described and discussed. First, the landscape heterogeneity is considered (Type I), and then demographic heterogeneity (Type II) in a homogeneous landscape is shown.

Fig. 1 Spatial patterns in host metapopulations: (**a**) Homogeneous lattice; (**b**) Lattice with **Type Ia** heterogeneity; (**c**) Lattice with **Type Ib** heterogeneity. Lattices with **Type Ic** heterogeneity—(**d**) without any passage, (**e**) with one passage, and (**f**) with three passages. The initial population densities of the host and the parasitoid at the occupied sites were the same in all cases

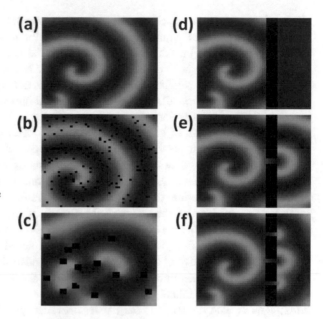

4.1 Type I: Landscape Heterogeneity

It was shown earlier [20] that in a homogeneous metapopulation, where all sites are occupied by H and P subpopulations with same life history parameters (r and β), the spatial dynamics shows complete synchrony in the majority ($>90\%$) of simulations done with different initial conditions. In very few cases, the lattice metapopulations showed slow evolution of large scale spatial patterns (e.g., spirals) in their population distribution (as shown in Fig. 1a). These spatial structures, when simulated long enough ($>10,000$ generations), slowly disappeared at the boundaries. This statistics changes completely when low levels of heterogeneity, in terms of randomly distributed 5% vacant sites, are present in the landscape (**Type Ia** heterogeneity). The presence of such heterogeneity in the landscape resists the process of spatial synchronization completely, and all lattice metapopulations (with different initial population distributions) showed spiral patterns in population distribution (Fig. 1b). However, these spatial structures in Fig. 1b not only evolved much faster than those seen in the homogeneous metapopulations (Fig. 1a), the spiral tips remain pinned to any of the vacant sites and the pattern was stable.

For **Type Ib** heterogeneity (clusters of 9 sites group of vacant patches distributed randomly), the metapopulation showed similar spatial patterns, i.e., complete spatial asynchrony with broken spirals. Here again the time taken for the pattern to evolve was faster than case of **Type Ia**, and the pattern was stable. It is obvious that low levels of landscape heterogeneity (**Type Ia,b**) lead to different numbers of neighbours to HP subpopulations in habitable patches. For **Type Ia** heterogeneity, depending on the distribution of these vacant sites, a subpopulation in a habitable site can in one extreme be solitary (i.e., all neighbours are vacant sites) or have all 8 neighbours. In general the number of habitable neighbours would be 1, 2, 3 to 7. This local asymmetry can introduce nucleation of spatial patterns once they arise and stabilize them as seen in **Type Ia**. The clustered pattern of landscape heterogeneity (**Type Ib**) introduces smaller numbers of larger vacant regions. This has stronger effect on breaking up the different parts of the spiral into smaller dynamical regions thereby introducing irregular spatial patterns.

Lattices with **Type Ic** of patterned landscape heterogeneity was studied for two cases—the vacant patches creating (1) an impermeable barrier and dividing the metapopulation into two disconnected domains of homogeneous landscapes, and, (2) a barrier of vacant line of sites with one or more passages through which limited population dispersal can occur between the domains as shown in Fig. 1d–f. In case of the impermeable barrier (Fig. 1d), the spatial patterns formed in the two separate metapopulations are similar to the ones as observed for homogeneous landscapes (Fig. 1a). In general spiral patterns evolve in these domains, but irrespective of the size, that domain which does not contain the spiral core shows spatial synchronization. Figure 1e and f show that the presence of one or more passages (i.e., habitable sites having a HP subpopulation that can disperse to neighbours in both domains) in the barrier prevents spatial synchronization. These connecting subpopulations act as continuous sources of dynamic heterogeneity in space and lead to generation of semi-circular travelling waves that get absorbed at the boundaries (Fig. 1d). For the case of more than one passages (Fig. 1f), the waves created at the different sources interfere with each other, and create complex wave patterns depending upon the domain size. For larger domains, the wave front can evolve to appear as if it was created by a single passage. However, if the domain size is small, each wavefront does not get sufficient space and time to develop into a single wave front. A detailed study of such barriers with breaks in real landscapes, in terms of major geographical features and transport links, was shown in spatial spread of the 2001 UK FMD epidemic [27].

4.2 Type II: Demographic Heterogeneity

It was shown earlier [20, 24] that the impact of demographic variability on the dynamics of the spatial HP system is similar to the landscape heterogeneity, i.e., it induces increasing number of cases of spatial asynchrony as parametric heterogeneity among the subpopulations increase. The **left plot** in Fig. 2a for $\beta = 4$ is similar to Fig. 1a, and is an example of the case where the metapopulation in a homogeneous landscape with no demographic heterogeneity shows spiral-like spatial patterns. The **right plot** in Fig. 2a for $\beta = 5$ (chaotic dynamics in single HP population) shows asynchrony with irregular pattern in population size distributions for majority of the cases. Here we show that the spatial patterns of the population abundance not only respond differently to different types of the demographic heterogeneity, but also to the predominant intrinsic dynamics of the

Fig. 2 Spatial patterns of
host metapopulations: (**a**)
Demographically
homogeneous landscape with
$\beta = 4$ (left) and $\beta = 5$
(right); (**b**) Lattices with
demographic heterogeneity of
Type IIa (left) and **Type IIb**
(right). (**c**) Lattices with
demographic heterogeneity of
Type IIc (left) and **Type IId**
(right)

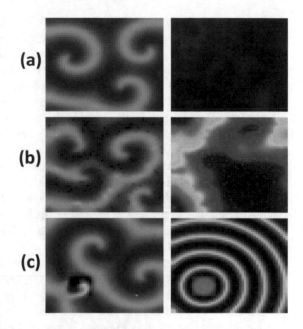

HP system. The **left plot** in Fig. 2b shows the spatial pattern in the metapopulations
with **Type IIa** heterogeneity (i.e., 5% randomly selected sites having parasitoid
populations with $\beta = 5$) while the rest have $\beta = 4$. The **right plot** shows **Type
IIb** heterogeneity (i.e., 5% randomly selected sites have parasitoid populations with
$\beta = 4$, while the rest have $\beta = 5$). Both plots in Fig. 2b appear more incoherent
and take longer time to reach any coherent patterns (such as spirals) than what was
needed for demographically homogeneous case (Fig. 2a). It was observed that these
spatial patterns were unstable even after 5×10^4 generations for small lattices as used
in this study.

The **Types IIc** and **IId** of demographic heterogeneity in metapopulations
represent cases where 4% of the total sites (i.e. 100 sites) form a single sub-lattice of
(10×10) sites, where the parasitoid populations have attack rate $\beta = 5$, while the
rest of the sites have $\beta = 4$ (**Type IIc**) and vice versa (**Type IId**). The spatial
patterns are shown in Fig. 2c. For **Type IIc**, the pattern is similar to **Type IIa**
(Fig. 2b left plot), except that the block of sites with $\beta = 5$ can pin the spiral core.
It was observed that if the sub-lattice placed randomly happens to cover the core
of a spiral wave, then spatial synchronization emerges finally. On the other hand,
the spatial pattern for **Type IId** heterogeneity is quite different. The block of sites
having HP subpopulations with $\beta = 4$ showed coherent patterns of spiral waves or
of concentric circular waves with their core fixed at the sub-lattice. The concentric
circular pattern was common ($\approx 80\%$ of cases).

The emergence of wave patterns (spiral or circular) in the case of these two
types of heterogeneity (**Type IIc** and **d**) can be explained as an outcome of
interaction between the two types of dynamics exhibited by the HP system for

$\beta = 4$ (quasiperiodic) and $\beta = 5$ (chaotic). Due to the slowly fluctuating and comparatively more stable dynamics in the sub-lattice for $\beta = 4$, it continues to maintain, on the average, a regular temporal pattern in the population density. The rest of the lattice (before the wave patterns set in) have wild temporal oscillations (for $\beta = 5$), which resist the formation of any coherent patterns. Once the transient dynamics die down, the sub-lattice acts as a source from which the migrant populations disperse out to the neighbouring sites with more regular periodic intensity. This regular rhythmic flow eventually helps in starting or pinning the core (centre), and subsequently in maintaining the spiral or circular wave patterns (right plot of Fig. 2c). In the reverse case (**Type IIc** and left plot of Fig. 2c), the sub-lattice sends out migrant populations at an erratic rate. As a consequence it fails to pin the core (centre) of the spiral wave. This failure in turn does not have any great effect on the existing spatial patterns in the rest of the lattice, which by and large remains unaffected, unless, as stated previously, the sub-lattice happens to be selected at the core of the spiral dominating the lattice.

When we used two sub-lattices of equal size (5×5), instead of a single (10×10) size, there were two cores (centres) acting as the source of spatial waves. The two sub-lattices can simultaneously be the core (centre) of the spiral (circular) wave pattern or one of them can act as the core while the other as centre. It was found that a minimum size for the sub-lattice is required below which it does not always induce such coherent spatial patterns. This minimum size was found to be (4×4) for both—for the single sub-lattice type, and the double sub-lattice type, in a 50×50 lattice metapopulation. This minimum size of the sub-lattice is required for pinning of the spiral core. The dependence of the sub-lattice size and the whole lattice size for the evolution of these different types of dynamic heterogeneity (stable, periodic, quasi-periodic, and chaotic) needs further work.

5   Conclusions

The (seemingly) trivial inclusion of spatial dimensions in the population models has shown the appearance of non-trivial and counter-intuitive spatial patterns [4], which is the outcome of the self-organizing process. The mechanism of self-organization has been found to operate in the functioning of many multi-component dynamical systems—where global pattern emerges from the collective interaction of the components following some simple local rules [28]. Self-organized patterns like *spiral waves* [7, 9, 14, 29] observed in the models have contributed immense insights in understanding the workings of natural systems. For example, it has explained why species interaction (host-parasitoid system or the system of species competing for a single resource) is inherently unstable and potentially capable of leading to a total collapse, do survive and persist in nature for long time [4, 7, 30]. It has also been argued that heterogeneity in landscape and phenotypic diversity in species clearly reduce the risk of population extinction by resisting spatial synchrony in dynamics, thereby indicating that ecological and demographic heterogeneity may have indirect adaptive value [20]. This argument, of course, remains valid till the loss or destruction of habitat patches remain below a threshold beyond which the habitat will get transformed into disconnected patches [4], which in turn will lead to the eventual collapse of the metapopulation due to lack of resources and space.

Although it has been shown that the emerging spiral wave patterns are very robust to environmental noise [8, 31], one of the intriguing questions that the ecologists are grappling with is the observation and verification of such spatial structures in field studies. Needless to say, any direct observation of the spiral wave patterns in the empirical data is difficult, even though attempts are made towards that [12, 32, 33]. The two major hurdles to uncover such patterns are—lack of relevant spatial data, and the long transient period that the space-time dynamics needs to settle down on any discernible geometrical patterns. This task becomes more difficult with the frequent anthropogenic activities, recognized as one of most causal factors for the habitat destructions and fragmentations [21]. The latter and its consequences are perceived to have challenged the ecologists with a daunting task of analysing and understanding the responses of various ecosystems to externally induced perturbations [34].

This work has shown the differences in spatial patterns exhibited by metapopulations for random single site variations and patterned heterogeneities. Getting an exact parallel of the effect of patterned vacancy on the spatial patterns in natural populations might be as difficult as the direct confirmation of the spiral waves (or spatial chaos) in natural ecosystems [12, 32]. However, there are many examples of human developmental activities that lead to the division of an otherwise single landscape into many domains separated by the patterned barriers. The presence of such barriers is seen everywhere [35]. As a specific example one can consider the construction of the Indira Gandhi Canal ($\approx$550 km long) in the late 1950s in the north-western districts of Rajasthan, India. The canal with its tributaries runs through a landscape that used to be a part (11%) of the great Indian desert. Needless to say, the ecological impact of the canal on the species abundances in the region is quite severe in terms of the loss of endemic biodiversity, as well as the emergence of new human diseases, among other things [36].

The existence of low levels of heterogeneity—in number of neighbouring patches to which migration can occur, or neighbours having different demographic parameters/traits—essentially creates local anisotropies in an otherwise homogeneous space. These act like *defects* inducing synchronization failure in lattices. A vacant site can act to *pin* the core of the spiral (like a quenched disorder), thereby making it robust. Thus spatial asynchrony tends to persist for longer time even in small lattices. Parametric inhomogeneity induces phase defects through difference in the local dynamics between neighbouring sites, and induce spatial asynchrony [37]. For demographically heterogeneous landscapes (**Type II**), at low levels of dispersal, phase differences introduced due to difference in intrinsic dynamics, even at few patches, will keep propagating through several hundreds or thousands of generations before the metapopulation finally gets spatially synchronized [38]. The ecological implication of the above observation throws up the question as to if the importance of transient spatial dynamics ever reduces in relevant ecological timescale. Then any effort to look for or anticipate stable spiral wave patterns in species abundances may prove to be futile. Such instances of long term transient dynamics were reported in other studies of spatially structured population models, and the cautionary conclusions were drawn that the transient dynamics should receive greater attention in order to fully understand the system's functional complexity than what it currently does [39].

In this study we have considered the presence of different types of isolated and patterned heterogeneities (migration barriers or genotypic coexistence) one at a time. However, in nature they are more likely to overlap. Identifying the natural barriers, predicting the interaction among physical landscapes, species dispersal, and gene flow, and using the information to manage ecological and epidemiological functions are important problems [40]. Needless to say, the spatial patterns in species populations will be exceedingly difficult to comprehend in the presence of the environmental heterogeneities that have such combinatorial forms. But ecologists continuously endeavour to find evidence(s) of stable patterns (spiral wave patterns) in natural populations [41, 42].

Acknowledgements BKS thanks Prof (Dr.) Paulien Hogeweg for stimulating discussions. SS thanks J C Bose Fellowship (DST), CPSDE (MHRD-CoE) and Indian National Science Academy for funding.

References

1. D. Mollison, Spatial contact models for ecological and epidemic spread. J. R. Stat. Soc. B **39**, 283–326 (1977)
2. M.C. Cross, P.C. Hohenberg, Pattern formations outside of equilibrium. Rev. Mod. Phys. **65**, 851–1112 (1993)
3. I. Hanski, *Metapopulation Ecology* (Oxford University Press, Oxford, 1999)
4. R.V. Sole, B. Goodwin, *Signs of Life: How Complexity Pervades Biology* (Basic Books, New York, 2000)
5. S.A. Levin, Dispersion and population interactions. Am. Nat. **108**, 207–228 (1974)
6. A. Hastings, Spatial heterogeneity and ecological models. Ecology **71**, 426–428 (1990)
7. M.P. Hassell, H.M. Comins, R.M. May, Spatial structure and chaos in insect population dynamics. Nature **353**, 255–258 (1991)
8. H.N. Comins, M.P. Hassell, R.M. May, The spatial dynamics of host-parasitoid systems. J. Anim. Ecol. **61**, 735–748 (1992)
9. R.V. Sole, J. Valls, J. Bascomte, Spiral waves, chaos and multiple attractors in lattice models of interacting populations. Phys. Lett. A **166**, 123–128 (1992)
10. A. Hastings, Complex interactions between dispersal and dynamics: lessons from coupled logistic equations. Ecology **74**, 1362–1372 (1993)
11. M.P. Hassell, H.M. Comins, R.M. May, Species coexistent and self-organising spatial dynamics. Nature **370**, 290–292 (1994)
12. M.P. Hassell, Host-parasitoid population dynamics. J. Anim. Ecol. **69**, 543–566 (2000)
13. P. Rohani, O. Miramontes, Host-parasitoid metapopulations, the consequences of parasitoid aggregation on spatial dynamics and searching efficiency. Proc. R. Soc. (Lond.) B **260**, 335–342 (1995)

14. N.J. Savill, P. Rohani, P. Hogeweg, Self-reinforcing spatial patterns enslave evolution in a host-parasitoid system. J. Theor. Biol. **188**, 11–20 (1997)
15. A. Moilanen, I. Hanski, Metapopulation dynamics, effects of habitat quality and landscape structure. Ecology **79**, 2503–2515 (1998)
16. A.J. Nicholson, V.A. Bailey, The balance of animal populations. Part 1. Proc. Zool. Soc. Lond. **3**, 551–598 (1935)
17. R.D. Holt, M.P. Hassell, Environmental heterogeneity and the stability of host-parasitoid interactions. J. Anim. Ecol. **62**, 89–100 (1993)
18. R.M. May, Host-parasitoid systems in patchy environments: a phenomenological model. J. Anim. Ecol. **47**, 833–843 (1978)
19. J.D. Reeve, Environmental variability, migration, and persistence in host-parasitoid systems. Am. Nat. **132**, 810–836 (1988)
20. B.K. Singh, J.S. Rao, R. Ramaswamy, S. Sinha, The role of heterogeneity on the spatiotemporal dynamics of host-parasite metapopulation. Ecol. Model. **180**, 435–443 (2004)
21. World Bank, *Sustainable Development in a Dynamic World: Transforming Institutions, Growth, and Quality of Life*, Chap. 1 (The World Bank, Washington, 2003)
22. P.L. Chesson, W.W. Murdoch, Aggregation of risk: relationships among host-parasitoid models. Am. Nat. **127**, 696–715 (1986)
23. T.H. Jones, M.P. Hassell, S.W. Pacala, Spatial heterogeneity and population dynamics of a host-parasitoid system. J. Anim. Ecol. **62**, 251–262 (1993)
24. B.K. Singh, Modelling infectious diseases: spatiotemporal dynamics and control. Ph.D. thesis, Jawaharlal Nehru University, New Delhi, 2001
25. B.K. Ehlers, C.F. Damgaard, F. Laroche, Intraspecific genetic variation and species coexistence in plant communities. Biol. Lett. **12**, 20150853 (2016). http://dx.doi.org/10.1098/rsbl.2015.0853
26. S. Cadel-Six, C. Peyraud-Thomas, L. Brient, N.T. Marsac, R. Rippka, A. Mejean, Different genotypes of anatoxin-producing cyanobacteria coexist in the Tarn river, France. Appl. Environ. Microbiol. **73**(23), 7605–7614 (2007). https://doi.org/10.1128/AEM.01225-07
27. N.J. Savill, D.J. Shaw, R. Deardon, M.J. Tildesley, M.J. Keeling, M.E.J. Woolhouse, S.P. Brooks, B.T. Grenfell, Topographic determinants of foot and mouth disease transmission in the UK 2001 epidemic. BMC Vet. Res. **2**, 3 (2006). https://doi.org/10.1186/1746-6148-2-3
28. P. Bak, C. Tang, K. Wiesenfeld, Self-organized criticality. Phys. Rev. A **38**, 364–374 (1988)
29. A.H. Hirzel, R.M. Nisbet, W.M. Murdoch, Host-parasitoid spatial dynamics in heterogeneous landscapes. Oikos **116**, 2082–2096 (2007). https://doi.org/10.1111/j.2007.0030-1299.15976.x
30. R.V. Sole, J. Bascompte, J. Valls, Nonequilibrium dynamics in lattice ecosystems: chaotic stability and dissipative structures. Chaos **2**, 387–395 (1992)
31. G.D. Ruxton, P. Rohani, The consequences of stochasticity for self-organized spatial dynamics, persistence and coexistence in spatially extended host-parasitoid communities. Proc. R. Soc. Lond. B **263**, 625–631 (1996)
32. H.N. Comins, M.P. Hassell, Persistence of multispecies host-parasitoid interactions in spatially distributed models with local dispersal. J. Theor. Biol. **183**, 19–28 (1996)
33. O.N. Bjornstad, J. Bascompte, Synchrony and second-order spatial correlation in host-parasitoid systems. J. Anim. Ecol. **70**, 924–933 (2001)
34. P.M. Kareiva, J.G. Kingsolver, R.B. Huey (eds.), *Biotic Interactions and Global Change* (Sinauer Associates, Sunderland, 1993)
35. M.P. Hassell, H.C.J. Godfray, H.N. Comins, Effects of global change on the dynamics of insect host-parasitoid interactions, in *Biotic Interactions and Global Change*, ed. by P.M. Kareiva, J.G. Kingsolver, R.B. Huey (Sinauer Associates, Sunderland, 1993)
36. I. Prakash, Biological invasion and loss of endemic biodiversity in the Thar desert. Resonance **6**, 76–85 (2001)
37. A. Holden, M. Markus, H.G. Othmer (eds.), *Non-linear Wave Processes in Excitable Media* (Plenum Press, New York, 1991)
38. B. Blasius, A. Hupert, L. Stone, Complex dynamics and phase synchronization in spatially extended ecological systems. Nature **399**, 354–359 (1999)

39. A. Hastings, K. Higgins, Persistence of transients in spatially structured ecological models. Science **263**, 1133–1136 (1994)
40. P. Caplat, P. Edelaar, R.Y. Dudaniec, A.J. Green, B. Okamura, J. Cote, J. Ekroos, P.R. Jonsson, J. Lndahl, S.VM. Tesson, E.J. Petit, Looking beyond the mountain: dispersal barriers in a changing world. Front. Ecol. Environ **14**(5), 261–268 (2016). https://doi.org/10.1002/fee.1280
41. M. Michel Baguette, S. Blanchet, D. Legrand, V.M. Stevens, C. Turlure, Individual dispersal, landscape connectivity and ecological networks. Biol. Rev. **88**, 310–326 (2013). https://doi.org/10.1111/brv.12000
42. K. Berthier, S. Piry, J.F. Cosson, P. Giraudoux, J.C. Foltête, R. Defaut, D. Truchetet, X. Lambin, Dispersal, landscape and travelling waves in cyclic vole populations. Ecol. Lett. **17**(1), 53–64 (2014). https://doi.org/10.1111/ele.12207

Dark States in Quantum Photosynthesis

S. V. Kozyrev and I. V. Volovich

1 Introduction

The effect of quantum photosynthesis is the experimental observation of quantum coherences (observation of photonic echo) in photosynthetic complexes. Effectiveness of exciton transport to reaction centers of photosynthetic complexes is also elevated, this phenomenon is discussed as a result of application of quantum coherencies by nature to improve the performance of photosynthetic systems. For discussion of quantum photosynthesis and other quantum effects in biological systems, see [1–4].

In this paper we model photosynthetic system in one exciton approximation by open three level quantum system with energy levels ε_0, ε_1, ε_2, $\varepsilon_0 < \varepsilon_1 < \varepsilon_2$, which interacts with three quantum fields (the reservoirs). The two lower levels $|0\rangle$, $|1\rangle$ are non-degenerate. The upper level ε_2 is degenerate. Energy level ε_0 describes photosynthetic system without excitons, energy level ε_1 is the state "exciton in the reaction center" and the degenerate energy level ε_2 describes excitons on chromophores. Three possible transitions between the energy levels are paired to different reservoirs: the transition between levels ε_2, ε_0 is paired to interaction with light, the transition between levels ε_2, ε_1 is related to interaction with phonons (vibrations of the protein matrix), the transition between levels ε_1, ε_0 is related to interaction with the special sink field with zero temperature.

We will use the method of dissipative dynamics of open quantum systems [5], namely the stochastic limit approach [6, 7]. The aim of the present paper is to prove that in this case (dissipative dynamics of sufficiently complex degenerate quantum

S. V. Kozyrev (✉) · I. V. Volovich
Steklov Mathematical Institute of the Russian Academy of Sciences, Moscow, Russia
e-mail: kozyrev@mi.ras.ru; volovich@mi.ras.ru

© Springer International Publishing AG, part of Springer Nature 2018
R. P. Mondaini (ed.), *Trends in Biomathematics: Modeling, Optimization and Computational Problems*, https://doi.org/10.1007/978-3-319-91092-5_2

13

systems) generation of the so-called dark states (or dark-state polaritons) known in quantum optics [8, 9] is possible. We conjecture that these dark states are related to the effect of quantum photosynthesis.

The scheme of experiments on quantum photosynthesis is as follows: first, one prepares a photosynthetic system and applies a laser field to the system (the frequency of the laser should be in resonance to transitions in the photosynthetic complex); second, one switches off the laser for short time; third, one applies the same laser field to the system and observes the photonic echo. Our aim is to reproduce this scheme using our model of dissipative dynamics in degenerate system.

Dynamics of the system density matrix in the model under consideration is described by a sum of three generators—the photonic generator, the phononic generator and the generator of absorption of excitons. We consider the following scheme of manipulation of quantum states of the system.

1. First, we compute the stationary state of the system taking into account only the photonic generator and neglecting the effects of transport and absorption (this can be considered as a strong light approximation). This state will contain quantum coherencies (off-diagonal part of the density matrix).
2. Second, we investigate joint action of generators of transport and absorption on the state prepared at the previous step. The observation is that a part of coherences will survive due to the effect of dark states.
3. Third, we apply the laser field again and show that the interaction of the prepared state with the laser field will be non-trivial, which gives the possibility of photonic echo for this state.

In non-degenerate case (when energy level ε_2 is non-degenerate) the considered in this paper non-equilibrium quantum system was discussed in [10] and the flow of excitons in the system was computed (earlier stochastic limit of nonequilibrium three level quantum open system was considered in [11]). Open quantum systems coupled to several reservoirs were considered in many papers, in particular in [12]. In paper [13] effects of degeneracy in exciton transport were investigated by the stochastic limit method and it was shown that it is possible to achieve quantum amplification of exciton transport (the supertransport effect). For discussion of supertransport in quantum photosynthesis, see also [14–17]. The possibility of excitation of non-decaying dark states in a degenerate system can be considered as a side effect of the supertransport.

In [18] dark states in photosynthetic systems were studied experimentally. In [19] dark states in photosynthesis were also considered but dark states in this paper were considered in a different way compared to the approach of present paper, in particular in this case the effect considered in the present paper does not take place.

In [5] application of quantum methods to computations and biology was discussed. Photosynthetic complex is an example of complex physical system, see [20] for the review of ultrametric approach to complex systems in physics and biology.

The exposition of the present paper is as follows. In Sect. 2 the Hamiltonian of light-harvesting system interacting with three reservoirs (photons, phonons and absorption) is described. In Sect. 3 the corresponding generators of dynamics of density matrix of the system are considered (there are three such generators which describe excitation, transport and absorption of excitons). In Sect. 4 bright, dark and off-diagonal states for the mentioned generators are discussed. In Sect. 5 manipulation of quantum states of the described degenerate system by dissipative dynamics is described.

2 Description of the Photosynthetic System

Let us consider the Hilbert space $\mathcal{H}_S = \mathbb{C}^{N+1}$ and use the Dirac notations. Let $\{|0\rangle, \ldots, |N\rangle\}$ denote an orthonormal basis in $\mathcal{H}_S$.

We consider a system with three energy levels $\varepsilon_0 < \varepsilon_1 < \varepsilon_2$, where the upper level is degenerate, with the Hamiltonian (operator in $\mathcal{H}_S$)

$$H_S = \varepsilon_0 |0\rangle\langle 0| + \varepsilon_1 |1\rangle\langle 1| + \varepsilon_2 \sum_{j=2}^{N} |j\rangle\langle j|. \tag{1}$$

This Hamiltonian describes a light-harvesting system, $|0\rangle$ corresponds to a state without excitons, $|1\rangle$ is a state "exciton in the reaction center" (or sink), $|j\rangle$ corresponds to one-exciton states of chromophores.

System interacts with three quantum fields (reservoirs) in a dipole way (the fields are in temperature states). Transitions between the levels with energies ε_0 and ε_2 (in particular creation of excitons) are related to interaction with light (Bose field in the Gibbs state with the temperature $\beta_{\mathrm{em}}^{-1} = 6000\,\mathrm{K}$, or laser field with the frequency $\varepsilon_2 - \varepsilon_0$), transitions between the levels ε_2 and ε_1 (transport of excitons to the reaction center) are related to interaction with phonons (described by the Bose field with the temperature $\beta_{\mathrm{ph}}^{-1} = 300\,\mathrm{K}$), and transitions between the levels ε_1 and ε_0 (absorption of excitons in the reaction center) are described by interaction with the sink reservoir in the Fock state (i.e. reservoir with the zero temperature).

Thus we have three reservoirs described by Hamiltonians of quantum Bose fields H_{em} (light, or electromagnetic field), H_{ph} (phonons, or vibrations of protein matrix), H_{sink} (sink, or absorption of excitons in the reaction center), each of the reservoir Hamiltonians has the form

$$H_R = \int_{\mathbb{R}^3} \omega_R(k) a_R^*(k) a_R(k)\, dk, \tag{2}$$

here $a_R^*(k)$, $a_R(k)$ are creation and annihilation operators, $[a_R(k), a_R^*(q)] = \delta(k - q)$, and the index $R = \mathrm{em}, \mathrm{ph}, \mathrm{sink}$ enumerates the reservoirs, ω_R is the dispersion of the Bose field a_R (some nonnegative function). Each of the reservoirs is in a mean zero Gaussian state with the quadratic correlator

$$\langle a_R^*(k) a_R(k') \rangle = N_R(k)\delta(k - k').$$

Here $N_R(k)$ (number of the field quanta with wave number k) for the temperature state is equal to

$$N_R(k) = \frac{1}{e^{\beta_R \omega_R(k)} - 1} \tag{3}$$

where β_R is the inverse temperature of the reservoir.

The full Hamiltonian of the system interacting with three reservoirs has the form

$$H = H_S + H_{\text{em}} + H_{\text{ph}} + H_{\text{sink}} + \lambda \left(H_{I,\text{em}} + H_{I,\text{ph}} + H_{I,\text{sink}} \right), \tag{4}$$

where λ is a positive constant (coupling constant) and the interaction Hamiltonians $H_{I,\text{em}}$, $H_{I,\text{ph}}$, $H_{I,\text{sink}}$ have dipole forms and are given by formulae (5)–(7) below. The Hamiltonian H is an operator in the Hilbert space $\mathcal{H}_S \otimes \mathcal{H}_{\text{em}} \otimes \mathcal{H}_{\text{ph}} \otimes \mathcal{H}_{\text{sink}}$ (in the following we omit the notation of tensor product).

Interaction of the system with light is described by the Hamiltonian

$$H_{I,\text{em}} = A_{\text{em}}|\chi\rangle\langle 0| + A_{\text{em}}^*|0\rangle\langle\chi|, \qquad A_{\text{em}}^* = \int_{\mathbb{R}^3} g_{\text{em}}(k)a_{\text{em}}^*(k)dk, \tag{5}$$

here function $g_{\text{em}}(k)$ is the form-factor of interaction with the field, and the bright photonic vector χ belongs to the level with energy ε_2, i.e. $H_S|\chi\rangle = \varepsilon_2|\chi\rangle$.

Transport of excitons to the reaction center is related to interaction of the system with phonons

$$H_{I,\text{ph}} = A_{\text{ph}}|\psi\rangle\langle 1| + A_{\text{ph}}^*|1\rangle\langle\psi|, \qquad A_{\text{ph}}^* = \int_{\mathbb{R}^3} g_{\text{ph}}(k)a_{\text{ph}}^*(k)dk, \tag{6}$$

where ψ is the bright phononic vector from the energy level ε_2.

Vectors ψ and χ belong to the same degenerate level with energy ε_2. Crucial feature of the model under consideration is the following—vectors ψ and χ are non-parallel.

Absorption of excitons in the reaction center is described by interaction with the additional field of sink (with the zero temperature)

$$H_{I,\text{sink}} = A_{\text{sink}}|1\rangle\langle 0| + A_{\text{sink}}^*|0\rangle\langle 1|, \qquad A_{\text{sink}}^* = \int_{\mathbb{R}^3} g_{\text{sink}}(k)a_{\text{sink}}^*(k)dk. \tag{7}$$

Remark Usually in papers on photosynthesis the system Hamiltonian H_S is taken non-diagonal and contains terms corresponding to the dipole interaction of excitons, and the Hamiltonian of interaction with phonons is also different. Here we consider the result of diagonalization of the system Hamiltonian H_S (transition to the so-called "global" basis), i.e. in our notations the states $|j\rangle$ of the system belong to the "global" basis. Relation between the "global" and "local" approaches in theory of open quantum systems was discussed in [21].

3   Generators of the Dissipative Dynamics

For investigation of the model we will use the method of quantum stochastic limit [6, 7]. In this limit dynamics of reduced density matrix of a system interacting with reservoir is generated by quantum dissipative operator in the Lindblad form, see formulae (10)–(12) below. This approach allows to take into account quantum decoherence, dissipation, transport and other thermodynamic phenomena. Quantum dynamics in the stochastic limit in presence of a laser field was discussed in [22]. Properties of antibunched light are discussed, for instance, in [23].

For the considered model with three reservoirs the dynamics will be generated by a sum of three generators (photonic, phononic and sink)

$$\frac{d}{dt}\rho(t) = \left(\theta_{\text{em}} + i[\cdot, H_{\text{eff}}] + \theta_{\text{ph}} + \theta_{\text{sink}}\right)(\rho(t)). \tag{8}$$

Creation and annihilation of excitons are described by the photonic generator equal to a sum of the dissipative Lindblad term θ_{em} and the term $i[\cdot, H_{\text{eff}}]$ related to interaction with a coherent field [6, 7]:

$$L_{\text{em}} = \theta_{\text{em}} + i[\cdot, H_{\text{eff}}], \quad H_{\text{eff}} = s(|\chi\rangle\langle 0| + |0\rangle\langle\chi|), \quad s \in \mathbb{R}. \tag{9}$$

$$\theta_{\text{em}}(\rho) = \|\chi\|^2 \Bigg[2\gamma_{\text{re,em}}^- \left(\langle\tilde{\chi}|\rho|\tilde{\chi}\rangle|0\rangle\langle 0| - \frac{1}{2}\{\rho, |\tilde{\chi}\rangle\langle\tilde{\chi}|\} \right) - i\gamma_{\text{im,em}}^-[\rho, |\tilde{\chi}\rangle\langle\tilde{\chi}|]$$

$$+ 2\gamma_{\text{re,em}}^+ \left(\langle 0|\rho|0\rangle|\tilde{\chi}\rangle\langle\tilde{\chi}| - \frac{1}{2}\{\rho, |0\rangle\langle 0|\} \right) + i\gamma_{\text{im,em}}^+[\rho, |0\rangle\langle 0|] \Bigg]. \tag{10}$$

The parameter s is the amplitude of the laser field. Here the normed bright photonic vector has the form

$$|\tilde{\chi}\rangle = \frac{|\chi\rangle}{\|\chi\|}.$$

Transport of excitons is described by the phononic generator

$$\theta_{\text{ph}}(\rho) = \|\psi\|^2 \Bigg[2\gamma_{\text{re,ph}}^- \left(\langle\tilde{\psi}|\rho|\tilde{\psi}\rangle|1\rangle\langle 1| - \frac{1}{2}\{\rho, |\tilde{\psi}\rangle\langle\tilde{\psi}|\} \right) - i\gamma_{\text{im,ph}}^-[\rho, |\tilde{\psi}\rangle\langle\tilde{\psi}|]$$

$$+ 2\gamma_{\text{re,ph}}^+ \left(\langle 1|\rho|1\rangle|\tilde{\psi}\rangle\langle\tilde{\psi}| - \frac{1}{2}\{\rho, |1\rangle\langle 1|\} \right) + i\gamma_{\text{im,em}}^+[\rho, |1\rangle\langle 1|] \Bigg]. \tag{11}$$

Here the normed bright phononic vector has the form

$$|\tilde{\psi}\rangle = \frac{|\psi\rangle}{\|\psi\|}.$$

Let us note that for large $\|\psi\|$ the transport of excitons will be amplified (the generator is multiplied by $\|\psi\|^2$). This corresponds to the supertransport phenomenon. The maximal possible amplification is equal to the degeneracy of the upper level.

Absorption of excitons is described by the sink generator

$$\theta_{\text{sink}}(\rho) = 2\gamma_{\text{re,sink}}^- \left(\langle 1|\rho|1\rangle |0\rangle\langle 0| - \frac{1}{2}\{\rho, |1\rangle\langle 1|\} \right) - i\gamma_{\text{im,sink}}^-[\rho, |1\rangle\langle 1|]. \tag{12}$$

The constants γ have the form (where P. is the principal part, or Cauchy principal value, generalized function)

$$\gamma_{\text{re},R}^+ = \pi \int |g_R(k)|^2 \delta(\omega_R(k) - \omega_R) N_R(k) dk, \tag{13}$$

$$\gamma_{\text{re},R}^- = \pi \int |g_R(k)|^2 \delta(\omega_R(k) - \omega_R)(N_R(k) + 1) dk, \tag{14}$$

$$\gamma_{\text{im},R}^+ = -\int |g_R(k)|^2 \text{ P.} \frac{1}{\omega_R(k) - \omega_R} N_R(k) dk, \tag{15}$$

$$\gamma_{\text{im},R}^- = -\int |g_R(k)|^2 \text{ P.} \frac{1}{\omega_R(k) - \omega_R}(N_R(k) + 1) dk, \tag{16}$$

$$\omega_{\text{em}} = \varepsilon_2 - \varepsilon_0, \qquad \omega_{\text{ph}} = \varepsilon_2 - \varepsilon_1, \qquad \omega_{\text{sink}} = \varepsilon_1 - \varepsilon_0.$$

Here the function $N_R(k)$ is given by (3), $\beta_{\text{em}}^{-1} = 6000\,\text{K}$ for the sun light, $\beta_{\text{ph}}^{-1} = 300\,\text{K}$ for protein matrix at room temperature, $\beta_{\text{sink}}^{-1} = 0\,\text{K}$.

For purely laser field $N_{\text{em}}(k) = 0$, this implies $\gamma_{\text{re,em}}^+ = \gamma_{\text{im,em}}^+ = 0$, but $\gamma_{\text{re,em}}^-, \gamma_{\text{im,em}}^- \neq 0$ (i.e. dissipative part of the generator is non-zero even for coherent field).

Combination of three generators (photonic, phononic and sink) in (8) provides a quantum thermodynamic machine—a quantum system with thermodynamic cycle. This machine operates by harvesting of photons, creation of excitons, transport of excitons to the reaction center and absorption of excitons. The current (or flow) of excitons [10] describes the effectiveness of this thermodynamic machine. In the model under consideration bright photonic and phononic vectors χ and ϕ belonging to the energy level ε_2 are not parallel. Thus the quantum thermodynamic machine is not perfectly developed and some quantum states may leak in the thermodynamic cycle. In this paper, see also [24], we show that this leads to excitation of quantum dark states with long lifetime. We conjecture that this phenomenon may describe the effect of quantum photosynthesis (existence of photonic echo in photosynthetic systems).

Remark Generators in (10), (11) are proportional to the squares $\|\chi\|^2$, $\|\psi\|^2$ of bright photonic and phononic vectors. This corresponds to the superradiance effect (actually superabsorption for (10) and supertransfer for (11))—effects of coherent amplification of quantum interaction [25]. In the superradiance effect the interaction between the field and a system of correlated oscillators is proportional to the square n^2 of the number of oscillators (not to the number n of oscillators as in the case of a system of oscillators without correlation). Supertransfer was discussed in relation to quantum photosynthesis [13–17].

4 Bright, Dark and Off-Diagonal Matrices

Lindblad generators (10)–(12) act on the space of matrices (which contains density matrices of the system). For each of these generators one can consider expansion of this space of matrices in a sum of orthogonal subspaces of bright, dark and off-diagonal matrices. Bright and dark states were extensively discussed in quantum optics, see [8, 9]. We use here the approach and notations of [13].

These subspaces depend on the generator and are different for different generators. Let us discuss the photonic generator θ_{em} given by (10). This generator describes creation and annihilation of excitons by interaction of chromophores and electromagnetic field. Bright matrices for this generator are linear combinations of matrices (projections in (10))

$$|0\rangle\langle 0|, \quad |\chi\rangle\langle\chi|.$$

Dark matrices B are matrices which give zero when multiplied by any bright matrix A, i.e.

$$AB = BA = 0.$$

For the generator θ_{em} dark matrices are linear combinations of matrices

$$|\phi\rangle\langle\phi'|, \quad |1\rangle\langle 1|, \quad |\phi\rangle\langle 1|, \quad |1\rangle\langle\phi|, \quad \phi\perp\chi, \phi'\perp\chi.$$

Off-diagonal matrices are matrices C orthogonal to all bright matrices A and all dark matrices B with respect to the scalar product $(\cdot, \cdot) = \mathrm{tr}(\cdot\cdot)$, i.e.

$$\mathrm{tr}(CA) = \mathrm{tr}(CB) = 0.$$

This subspace contains matrices corresponding to transitions between bright and dark subspaces, and between levels in the bright subspace, for the generator θ_{em} the off-diagonal subspace is generated by matrices

$$|\chi\rangle\langle 0|, \quad |\chi\rangle\langle\phi|, \quad |\chi\rangle\langle 1|, \quad |1\rangle\langle 0|, \quad |\phi\rangle\langle 0|, \quad \phi\perp\chi$$

and conjugated.

Dark matrices described above are stationary with respect to the dynamics generated by θ_{em}. The bright space corresponds to processes of creation and annihilation of excitons (for the generator θ_{em}) and to process of transport of excitons (for the generator θ_{ph}). Off-diagonal matrices decay exponentially which corresponds to the decoherence phenomenon. In the presence of a coherent field (generator $i[\cdot, H_{eff}]$) off-diagonal matrices can be created.

Described expansion of the state of matrices depends on the generator and for the phononic generator θ_{ph} will be completely different. In particular matrix $|1\rangle\langle 1|$ is dark for θ_{em} and bright for θ_{ph}, and matrix $|\chi\rangle\langle\chi|$ is bright for θ_{em} and contains a combination of bright, dark and off-diagonal terms for θ_{ph}. This non-coincidence of expansions of a state of the system in a sum of bright, dark and off-diagonal matrices due to non-parallel photonic and phononic bright vectors χ and ψ may lead to interesting phenomena of possibility of manipulations with quantum states discussed in the next section.

Expansion of the space of matrices in a sum of bright, dark and off-diagonal subspaces for the phononic generator θ_{ph} given by (11) is as follows. Bright matrices for θ_{ph} are linear combinations of matrices

$$|1\rangle\langle 1|, \quad |\psi\rangle\langle\psi|.$$

The space of dark matrices is generated by matrices

$$|\eta\rangle\langle\eta'|, \quad |0\rangle\langle 0|, \quad |\eta\rangle\langle 0|, \quad |0\rangle\langle\eta|, \quad \eta\perp\psi, \eta'\perp\psi.$$

The off-diagonal space is a linear span of matrices

$$|\psi\rangle\langle 1|, \quad |\psi\rangle\langle\eta|, \quad |\psi\rangle\langle 0|, \quad |0\rangle\langle 1|, \quad |\eta\rangle\langle 1|, \quad \eta\perp\psi$$

and conjugated.

5   Manipulation of Quantum States

The scheme of experiments on quantum photosynthesis is as follows. At the first step photosynthetic system is excited by a laser pulse, photons are absorbed and excitons are created. Then the system performs decoherence and transport of excitons in absence of light (in the time period of order of one microsecond). At the third step of the experiment the system interacts with another laser pulse. Observation of photonic echo at the third step of this experiment was called the effect of quantum photosynthesis, this phenomenon attracts a lot of attention [1–3, 18].

In this section we will reproduce this phenomenon by investigation of quantum dynamics of the discussed in the present paper degenerate system interacting with nonequilibrium environment in presence of coherent field (4), (1), (2), (5)–(7). We will consider a three step manipulation by quantum states of the system given by application of quantum dissipative dynamics generated by different combinations of operators (9)–(12).

5.1 Application of a Light Pulse

Let us discuss the process of manipulation of quantum states of excitons. The initial state is the density matrix without excitons

$$\rho_0 = |0\rangle\langle 0|.$$

Let us apply to this matrix the dynamics given by the photonic generator $L_{em} = \theta_{em} + i[\cdot, H_{eff}]$ of the form (9). One can say that we use the approximation of strong light and ignore the contributions to the dynamics from the phononic generator θ_{ph} and the sink generator θ_{sink} in (8). In the limit $t \to \infty$ this dynamics puts the system in a stationary state of the form

$$\rho_1 = \rho_{00}|0\rangle\langle 0| + \rho_{\chi\chi}|\tilde{\chi}\rangle\langle\tilde{\chi}| + \rho_{\chi 0}|\tilde{\chi}\rangle\langle 0| + \rho_{0\chi}|0\rangle\langle\tilde{\chi}|, \tag{17}$$

where

$$\rho_{00} = \frac{\gamma_{re,em}^- - \frac{s^2}{\|\chi\|^2}\mathrm{Re}\left(\frac{1}{\mu_{\chi 0}}\right)}{\gamma_{re,em}^+ + \gamma_{re,em}^- - 2\frac{s^2}{\|\chi\|^2}\mathrm{Re}\left(\frac{1}{\mu_{\chi 0}}\right)}, \tag{18}$$

$$\rho_{\chi\chi} = \frac{\gamma_{re,em}^+ - \frac{s^2}{\|\chi\|^2}\mathrm{Re}\left(\frac{1}{\mu_{\chi 0}}\right)}{\gamma_{re,em}^+ + \gamma_{re,em}^- - 2\frac{s^2}{\|\chi\|^2}\mathrm{Re}\left(\frac{1}{\mu_{\chi 0}}\right)}, \tag{19}$$

$$\rho_{\chi 0} = \frac{is}{\|\chi\|\mu_{\chi 0}} \frac{\gamma_{re,em}^- - \gamma_{re,em}^+}{\gamma_{re,em}^+ + \gamma_{re,em}^- - 2\frac{s^2}{\|\chi\|^2}\mathrm{Re}\left(\frac{1}{\mu_{\chi 0}}\right)}, \tag{20}$$

$$\rho_{0\chi} = -\frac{is}{\|\chi\|\mu_{0\chi}} \frac{\gamma_{re,em}^- - \gamma_{re,em}^+}{\gamma_{re,em}^+ + \gamma_{re,em}^- - 2\frac{s^2}{\|\chi\|^2}\mathrm{Re}\left(\frac{1}{\mu_{\chi 0}}\right)}, \tag{21}$$

$$\mu_{\chi 0} = \mu_{0\chi}^* = -\gamma_{re,em}^- - \gamma_{re,em}^+ + i\gamma_{im,em}^- + i\gamma_{im,em}^+. \tag{22}$$

In particular, if there is no coherent field (i.e. $s = 0$), then off-diagonal elements of the matrix ρ_1 vanish $\rho_{\chi 0} = \rho_{0\chi} = 0$ and diagonal elements ρ_{00}, $\rho_{\chi\chi}$ will be given by the Gibbs state satisfying

$$\frac{\rho_{\chi\chi}}{\rho_{00}} = e^{-\beta_{\text{em}}(\varepsilon_2 - \varepsilon_0)}.$$

For the photonic generator $L_{\text{em}} = \theta_{\text{em}} + i[\cdot, H_{\text{eff}}]$ there exist also dark stationary matrices described in previous section, i.e. addition of a coherent field does not change the space of stationary dark matrices.

5.2 Transport and Absorption of Excitons

At the second step of manipulation of quantum states of excitons we switch off the light, i.e. we use the obtained at the previous step state ρ_1 given by (17)–(21) as the initial state for dynamics with the generator $\theta_{\text{ph}} + \theta_{\text{sink}}$ (sum of the phononic and the sink generators).

Let us consider the expansion of the bright photonic vector $\widetilde{\chi}$

$$\widetilde{\chi} = \widetilde{\chi}_0 + \widetilde{\chi}_1, \qquad \widetilde{\chi}_0 \| \widetilde{\psi}, \quad \widetilde{\chi}_1 \perp \widetilde{\psi}$$

in the sum of contributions parallel and orthogonal to the bright phononic vector $\widetilde{\psi}$

$$|\widetilde{\chi}_0\rangle = \langle \widetilde{\psi}, \widetilde{\chi} \rangle |\widetilde{\psi}\rangle = |\widetilde{\psi}\rangle \langle \widetilde{\psi} \| \widetilde{\chi}\rangle, \qquad |\widetilde{\chi}_1\rangle = (1 - |\widetilde{\psi}\rangle\langle\widetilde{\psi}|)|\widetilde{\chi}\rangle.$$

Let us substitute this expansion in expression (17)–(21) for the stationary photonic state ρ_1, apply the generator $\theta_{\text{ph}} + \theta_{\text{sink}}$ and consider the corresponding dynamics. Discussion of Sect. 4 of dynamics in spaces of bright, dark and off-diagonal matrices implies that all terms in the expansion which contain χ_0 will decay since the transport generator θ_{ph} will transfer excitons to the reaction center where excitons will be absorbed. Therefore the system density matrix will tend to the stationary state ρ_2 of the form

$$\rho_2 = \rho_{00}|0\rangle\langle 0| + \rho_{\chi\chi}|\widetilde{\chi}_1\rangle\langle\widetilde{\chi}_1| + \rho_{\chi 0}|\widetilde{\chi}_1\rangle\langle 0| + \rho_{0\chi}|0\rangle\langle\widetilde{\chi}_1|, \qquad (23)$$

where $\rho_{\chi\chi}$, $\rho_{\chi 0}$, $\rho_{0\chi}$ are given by (18)–(21) and ρ_{00} is given by the condition that trace of density matrix is equal to one

$$\rho_{00} = 1 - \|\widetilde{\chi}_1\|^2 \rho_{\chi\chi}.$$

We see that application of expansion of the space of matrices on the system Hilbert space in a sum of bright, dark and off-diagonal subspaces allows to describe the dynamics generated by dissipative Lindblad generators for degenerate quantum system with complex interaction (formula (23) was obtained from (17) by substitution of $\widetilde{\chi}$ by $\widetilde{\chi}_1$).

In the model under consideration the obtained state (23) can exist infinite time (in reality the lifetime should be finite but long). If the bright vectors for photons and phonons are parallel (i.e. $|\chi_1\rangle = 0$) then we will get $\rho_2 = |0\rangle\langle 0|$. In general case of non-parallel ψ, χ the obtained state ρ_2 will contain non-decaying dark part. This part can be observed by interaction with laser pulse, see the next subsection.

5.3 Interaction with Laser

At the third step of manipulation of quantum states of excitons we subject the obtained at the previous step state ρ_2 of the form (23) to interaction with coherent light. We ignore transport and absorption (for example, the light is strong and we consider small times). We will show that in this situation the interaction will be non-trivial, therefore it would be possible to observe photonic echo.

Spectroscopy is related to application of dynamics with the generator $i[\cdot, H_{\mathrm{eff}}]$ to the off-diagonal part of the density matrix. Thus we are interested in the off-diagonal term in (23) of the form

$$\rho_{\chi 0}|\widetilde{\chi}_1\rangle\langle 0| + \rho_{0\chi}|0\rangle\langle\widetilde{\chi}_1|, \tag{24}$$

where the matrix elements $\rho_{\chi 0}$, $\rho_{0\chi}$ are given by (20). In particular this term can be non-zero only if $s \neq 0$, i.e. the state (17) should be excited by light with coherent component.

Let us recall that $\widetilde{\chi}_1$ is the orthogonal complement to projection of $\widetilde{\chi}$ (bright photonic vector) to $\widetilde{\psi}$ (bright phononic vector). We define $\widetilde{\chi}_2$ as projection of $\widetilde{\chi}_1$ to $\widetilde{\chi}$ (i.e. $\widetilde{\chi}_2$ is parallel to $\widetilde{\chi}$), then

$$|\widetilde{\chi}_2\rangle = \left(1 - |\langle\widetilde{\psi}, \widetilde{\chi}\rangle|^2\right)|\widetilde{\chi}\rangle.$$

Non-trivial contribution to spectroscopy can be given by application of $[\cdot, H_{\mathrm{eff}}]$ to (24), i.e. by

$$\rho_3 = i[\rho_{\chi 0}|\widetilde{\chi}_2\rangle\langle 0| + \rho_{0\chi}|0\rangle\langle\widetilde{\chi}_2|, H_{\mathrm{eff}}].$$

Let us assume that the state ρ_1 is prepared by application of laser field with the amplitude s, and that for spectroscopy we use laser field with the same amplitude. In the limit $s \to \infty$ (i.e. for strong laser fields) we get

$$\lim_{s\to\infty} \rho_3 = -\frac{1}{2}\|\chi\|^2\pi \int |g_{\mathrm{em}}(k)|^2\delta(\omega_{\mathrm{em}}(k) - \varepsilon_2 + \varepsilon_0)dk$$

$$\cdot\left(1 - |\langle\widetilde{\psi}, \widetilde{\chi}\rangle|^2\right)(|0\rangle\langle 0| - |\widetilde{\chi}\rangle\langle\widetilde{\chi}|). \tag{25}$$

Let us note that

$$1 - |\langle\widetilde{\psi}, \widetilde{\chi}\rangle|^2 = \sin^2\alpha,$$

where α is the angle between bright photonic and phononic vectors χ and ψ. If $\alpha \neq 0$, then the above expression (25) is non-zero. In this case one should get the photonic echo in spectroscopic experiments. We conjecture that this effect could explain the phenomenon of quantum photosynthesis, i.e. existence of quantum coherences with long lifetime observed in photosynthetic systems.

Summary We have shown that the model of light-harvesting complex as a degenerate system with absorption and interaction with photons and phonons describes excitation of dark states which will have long lifetime and will be visible in spectroscopic experiments.

Earlier it was shown [13–17] that degeneracy can be used for quantum amplification of exciton transport (the supertransport effect). Discussion of the present paper shows that in this case, as a side effect of the supertransport, we will obtain coherent dark states with long lifetime. This result can be discussed in relation to the phenomenon of quantum photosynthesis [1, 2] and experimental observation of dark states in photosynthetic systems [18].

Remark We have discussed manipulations with quantum states, in particular with dark states: excitation of the states by coherent fields and using of Lindblad dissipation for projections of the states to some subspaces of matrices; since the bright vectors for different fields (χ and ψ in the present model) can be non-parallel the manipulations under consideration are nontrivial. Manipulations of quantum states can be used for quantum computations [5].

Long lifetime of dark states could help to avoid decoherence in quantum computers. For another approaches to reduce decoherence in quantum computers and to manipulation of quantum states to quantum control, see [26–31]. Different ideas of how to use photosynthetic quantum effects for computations and quantum states manipulations are discussed in [1, 3, 24, 32]. In [33, 34] problems of quantum information theory in particular quantum channel capacity were considered.

In [35] a new approach to investigation of quantum photosynthesis based on the so-called holographic approach used earlier in high energy physics [36–38] was proposed.

References

1. G.S. Engel, T.R. Calhoun, E.L. Read, T.K. Ahn, T. Mancal, Y.C. Cheng, R.E. Blankenship, G.R. Fleming, Evidence for wavelike energy transfer through quantum coherence in photosynthetic systems. Nature **446**, 782–786 (2007)
2. G.D. Scholes, G.R. Fleming, A. Olaya-Castro, R. van Grondelle, Lessons from nature about solar light harvesting. Nat. Chem. **3**, 763–774 (2011)
3. M. Mohseni, P. Rebentrost, S. Lloyd, A. Aspuru-Guzik, Environment-assisted quantum walks in photosynthetic energy transfer. J. Chem. Phys. **129**, 174106 (2008)
4. M. Mohseni, Y. Omar, G. Engel, M.B. Plenio (eds.), *Quantum Effects in Biology* (Cambridge University Press, Cambridge, 2014)

5. M. Ohya, I. Volovich, *Mathematical Foundations of Quantum Information and Computation and Its Applications to Nano- and Bio-systems* (Springer, Berlin, 2011)
6. L. Accardi, L.Y. Gang, I. Volovich, *Quantum Theory and Its Stochastic Limit* (Springer, Berlin, 2002)
7. L. Accardi, S.V. Kozyrev, Lectures on quantum interacting particle systems, in *QP-PQ: Quantum Probability and White Noise Analysis - Vol. 14 Quantum Interacting Particle Systems* (World Scientific Publishing, Singapore, 2002)
8. M.O. Scully, M.S. Zubairy, *Quantum Optics* (Cambridge University Prress, Cambridge, 1997)
9. M. Fleischhauer, M.D. Lukin, Dark-State polaritons in electromagnetically induced transparency, Phys. Rev. Lett. **84**, 5094 (2000). arXiv:quant-ph/0001094
10. S.V. Kozyrev, A.A. Mironov, A.E. Teretenkov, I.V. Volovich, Flows in nonequilibrium quantum systems and quantum photosynthesis. Infin. Dimens. Anal. Quantum. Probab. Relat. Top. **20**, 175002 (2017). arXiv:1612.00213 [quant-ph]
11. L. Accardi, K. Imafuku, S.V. Kozyrev, Interaction of 3-level atom with radiation. Opt. Spectrosc. **94**(6), 904–910 (2003)
12. C.-K. Chan, G.-D. Lin, S.F. Yelin, M.D. Lukin, Quantum interference between independent reservoirs in open quantum systems. Phys. Rev. **A89**, 042117 (2014)
13. I.Y. Aref'eva, I.V. Volovich, S.V. Kozyrev, Stochastic limit method and interference in quantum many-particle systems. Theor. Math. Phys. **183**(3), 782–799 (2015)
14. R. Monshouwer, M. Abrahamsson, F. van Mourik, R. van Grondelle, Superradiance and exciton delocalization in bacterial photosynthetic light-harvesting systems, J. Phys. Chem. **B101**, 7241–7248 (1997)
15. A. Olaya-Castro, C.F. Lee, F.F. Olsen, N.F. Johnson, Efficiency of energy transfer in a light-harvesting system under quantum coherence. Phys. Rev. **B78**, 085115 (2008)
16. S. Lloyd, M. Mohseni, Symmetry-enhanced supertransfer of delocalized quantum states. New J. Phys. **12**, 075020 (2010). arXiv:1005.2579
17. D.F. Abasto, M. Mohseni, S. Lloyd, P. Zanardi, Exciton diffusion length in complex quantum systems: the effects of disorder and environmental fluctuations on symmetry-enhanced supertransfer. Philos. Trans. R. Soc. **A370**, 3750–3770 (2012)
18. M. Ferretti, R. Hendrikx, E. Romero, J. Southall, R.J. Cogdell, V.I. Novoderezhkin, G.D. Scholes, R. van Grondelle, Dark states in the light-harvesting complex 2 revealed by two-dimensional electronic spectroscopy. Sci. Rep. **6**, 20834 (2016)
19. H. Dong, D.-Z. Xu, J.-F. Huang, C.-P. Sun, Coherent excitation transfer via the dark-state channel in a bionic system. Light Sci. Appl. **1**, e2 (2012)
20. S.V. Kozyrev, Ultrametricity in the theory of complex systems. Theor. Math. Phys. **185**(2), 346–360 (2015)
21. A.S. Trushechkin, I.V. Volovich, Perturbative treatment of inter-site couplings in the local description of open quantum networks. Europhys. Lett. **113**(3), 30005 (2016)
22. L. Accardi, S.V. Kozyrev, A.N. Pechen, Coherent quantum control of Λ-atoms through the stochastic limit, in *Quantum Information and Computing, QP-PQ: Quantum Probability and White Noise Analysis - Vol. 19*, ed. by L. Accardi, M. Ohya, N. Watanabe (World Scientific, Singapore, 2006), pp. 1–17. arXiv:quant-ph/0403100
23. I.V. Volovich, Cauchy–Schwarz inequality-based criteria for non-classicality of sub-Poisson and antibunched light. Phys. Lett. **A380**(1–2), 56–58 (2016)
24. I.V. Volovich, S.V. Kozyrev, Manipulation of states of a degenerate quantum system. Proc. Steklov Inst. Math. **294**, 241–251 (2016)
25. R.H. Dicke, Coherence in spontaneous radiation processes. Phys. Rev. **93**, 99 (1954)
26. I.V. Volovich, Models of quantum computers and decoherence problem, in *Quantum Information (Nagoya, 1997)* (World Scientific Publishing, River Edge, 1999), pp. 211–224. arXiv:quant-ph/9902055
27. A.N. Pechen, N.B. Il'in, Existence of traps in the problem of maximizing quantum observable averages for a qubit at short times. Proc. Steklov Inst. Math. **289**, 213–220 (2015)
28. A. Pechen, A. Trushechkin, Measurement-assisted Landau-Zener transitions. Phys. Rev. **A91**(5), 52316 (2015)

29. A.N. Pechen, N.B. Il'in, On the problem of maximizing the transition probability in an n-level quantum system using nonselective measurements. Proc. Steklov Inst. Math. **294**, 233–240 (2016)
30. A.N. Pechen, N.B. Il'in, Control landscape for ultrafast manipulation by a qubit. J. Phys. **A50**(7), 75301 (2017)
31. G.G. Amosov, S.N. Filippov, Spectral properties of reduced fermionic density operators and parity superselection rule. Quantum Inf. Process **16**(1), 2 (2017)
32. G. Vattay, S.A. Kauffman, Evolutionary design in biological quantum computing (2013). arXiv:1311.4688 [cond-mat.dis-nn]
33. A.S. Holevo, Gaussian optimizers and the additivity problem in quantum information theory. Russ. Math. Surv. **70**(2), 331–367 (2015)
34. M.E. Shirokov, On quantum zero-error capacity. Russ. Math. Surv. **70**(1), 176–178 (2015)
35. I. Aref'eva, I. Volovich, Holographic photosynthesis (2016). arXiv:1603.09107
36. I.Y. Aref'eva, Formation time of quark–gluon plasma in heavy-ion collisions in the holographic shock wave model. Theor. Math. Phys. **184**(3), 1239–1255 (2015). arXiv:1503.02185
37. I. Aref'eva, Multiplicity and thermalization time in heavy–ions collisions, in *19-th International Seminar on High Energy Physics (QUARKS–2016)*, Peterburg, May 29–June 4 2016. EPJ Web of Conferences, vol. 125 (2016), p. 1007. doi:10.1051/epjconf/201612501007
38. I.Y. Aref'eva, M.A. Khramtsov, AdS/CFT prescription for angle–deficit space and winding geodesics, J. High Energy Phys. **4**, 121 (2016). arXiv:1601.0200

Boundary Control Problems in Hemodynamics

Adélia Sequeira, Jorge Tiago, and Telma Guerra

1 Introduction

Cardiovascular diseases, such as heart attack and strokes, are the major causes of death in developed countries, with a significant impact in the cost and overall status of healthcare. Understanding the fundamental mechanisms of the pathophysiology and treatment of these diseases are matters of the greatest importance around the world. This gives a key impulse to the progress in mathematical and numerical modeling of the associated phenomena governed by complex physical laws, using adequate and fully reliable in silico settings.

The increasing collaboration between scientists working in multidisciplinary areas such as medical researchers and clinicians, mathematicians and bioengineers has contributed to data information exchange that can be used in the numerical simulations providing more realistic results. The final goal is to set up patient-specific models and simulations incorporating data and measurements taken from each single patient, that will be able to predict results of medical diagnosis and therapeutic planning with reasonable accuracy and using non-invasive means.

Techniques based on the inclusion of data measurements in the numerical simulations to improve the computational solution are known, in the literature, as Data Assimilation (DA) techniques. They allow for the reconstruction or improvement of the numerical solution of a given model that should match the data at the locations where they were observed and include different types of approaches. This

A. Sequeira (✉) · J. Tiago
Department of Mathematics and CEMAT, Instituto Superior Técnico, Universidade de Lisboa, Lisboa, Portugal
e-mail: adelia.sequeira@tecnico.ulisboa.pt; jftiago@tecnico.ulisboa.pt

T. Guerra
Escola Superior de Tecnologia do Barreiro, Instituto Politécnico de Setúbal, Lavradio, Portugal
e-mail: telma.guerra@estbarreiro.ips.pt

© Springer International Publishing AG, part of Springer Nature 2018   27
R. P. Mondaini (ed.), *Trends in Biomathematics: Modeling, Optimization and Computational Problems*, https://doi.org/10.1007/978-3-319-91092-5_3

methodology has already been used in other engineering fields like geophysics and meteorology (for an overview, see [1] and the references therein). More recently, DA variational approaches were also used in hemodynamics, namely in model parameter estimation, including material properties needed to properly define FSI models and improve blood flow simulations, e.g. [2, 3]. Moreover, Kalman filter techniques were suggested to perform parameter estimation in cardiovascular modeling (see, e.g., [4, 5] and [6] for an overview).

In [2], several approaches were compared, namely the domain splitting method, the matrix updating technique and also the variational approach which was shown to give the best results among the three techniques. This approach consists in minimizing the misfit between the observed data and the solution of the blood flow modeled by Navier-Stokes equations, by controlling certain free parameters. The authors assumed the value of the pressure on the inlet boundary as the control parameter. In [3] this approach was reformulated to include the possibility of controlling the blood inflow velocity profile. The authors performed a parameter fitting for the cost function and verified the robustness of the approach with respect to noise reduction in a 2D idealized stenosis. Another application of the DA techniques was used in [7] to improve the accuracy of computational domains reconstructed from medical images.

The existence of solution for these mathematical approaches to the DA problem was not established yet. However, numerical solutions based on a Discretize then Optimize (DO) direct approach were shown to be successful in several cases. (DO) consists in first discretizing the variational (optimal control) problem describing the DA method, and then in solving the resulting nonlinear finite-dimensional optimization problem. An alternative method is the adjoint (indirect) or Optimize then Discretize (OD) approach. Some authors [8, 9] suggested that the (DO) approach may be preferred. In [10, 11] it was shown that, in the case of nonlinear problems, such as in fluid control problems, (OD) could result in a discrete optimal solution failing to be optimal for the continuous problem. However, for stabilized advection equations both approaches can lead to different solutions but, in certain cases, the (OD) has better asymptotic convergence properties [12, 13]. For the Navier-Stokes equations, different perspectives were suggested (see [14] and [15] using (DO) approaches of boundary control problem and distributed control problems, respectively). It appears that, at the present stage, no general answer can be given. In particular, for the problem studied here (see (1)–(6), Sect. 2), this question remains open. In [2], where a pressure type control was considered, the authors obtained a better performance of the (DO), in terms of accuracy of the controlled solution. Based on these results, we have adopted here the (DO) approach. Nevertheless, a detailed comparison of these two approaches still deserves some attention.

In this paper we present a review of some results obtained for DA problem to control the velocity inflow profile in 3D domains, based on the previous studies using the (DO) approach, namely in [3, 16]. To simplify, the model is assumed to be stationary, in order to neglect the fluid interaction with the vessel walls. The assimilation of time dependent data to apply in clinical cases where the pulsatile nature of blood flow is determinant will be the subject of future work. The methodology adopted in this work includes the usage of the Sequential Quadratic Programming method [17] to solve the discretized optimization problem. As a result we obtain a large scale finite dimensional, nonlinear optimization problem. Here, we consider different idealized and realistic geometries, reconstructed from medical images, to investigate the robustness of the method in such domains and analyze the influence of the location of data measurements.

The paper is organized as follows. In Sect. 2 we introduce the 3D mathematical blood flow model, considered as a shear-thinning non-Newtonian fluid, and describe the variational approach for the DA problem. The (DO) methodology to solve the control problem is also described and the optimization algorithm is presented. Section 3 is devoted to numerical results obtained for different 3D geometries, where blood is modeled as a Newtonian or a Generalized Newtonian fluid. This section includes also a discussion of the obtained results.

2   Mathematical Models and Methods

Blood is a concentrated heterogeneous suspension of several formed cellular elements, the *blood cells* or *hematocytes*, red blood cells (RBCs or *erythrocytes*), white blood cells (WBCs or *leukocytes*) and platelets (*thrombocytes*), in an aqueous polymeric and ionic solution (mainly Na^+, K^+, Ca^{2+} and Cl^-), the *plasma*. Plasma represents $\sim 55\%$ of the blood volume and is composed of $\sim 92\%$ water and $\sim 3\%$ particles, namely, electrolytes, organic molecules, numerous proteins (albumin, globulins and fibrinogen) and waste products. Its central physiological function is to transport these dissolved substances, nutrients, wastes and the formed cellular elements throughout the circulatory system.

Experimental studies over many years have shown that blood flow exhibits non-Newtonian characteristics such as shear-thinning, viscoelasticity, yield stress and thixotropy. The complex rheology of blood is influenced by numerous factors including plasma viscosity, hematocrit (volume percentage of RBCs in blood) and in particular, the ability of RBCs to form aggregates when at rest or at low shear rates and to deform at high shear rates, storing and releasing energy (see, e.g., [18] and the references cited therein). Hemodynamic analysis of blood flow in vascular beds and prosthetic devices requires the rheological behavior of blood to be characterized by phenomenological constitutive equations relating the stress to the rate of deformation and flow.

Blood is a non-Newtonian fluid, but it can be regarded as Newtonian depending on the size of the blood vessels and the flow behavior, as in arteries with diameters larger than $100\,\mu$m where measurements of the apparent viscosity show that it ranges from 0.003 to 0.004 Pa s and the typical Reynolds number is about 0.5.

In one of the case studies presented in this paper we account for the shear-thinning viscosity behavior of blood at low shear rates and consider a Generalized Newtonian model for blood. To simplify, blood flow is supposed to be steady and

Fig. 1 Example: two-dimensional domain

the interaction with the vessel wall is neglected. The model reads as follows: let the vector function **u** and the scalar function p represent the blood velocity and pressure, respectively. Both quantities satisfy the momentum and mass balance equations:

$$\begin{cases} -div\,\tau(D\mathbf{u}) + \rho(\mathbf{u}\cdot\nabla)\mathbf{u} + \nabla p = \mathbf{f} & \text{in } \Omega \\ div\,\mathbf{u} = 0 & \text{in } \Omega \\ \mathbf{u} = 0 & \text{on } \Gamma_{wall} \\ \mathbf{u} = \mathbf{g} & \text{on } \Gamma_{in} \\ \sigma\cdot\mathbf{n} = 0 & \text{on } \Gamma_{out}. \end{cases} \tag{1}$$

Here Ω represents part of an artery truncated by two artificial sections, the inflow and outflow boundaries (Fig. 1). The vector function **g** describes the velocity profile at the inflow boundary Γ_{in}. The Cauchy stress tensor is represented by σ and **n** is the unitary vector normal to the outflow boundary surface Γ_{out}.

We consider a homogeneous Dirichlet boundary condition on the vessel wall Γ_{wall} and a homogeneous Neumann boundary condition on the outflow boundary Γ_{out}. The constant parameter ρ represents blood density and **f** is the body force which we neglect by taking $\mathbf{f} = \mathbf{0}$. The tensor of viscous stresses is represented by

$$\tau = 2\mu(\dot{\gamma})D\mathbf{u}, \tag{2}$$

where μ is the dynamic viscosity, $\dot{\gamma}$ refers to the shear rate

$$\dot{\gamma} = \sqrt{\frac{1}{2}(\nabla\mathbf{u} + (\nabla\mathbf{u})^T):(\nabla\mathbf{u} + (\nabla\mathbf{u})^T)} = \sqrt{2}|D\mathbf{u}|$$

(the symbol ":" represents the inner product of two second-order tensors) and D is the strain rate tensor

$$D\mathbf{u} = \frac{1}{2}(\nabla\mathbf{u} + (\nabla\mathbf{u})^T),$$

corresponding to the symmetric part of the velocity gradient.

When μ is constant (and does not depend on the shear rate), the extra-stress tensor τ defined by (2) is proportional to the strain rate tensor D, and (1) becomes the incompressible Navier-Stokes system, modeling blood flow in large vessels and healthy conditions.

Viscosity functions with bounded and non-zero limiting values of viscosity can be written in the general form

$$\mu(\dot{\gamma}) = \mu_\infty + (\mu_0 - \mu_\infty)F(\dot{\gamma}). \tag{3}$$

where $F(\dot{\gamma})$ is a shear dependent function, satisfying the limit conditions

$$\lim_{\dot{\gamma} \to 0} F(\dot{\gamma}) = 1 \quad \text{and} \quad \lim_{\dot{\gamma} \to +\infty} F(\dot{\gamma}) = 0.$$

Hence, the viscosity ranges from μ_0, when $\dot{\gamma}$ tends to zero, to μ_∞ when $\dot{\gamma}$ goes to infinity.

Different choices of $F(\dot{\gamma})$, with μ_0 higher than μ_∞ correspond to different shear-thinning models for blood flow, with material constants quite sensitive and depending on a number of factors including hematocrit, temperature, plasma viscosity, age of RBCs, exercise level, gender or health conditions [18]. The most common is the so-called Generalized Cross model, where

$$F(\dot{\gamma}) = \frac{1}{(1 + (\lambda \dot{\gamma})^b)^a}, \quad a, b, \lambda > 0 \tag{4}$$

and the Carreau model, corresponding to

$$F(\dot{\gamma}) = (1 + (\lambda \dot{\gamma})^2)^{\frac{n-1}{2}}, \quad \lambda > 0, \quad 0 < n < 1. \tag{5}$$

Here we only consider the Carreau viscosity model (see [3] for optimal control results using the Generalized Cross model).

We follow in this work the DA approach proposed in [3] for the 2D case. It consists in looking for the control function $\mathbf{g}$ such that the following cost functional

$$J(\mathbf{u}, \mathbf{g}) = \beta_1 \int_{\Omega_{part}} |\mathbf{u} - \mathbf{u}_d|^2 \, dx + \beta_2 \int_{\Gamma_{in}} |\nabla_s \mathbf{g}|^2 \, ds, \tag{6}$$

will be minimized. Here $\mathbf{u}$ is the solution of (1) corresponding to $\mathbf{g}$, and $\mathbf{u}_d$ represents the data available only on a part of the domain called Ω_{part}. The first term in J is the kernel term to be minimized, representing a misfit between the solution $\mathbf{u}$ and the data $\mathbf{u}_d$; the second term in J is the regularizing term that adds some convexity to the cost function with respect to the dependence on the control parameter represented by $\mathbf{g}$, defined on Γ_{in}. By fixing parameters β_1 and β_2, it is possible to decide whether the minimization of J should emphasize a good approximation of the velocity vector to $\mathbf{u}_d$ or a smoother control measured by the norm of the tangential derivative $\nabla_s(.)$.

The above problem is a particular case of the broader class of variational problems consisting of different choices of the functional J. We remark that in [2, 19], for the Newtonian case, a Neumann control of the type

$$[-p\mathbf{I} + \mu(\nabla\mathbf{u} + (\nabla\mathbf{u})^T)]\mathbf{n} = -g\mathbf{n} \tag{7}$$

was considered on Γ_{in}. In Sect. 3 we shall present, for the non-Newtonian case, results where the Dirichlet velocity control is compared with the Neumann pressure control.

2.1 Mathematical Analysis

The well-posedness of several control problems has been studied for shear-thinning models (see, e.g., [20, 21]). In the Newtonian case some results have also been proved for problem (1)–(6) (see [22]).

Let $\Gamma \subset \partial\Omega$ and

$$\mathbf{H}_0^1(\Gamma) = \left\{ \mathbf{v} \in \mathbf{L}^2(\Gamma) \mid \nabla_s\mathbf{v} \in \mathbf{L}^2(\Gamma), \ \gamma_{\partial\Gamma}\mathbf{v} = \mathbf{0} \right\}.$$

At the inflow boundary we consider a vector function $\mathbf{g} \in \mathcal{U}$ where

$$\mathcal{U} = \left\{ \mathbf{g} \in \mathbf{H}_0^1(\Gamma_{in}) : \text{such that (1) has a unique weak solution} \right\}.$$

We remark that $\mathcal{U}$ is not an empty set, since we can consider, for instance, $\mathbf{g}$ such that $\|\mathbf{g}\|_{H_0^1(\Gamma)} \leq \delta$, for certain δ small enough.

Now consider $(\Omega_{p_i})_i$ to be a monotone sequence of subsets of Ω, such that

$$\Omega_{p_1} \subset \Omega_{p_2} \ldots \subset \Omega_{p_m} \subset \Omega. \tag{8}$$

In addition, assume also that for each $i \in \{1, \ldots, m\}$, we have

$$\partial\Omega_{p_i} = \Gamma_{in} \cup \Gamma_{wall_i} \cup \Gamma_{out_i}$$

where Γ_{out_i} are disjoint surfaces corresponding to cross sections of Ω, and Γ_{wall_i} are nonempty wall segments such that $\Gamma_{wall_i} \subset \Gamma_{wall}$. Note that the construction of each Ω_{p_i} in this way ensures that the inclusions (8) are verified and that each Ω_{p_i} represents a part of the vessel Ω. Therefore, each Γ_{out_i} is in fact a cross section of Ω.

We can now state the following result, as a consequence of Theorem 4.5 [22]:

Corollary 2.1 *Let $\beta_1, \beta_2 > 0$ and assume that the data $\mathbf{u}_d$ is known in a part of the domain given by $\Omega_{part} = \cup_{i=1}^m s_i$ where $s_i = \Gamma_{out_i}$, for all $i \in \{1, \ldots, m\}$. Then there is an optimal solution $(\mathbf{u}, \mathbf{g}) \in \mathbf{H}^1(\Omega) \times \mathcal{U}$ to problem (1)–(6).*

2.2 Numerical Approximation

In this section we describe the numerical algorithm to solve problem (1)–(6). It is based on the Discretize then Optimize (DO) direct approach which consists in first discretizing the optimal control problem and then solving the finite dimensional optimization problem resulting from the discretization. Instead of the (DO) approach it is often used the Optimize then Discretize (OD) approach, but it is not yet clear which one of them is the most suitable for fluid control problems, as already referred.

Assume that we look for $\mathbf{u} \in \mathbf{H}^1(\Omega)$ and $p \in L_0^2(\Omega)$. We consider $\mathbf{V} = \{\mathbf{v} \in \mathbf{H}^1(\Omega) : \mathbf{v}|_{\partial\Omega_D} = 0\}$, where $\partial\Omega_D = \Gamma_{in} \cup \Gamma_{wall}$, $\mathbf{Q} = L_0^2(\Omega)$, the spaces of test functions corresponding to $\mathbf{u}$ and p, respectively. The weak formulation of (1) can be formally obtained by multiplying both equations by suitable test functions and integrating by parts, as follows:

$$\begin{cases} \int_\Omega \tau(D\mathbf{u}) : \nabla\mathbf{v}\,dx + \int_\Omega (\rho(\mathbf{u} \cdot \nabla)\mathbf{u}) \cdot \mathbf{v}\,dx - \int_\Omega p\,div\,\mathbf{v}\,dx = \int_\Omega \mathbf{f} \cdot \mathbf{v}\,dx \\ \int_\Omega q\,div\,\mathbf{u}\,dx = 0 \end{cases}$$

(9)

for all $\mathbf{v} \in \mathbf{V}$ and $q \in \mathbf{Q}$.

For the discretization of these equations we consider the finite dimensional approximations

$$\mathbf{u}_h = \sum_{j=1}^{N_u} u_j\phi_j \in \mathbf{V}_h, \quad p_h = \sum_{k=1}^{N_p} p_k\psi_k \in \mathbf{Q}_h$$

(10)

Here, $\mathbf{V}_h$ and $\mathbf{Q}_h$ are finite dimensional subspaces of $\mathbf{V}$ and $\mathbf{Q}$, such that $dim(\mathbf{V}_h) = N_u$ and $dim(\mathbf{Q}_h) = N_p$, ϕ_j and ψ_k are the shape functions, basis of $\mathbf{V}_h$ and $\mathbf{Q}_h$, respectively, and u_j and p_k are the corresponding unknown coefficients. $\mathbf{V}_h$ and $\mathbf{Q}_h$ are Lagrange type Finite Elements spaces, associated to a partition $\mathcal{T}_h$ of $\overline{\Omega}$, where $h > 0$ is the discretization parameter.

The discrete problem is written as: find $\mathbf{u}_h \in \mathbf{V}_h$ and $p_h \in \mathbf{Q}_h$ such that

$$\begin{cases} \int_\Omega \tau(D\mathbf{u}_h) : \nabla\mathbf{v}_h\,dx + \int_\Omega (\rho(\mathbf{u}_h \cdot \nabla)\mathbf{u}_h) \cdot \mathbf{v}_h\,dx - \int_\Omega p_h\,div\,\mathbf{v}_h\,dx \\ = \int_\Omega \mathbf{f} \cdot \mathbf{v}_h\,dx, \quad \forall v_h \in V_{0,h} \\ \int_\Omega q_h\,div\,\mathbf{u}_h\,dx = 0, \quad \forall q_h \in Q_h. \end{cases}$$

(11)

Since this is a convected dominated problem, a GLS (Galerkin-Least-Squares) stabilization is adopted. We refer to [23, 24] for details on the numerical analysis of the finite element discrete problem (11).

Let us consider the bilinear forms

$$a(\mathbf{u}_h, \mathbf{v}_h) = \int_\Omega \tau(D\mathbf{u}_h) : \nabla \mathbf{v}_h + \int_\Omega (\rho(\mathbf{u}_h \cdot \nabla)\mathbf{u}_h) \cdot \mathbf{v}_h - \int_\Omega p_h \, div \, \mathbf{v}_h$$

and

$$b(\mathbf{u}_h, q_h) = \int_\Omega q_h \, div \, \mathbf{u}_h.$$

Under this notation, the stabilization of system (11) consists in finding $\mathbf{u}_h \in \mathbf{V}_h$ and $p_h \in \mathbf{Q}_h$ by adding new terms $\mathcal{L}_h^1$ and $\mathcal{L}_h^2$ such that

$$\begin{cases} a(\mathbf{u}_h, \mathbf{v}_h) + \mathcal{L}_h^1(\mathbf{u}_h, \mathbf{f}, \mathbf{v}_h) = (\mathbf{f}, \mathbf{v}_h) \\ b(\mathbf{u}_h, q_h) = \mathcal{L}_h^2(p_h, q_h) \end{cases} \tag{12}$$

where

$$\mathcal{L}_h^1(\mathbf{u}_h, \mathbf{f}, \mathbf{v}_h) = \sum_{K \in \tau_h} (L(\mathbf{u}_h, p_h) - \mathbf{f}, \varphi(\mathbf{u}_h, \mathbf{v}_h))$$

and

$$\mathcal{L}_h^2(p_h, q_h) = \left(-\frac{1}{\lambda} p_h, q_h\right).$$

so that $\mathcal{L}_h^1$ verifies

$$\mathcal{L}_h^1(\mathbf{u}_h, \mathbf{f}, \mathbf{v}_h) = 0. \tag{13}$$

Here, λ is a penalty parameter (see [25]) and L and φ are given by

$$L(\mathbf{u}, p) = -div \, \tau(D(\mathbf{u})) + (\mathbf{u} \cdot \nabla)\mathbf{u} + \nabla p$$

$$\varphi(\mathbf{u}_h, \mathbf{v}_h) = \delta((\mathbf{u}_h \cdot \nabla)\mathbf{v}_h + div \, \tau(D(\mathbf{v}_h))).$$

The parameter δ should be suitably chosen. In this work, the parameter is taken from [26] (see also [27] for more details) and can be optimized in the frame of optimal control problems [12].

In order to solve system (12), we first look to the diffusion term of $a(\mathbf{y_h}, \mathbf{v_h})$ and its counterpart in $\mathcal{L}_h^1(\mathbf{u}_h, \mathbf{f}, \mathbf{v}_h)$

$$\int_\Omega \tau(D\mathbf{u}_h) : \nabla\mathbf{v}_h + \sum_{K\in\mathcal{T}_h} \int_K -div\,\tau(D\mathbf{u}_h) \cdot \varphi(\mathbf{u}_h, \mathbf{v}_h).$$

Using the approximations (10) we can write this term in the matrix form, as follows (see [3]):

$$\left(\int_\Omega 2\mu \left(\left| \sum_{j=1}^{N_u} u_j D\phi_j \right| \right) \sum_{k=1}^{N_u} u_k D\phi_k : \nabla\phi_i \right)$$

$$+ \sum_{K\in\mathcal{T}_h} \int_K -div\,\tau \left(D \left(\sum_{j=1}^{N_u} u_j D\phi_j \right) \right) \cdot \left(\delta \sum_{i=1}^{N_u} u_i\phi_i \cdot \nabla\phi_i \right)$$

$$= (Q + \mathcal{Q})(\mathbf{U}), \quad \forall i, j = 1 \dots N_u$$

where $\mathbf{U} = (u_1, \dots, u_{N_u})^T$, and Q, $\mathcal{Q}$ are square matrices of order N_u.

Let us now consider the convective term and its counterpart in $\mathcal{L}_h^1(\mathbf{u}_h, \mathbf{f}, \mathbf{v}_h)$:

$$\int_\Omega ((\mathbf{u}_h \cdot \nabla)\mathbf{u}_h) \cdot \mathbf{v}_h + \sum_{K\in\mathcal{T}_h} \int_K ((\mathbf{u}_h \cdot \nabla)\mathbf{u}_h) \cdot \varphi(\mathbf{u}_h, \mathbf{v}_h).$$

Using again approximations (10), the above expression can be written as

$$\left(\sum_{j=1}^{N_u} u_j \sum_{k=1}^{N_u} u_k \int_\Omega (\phi_j \cdot \nabla)\phi_k \cdot \phi_i \right)_{i=1,\dots,N_u} + \left(\sum_{K\in\tau_h} \sum_{j=1}^{N_u} u_j \sum_{k-1}^{N_u} u_k \int_K (\phi_j \cdot \nabla)\phi_k \cdot \right.$$

$$\left. \left(\delta \left(\sum_{l=1}^{N_u} u_l\phi_l \cdot \nabla\phi_i + div\,\tau(D(\phi_i)) \right) \right) \right)_{i=1,\dots,N_u} = (N(\mathbf{U}) + \mathcal{N}(\mathbf{U}))\mathbf{U},$$

where $N(\mathbf{U})$ and $\mathcal{N}(\mathbf{U})$ are square matrices of order N_u.

We act in a similar way on the pressure and define the pressure term $(B^T + \mathcal{B})\mathbf{P}$ where

$$[B^T]_{i,j} = \int_\Omega \psi_j div\,\phi_i \quad \forall i = 1, \dots, N_u; \ j = 1, \dots, N_p$$

$$[\mathcal{B}]_{i,j} = \sum_{K \in \mathcal{T}_h} \int_K \nabla \psi_j \cdot \left(\delta \left(\int_\Omega \psi_j \sum_{l=1}^{N_u} u_l \phi_l \cdot \nabla \phi_i + div\, \tau(D(\phi_i)) \right) \right)$$

$$\forall i = 1, \ldots, N_u; \ j = 1, \ldots, N_p$$

Hence, adding the discretized terms of $\mathcal{L}_1$ and $\mathcal{L}_2$, system (12) becomes

$$\begin{cases} (Q + \mathcal{Q})(\mathbf{U}) + (N(U) + \mathcal{N}(\mathbf{U}))\mathbf{U} + (B^T + \mathcal{B})P = F \\ B\mathbf{U} = \mathcal{B}_1 P + \text{boundary conditions} \end{cases} \quad (14)$$

where

$$[\mathcal{B}_1]_{i,j} = - \sum_{K \in \tau_h} \int_K \frac{1}{\lambda} \psi_i \psi_j \quad \forall i, j = 1, \ldots, N_p.$$

As for the cost function (6) we assume that $\mathbf{V}_h$ is associated to a partition $\mathcal{T}_h$ of the domain Ω so that some of the basis functions $(\phi_i)_{i=1\ldots N_o}$ can be associated to the nodes on Ω_{part} and others to the nodes on Γ_{in}, which we denote by $(\phi_i)_{i=1\ldots N_g}$. Hence, similarly to $\mathbf{u}_h$, the approximated control function is defined by

$$\mathbf{g}_h = \sum_{j=1}^{N_g} g_j \phi_j,$$

where the coefficients g_j correspond to the velocity coefficients u_j associated to Γ_{in}.

To discretize the first term we replace both $\mathbf{u}$ and $\mathbf{u}_d$ by their respective finite dimensional approximations $\mathbf{u}_h$ and $\mathbf{u}_{d,h}$. The latter is given by

$$\mathbf{u}_{d,h} = \sum_{i=1}^{N_o} u_{d_i} \phi_i.$$

Then, we obtain

$$\int_{\Omega_{part}} \left\langle \sum_i^{N_o} (u_i - u_{di}) \phi_i, \sum_j^{N_o} (u_j - u_{dj}) \phi_j \right\rangle dx$$

$$= \int_{\Omega_{part}} \sum_i^{N_o} (u_i - u_{di}) \sum_j^{N_o} (u_j - u_{di}) \langle \phi_i, \phi_j \rangle dx$$

$$= \sum_{i}^{N_o}(u_i - u_{di}) \sum_{j}^{N_o}(u_j - u_{dj}) \int_{\Omega_{part}} \phi_i \phi_j \, dx$$

$$= (\mathbf{U} - \mathbf{U}_d)^T \mathbf{M}(\mathbf{U} - \mathbf{U}_d) = \langle (\mathbf{U} - \mathbf{U}_d), \mathbf{M}(\mathbf{U} - \mathbf{U}_d) \rangle$$

$$= (\mathbf{U} - \mathbf{U}_d, \mathbf{U} - \mathbf{U}_d)_M = \|\mathbf{U} - \mathbf{U}_d\|_{N_o}^2 \tag{15}$$

where $\| \cdot \|_{N_o}$ is the norm induced by the inner product $(\cdot, \cdot)_M$ and $\mathbf{M}$ is a symmetric $N_o \times N_o$ matrix with general term given by

$$m_{ij} = \int_{\Omega_{part}} \phi_i \phi_j \, dx \,.$$

Finally, for the regularization term we have

$$\int_{\Gamma_{in}} \left| \sum_{i}^{N_g} g_i \nabla \phi_i \right|^2 dx = \int_{\Gamma_{in}} \left\langle \sum_{i}^{N_g} g_i \nabla \phi_i, \sum_{j}^{N_g} g_j \nabla \phi_j \right\rangle dx$$

$$= \sum_{i}^{N_g} g_i \sum_{j}^{N_g} g_j \int_{\Gamma_{in}} \nabla \phi_i : \nabla \phi_j = \mathbf{G}^T \mathbf{A} \mathbf{G}$$

$$= \langle \mathbf{G}, \mathbf{A}\mathbf{G} \rangle = (\mathbf{G}, \mathbf{G})_A = \|\mathbf{G}\|_{N_g}^2 \tag{16}$$

where $\| \cdot \|_{N_g}$ is the norm induced by the inner product $(\cdot, \cdot)_A$ and $\mathbf{A}$ is a symmetric $N_g \times N_g$ matrix whose elements are defined by

$$a_{ij} = \int_{\Gamma_{in}} \nabla \phi_i : \nabla \phi_j \, dx \,.$$

Taking into account (14)–(16), the discrete version of the control problem (6)–(1) consists in minimizing the following discrete cost function:

$$J(\mathbf{U}, \mathbf{G}) = \beta_1 \|\mathbf{U} - \mathbf{U}_d\|_{N_o}^2 + \beta_2 \|\mathbf{G}\|_{N_g}^2 \tag{17}$$

subject to (14).

We remark that vector $\mathbf{U} = (\mathbf{U}_g, \mathbf{G})$ includes the controlled velocity coefficients $\mathbf{G}$ and the uncontrolled ones $\mathbf{U}_g$, which also depend on $\mathbf{G}$. Therefore, the stabilized problem can be written in the general form

$$\min_{\mathbf{G}} F(G) = J(\mathbf{U}(G), \mathbf{G}) \tag{18}$$

$$C(\mathbf{G}) \geq 0, \tag{19}$$

where (19) represents the problem constraints (14), including boundary conditions. System (18)–(19) represents a large scale finite dimensional optimization problem with nonlinear constraints and a quadratic cost. To solve this problem, we use the Sequential Quadratic Programming algorithm, as described in [17]. The algorithm is available in the SNOPT library [28] and was tested in several benchmark large scale problems. The iterative procedure requires the evaluation of $F(\mathbf{G})$ which, in turn, implies solving the nonlinear system (14), which is done by the damped Newton method, as described in [29].

We will now briefly describe the algorithm and refer to [17], for more details.

Let us assume that the solution $\mathbf{G}$ of (18)–(19) verifies the Karush-Kuhn-Tucker (KKT) optimality conditions

$$\mathcal{D}C(\mathbf{G})^T \lambda = \mathcal{D}F(\mathbf{G})$$
$$C(\mathbf{G})^T \lambda = 0$$
$$C(\mathbf{G}) \geq 0$$
$$\lambda \geq 0$$

where $\mathcal{D}F$ and $\mathcal{D}C$ are the gradients of F and C, respectively, and λ is the vector of the Lagrange multipliers.

If one is able to find a good initial estimate $\mathbf{G}_0$ (and the corresponding λ_0), close enough to the optimal $\mathbf{G}$, the following algorithm produces a sequence that is globally convergent [17].

We remark that step 2 of Algorithm 1, which concerns the solution of the linear quadratic problem, is implemented using the library SQOPT [28].

3 Results and Discussion

3.1 Controlling the Pressure Versus the Velocity Field at the Inlet

As mentioned in the previous section, it is interesting to compare our results obtained for the Dirichlet velocity control (approach (P1) of a vector function) with those of [2] for the Neumann pressure control of type (7) (approach (P2) of a scalar function). To this end, we first reproduce the results presented in [2] for an idealized 2D straight channel $\Omega = [0, 5] \times [-0.5, 0.5]$ with $\Gamma_{in} = \{0\} \times [-0.5, 0.5]$ and $\Gamma_{out} = \{5\} \times [-0.5, 0.5]$ (Fig. 1). The observations were assumed to correspond to the sections $\{1\} \times [-0.5, 0.5]$, $\{2.5\} \times [-0.5, 0.5]$ and $\{4\} \times [-0.5, 0.5]$. Taking $\mu = 1$, we consider the ground truth solution to be known exactly and given by $\mathbf{u} = 1 - 4y^2$ (in particular $\mathbf{u} = 0$ on $\Gamma_{wall} = \partial\Omega \setminus (\Gamma_{in} \cup \Gamma_{out})$).

To solve (P2), the cost function needs to be properly rewritten, and Algorithm 1 (presented in the previous section) can be applied in a similar way. In [2], the

---

Algorithm 1: SNOPT

---

while Optimality tolerance of KKT less than threshold **do**

    1- Determine a quasi-Newton approximation $\mathcal{H}_k$ for the Hessian of the modified Lagrangian

$$\mathcal{L}(\mathbf{G}, \mathbf{G}_k, \lambda_k) = F(\mathbf{G}) - \lambda_k^T [C(\mathbf{G}_k) - \mathbf{G}_k - \mathcal{D}C(\mathbf{G}_k)(\mathbf{G} - \mathbf{G}_k)].$$

    2- Solve the auxiliary Linear Quadratic problem

$$\min_C \mathcal{Q}(\mathbf{G}, \mathbf{G}_k, \lambda_k) = F(\mathbf{G}_k) + \mathcal{D}F^T(\mathbf{G}_k)(\mathbf{G} - \mathbf{G}_k) - \frac{1}{2}(\mathbf{G} - \mathbf{G}_k)^T \mathcal{H}_k(\mathbf{G} - \mathbf{G}_k)$$

$$\mathbf{G}_k + \mathcal{D}C(\mathbf{G}_k)(\mathbf{G} - \mathbf{G}_k) \geq 0 \qquad\qquad (20)$$

to obtain the intermediate iterate $(\bar{\mathbf{G}}_k, \bar{\lambda}_k, \bar{s}_k)$, where $\hat{s}_k$ is the vector of the slack variables associated to the linear constraints in (20).

    3- Compute $\alpha_{k+1} \in (0, 1]$ as the minimizer of the merit function

$$M_\gamma(\mathbf{G}, \lambda, s) = F(\mathbf{G}) + \lambda^T(C(\mathbf{G}) - s) + \frac{1}{2} \sum_{i=1}^m \gamma_i (C_i(\mathbf{G}) - s_i)^2$$

along the line

$$d(\alpha) = (\mathbf{G}_k, \lambda_k, s_k) + \alpha[(\bar{\mathbf{G}}_k, \bar{\lambda}_k, \bar{s}_k) - (\mathbf{G}_k, \lambda_k, s_k)],$$

where s_i, for $i = 1 \ldots m$, are the components of s and γ is a vector of penalty parameters (see [3] for details on how to choose γ).

    4- Set $(\mathbf{G}_{k+1}, \lambda_{k+1}, s_{k+1}) = d(\alpha_{k+1})$.

    5- Compute the optimality tolerance for the KKT conditions.

end while

---

weights in the cost function were set to be $\beta_1 = \frac{1}{2}$ and $\beta_2 = \frac{10^{-9}}{2}$, according to the Morozov Discrepancy Principle associated to a certain fixed signal-to-noise ratio (see, for instance, [30]).

We reproduced the results found in [2] to conclude that the controlled solution given by (P2) approximates the exact solution with an acceptable relative error of 0.00112, that is, of order $\approx 0.1\%$. For this reason, we used this percentage as a reference relative error to fix the weights β_1 and β_2 to compare results obtained with (P1) and (P2).

Let us extend the previous domain to obtain the curved vessel represented in Fig. 2 (left) and referred to as the ground truth domain.

We consider a parabolic profile at the inlet and solve system (1) associated to the incompressible Navier-Stokes equations to obtain the ground truth solution $\mathbf{u}_d$.

An unstructured mesh corresponding to 43K degrees of freedom (max $h = 1/20$) has been used. The nonlinear system was solved using the damped Newton's method, as mentioned in Sect. 2.

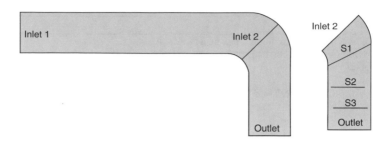

Fig. 2 Ground truth domain (left); working domain Ω with $\Omega_{part} = S_1 \cup S_2 \cup S_3$ (right)

Table 1 Relative errors $\mathrm{RE}_{\Omega_{part}, \beta_2}$, for both (P1) and (P2) approaches

β_2	$(P1)$	β_2	$(P2)$
0.5×10^{-2}	0.02992	0.5×10^{-5}	0.02350
0.5×10^{-3}	0.01005	0.5×10^{-6}	0.01757
0.5×10^{-4}	0.00219	0.5×10^{-7}	0.00978
2.5×10^{-5}	0.00118	0.5×10^{-8}	0.00412
0.5×10^{-5}	0.000259	0.5×10^{-9}	0.00129

We also considered a more realistic situation where the unknown inlet boundary condition did not correspond to a parabolic velocity profile, normal at Γ_{in}, nor to a pressure profile, that could be assumed axial dependent. For this reason, we truncated the channel on the section labeled *inlet 2*, which became the new artificial inlet Γ_{in} of the smaller domain Ω represented in Fig. 2 (right). Therefore, we want to fix a boundary condition at Γ_{in} so that the solution in Ω would match, as much as possible, with the ground truth solution $\mathbf{u}_d$. For the observations, we assumed to have measured exactly the velocity profiles of $\mathbf{u}_d$ at $\Omega_{part} = S1 \cup S2 \cup S3$, where $S1$, $S2$ and $S3$ are lines that were chosen arbitrarily inside Ω (Fig. 2, left). Before solving both problems (P1) and (P2), it was necessary to set β_1 and β_2. According to the example in [2], we fixed $\beta_1 = \frac{1}{2}$ and looked for β_2 so that the relative error

$$\mathrm{RE}_{\Omega_{part}, \beta_2} = \frac{\|\mathbf{u}_{\beta_2} - \mathbf{u}_d\|_{\mathbf{L}^2(\Omega_{part})}}{\|\mathbf{u}_d\|_{\mathbf{L}^2(\Omega_{part})}},$$

is ≈ 0.00112. In the expression of the relative error, $\mathbf{u}_{\beta_2}$ represents the solution of the control problem associated to β_2. This has been done by heuristically fixing a value for β_2 and evaluating the corresponding relative errors using Algorithm 1, with an optimality tolerance of 10^{-6}. The results are shown in Table 1.

From these conclusions $\beta_2 = 2.5 \times 10^{-5}$ has been fixed for (P1) and $\beta_2 = 0.5 \times 10^{-9}$ for (P2).

The ground truth solution is represented in the first row of Fig. 3. As expected, we can see that the pressure contours are not parallel to the cross sections in the curved part of the truncated channel, and the velocity profile loses the parabolic shape on those cross sections. The second and third rows (Fig. 3) represent the solutions obtained for (P1) and (P2), respectively.

Fig. 3 First row: ground truth velocity magnitude (m/s) (left) and ground truth pressure (right). Second row: controlled solution (P1)—velocity magnitude (m/s) (left) and pressure (Pa) (right). Third row: controlled solution (P2)—velocity magnitude (m/s) (left) and pressure (Pa) (right)

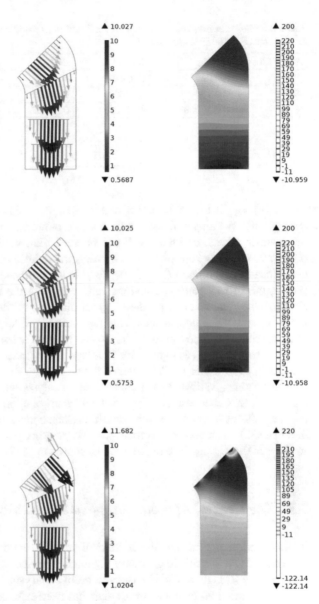

The results show that the solution obtained with the velocity control (P1) is qualitatively closer to the ground truth solution, represented in the first row of Fig. 3. To quantify these different performances, we use the relative error of the controlled solutions with respect to $\mathbf{u}_d$, evaluated at different sites. In Table 2 we present the values for

Table 2 Relative errors, final value of the cost function (J), and number of objective evaluations NE for both (P1) and (P2) approaches

Approach	RE_Ω	$RE_{\Gamma_{in}}$	$RE_{\Omega_{part}}$	Cost	NE
(P1)	0.00517	0.02286	0.00118	0.00953	230
(P2)	0.13443	0.57843	0.00129	$7.40043e^{-4}$	126
(P2r)	0.10068	0.47189	0.01757	0.03435	68

$$RE_\Omega = \frac{\|\mathbf{u} - \mathbf{u}_d\|_{\mathbf{L}^2(\Omega)}}{\|\mathbf{u}_d\|_{\mathbf{L}^2(\Omega)}},$$

where $\|\cdot\|_{\mathbf{L}^2(\Omega)}$ is the $\mathbf{L}^2(\Omega)$ norm, and for $RE_{\Gamma_{in}}$ and $RE_{\Omega_{part}}$, which are computed analogously. We also indicate the final value obtained for the cost functional and the number of cost evaluations. It can be seen that, while a relative error on the observations site is kept of the same order, the solution of (P1) is globally closer to $\mathbf{u}_d$ than the solution of (P2). Actually, looking closer to the later pressure profile (Fig. 3, third row), some oscillations can be seen at the inlet. This indicates that, although the relative error on the observations was of the order 0.1%, the weight $\beta_2 = 0.5 \times 10^{-9}$ almost canceled the regularizing effect of the second term in the cost function. An increase in β_2 improves the regularizing effect, but the desired relative error in Ω_{part} is higher. We illustrate this behavior by considering the case (P2r) with $\beta_2 = 0.5 \times 10^{-6}$, for which the results are shown in Table 2.

These values, validated by a convergence analysis with respect to mesh refinement, indicate better results using a velocity control approach. A higher number of cost evaluations might however be required. Such conclusion does not invalidate the fact that (P2) has a good performance, when pressure contours align with the cross sections in the region close to the inlet, as shown in [2, 31].

3.2 The Control Approach Applied to a Realistic Domain

In this section we present the numerical results found when applying the DA approach (P1) to a realistic geometry obtained from the segmentation of Computed Tomography (CT) data sets of a saccular brain aneurysm.

As in the previous example, the domain has been extended to compute the ground truth solution $\mathbf{u}_d$ used both to select the measured data and to estimate the accuracy of the method. The ground truth domain is represented in Fig. 4 (left).

As model parameters, we considered $\mu = \frac{\nu}{\rho}$ with $\nu = 3.67 \times 10^{-3}$ Pa s, a value within the range suggested in [32]. We also set $\rho = 1050$ kg/m^3 and fixed a laminar inflow profile—normal to the inlet—which corresponds to a flow rate of $Q = 4 \times 10^{-6}$ m^3/s. Again, these are typical parameters used for blood flow simulations in [32]. At the inlet these values imply a physiological Reynolds number $Re = 367$.

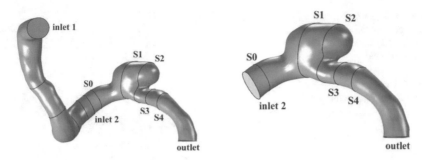

Fig. 4 Ground truth geometry (left); working geometry Ω (right)

No slip boundary conditions were imposed on the vessel wall and a homogeneous Neumann (zero normal stress) condition was fixed on the outflow boundary.

The discrete Navier-Stokes system (particular case of (11)) has been solved using P1-P1 finite elements and a GLS stabilizing method. First, we analyzed the case when the same degrees of freedom were used both to generate $\mathbf{u}_d$ and for the DA procedure. Subsequently, in order to avoid the so-called *inverse crime* problem, the ground truth solution was generated using a finer mesh. This is the usual strategy when the same model and discretization are used both to generate the synthetic data, from which the observations are chosen, and to solve the control (inverse) problem.

We consider Ω to be the subdomain starting in section *inlet 2*, which is shown in Fig. 4, on the right. We identify this section with Γ_{in} in problem (1)–(6). The goal is again finding a velocity boundary condition to use at this section in such a way that the corresponding solution matches $\mathbf{u}_d$. Additionally, we assume to have exact measurements of the velocity on $\Omega_{part} = S1 \cup S2 \cup S3 \cup S4$ where $S1$, $S2$, $S3$ and $S4$ are the sections represented in Fig. 4. The later assumption, concerning the exactness of the measurements, will be relaxed in the next study case.

Concerning the choice of the weights for the cost function, they are set to $(\beta_1, \beta_2) = (10^5, 10^{-3})$. This choice can be justified when the presence of noise on the observations is considered (see [16]).

To obtain the finite dimensional problem (18)–(19) the same type of finite elements has been used. The control problem was solved using Algorithm 1 with an optimality tolerance of 10^{-5}.

Table 3 shows in the first row the relative error of the controlled solution $\mathbf{u}$, with respect to the ground truth solution $\mathbf{u}_d$, evaluated in different parts of the domain. The relative error on Γ_{in} allows to measure the significant difference between the control vector and the ground truth solution at the artificial boundary, when compared to the relative errors RE_Ω and $RE_{\Omega_{part}}$ that show a very good accuracy in the working domain, and almost a perfect match in Ω_{part}.

To emphasize the gain achieved by the DA approach, we computed an alternative solution, $\mathbf{u}_Q$, based on the assumption that on Ω_{part} we can measure the exact flow rate instead of the velocity profile. The corresponding solution was obtained similarly to $\mathbf{u}_d$. The second row of Table 3 shows the relative errors of this alternative solution with respect to the true solution $\mathbf{u}_d$. It can be found that the

Table 3 Relative errors and final value of the cost function (J) for (P1) in the realistic domain

Solution	RE_Ω	$RE_{\Gamma_{in}}$	$RE_{\Omega_{part}}$	Cost
$\mathbf{u}$	0.019522	0.176382	0.002823	0.001037
$\mathbf{u}_Q$	0.225058	0.40286	0.25353	0.04693

Fig. 5 Bypass: computational domain

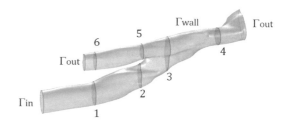

DA approach, resulting in $\mathbf{u}$, undergoes an error reduction, in the whole domain, from 22% to less than 2% when compared to the idealized solution $\mathbf{u}_Q$. In [16] a representation of the relative error, for the WSS magnitude, of both $\mathbf{u}$ and $\mathbf{u}_Q$ with respect to the ground truth solution $\mathbf{u}_d$, has also been shown. While for the solution $\mathbf{u}_Q$, based on the idealized laminar profile, the relative error is frequently above 40% and sometimes above 60%, the relative error associated to the controlled solution only reaches 10% close to the inlet, where we have chosen a coarse mesh. This indicates an important potential gain of the DA approach in reducing the error associated to WSS in silico measurements.

The robustness of the approach can be accessed with respect to the presence of noise in the data, increasing Reynolds numbers, or observation sites available. These aspects have been treated in [7].

3.3 Application to a Second Clinical Case

In this section, we present the results obtained with the Carreau viscosity model (5), using the parameters $\mu_0 = 0.0456\,\mathrm{Pa\,s}$; $\mu_\infty = 0.0032\,\mathrm{Pa\,s}$; $\lambda = 10.03\,\mathrm{s}$; $n = 0.344$ (see [18]).

We apply the proposed Data Assimilation approach to a computational domain representing the region of an artery including a bypass junction. The geometry was obtained from a Magnetic Resonance Imaging (MRI) data set. After segmentation, a NURBS based parametrization was used to generate the domain represented in Fig. 5.

To obtain the synthetic solution, we used stabilized P1-P1 finite elements, corresponding to 372,636 DOFS. As in the previous examples, at the inlet we imposed a parabolic profile corresponding to a flow rate $Q = 4 \times 10^{-6}\,\mathrm{m^3/s}$, a Reynolds number $Re = 355$ and a maximum velocity $U_0 = 0.378788\,\mathrm{m/s}$. On Γ_{out}, consisting of two boundaries, we have still imposed a homogeneous Neumann boundary condition. This choice is clearly not representative of the pulsatile flow arising in such type of pathological situation, but it serves the purpose of testing our method in different 3D models, with realistic Reynolds numbers and computational domains.

For the DA approach, we considered Ω_{part} as the set of cross sections represented in Fig. 5. The discretization of the control problem is based on the same finite elements used for the synthetic data. The control variable was discretized with 382 DOFS. Numerical tests have been performed for different pairs of parameters (β_1, β_2) of the cost functional. Here, we present the results for two significant pairs, $p_1 = (10^3, 10^{-3})$ and $p_2 = (10^4, 10^{-3})$. The objective is to highlight the improvement of the results as the weight assigned to the data misfit term β_1 increases from 10^3 to 10^4.

As in the previous cases, we also compare the controlled solution with the solution $\mathbf{u}_Q$, obtained by solving (14) using a constant velocity (averaged from the flow rate Q) at the inlet. Therefore, when comparing $\mathbf{u}$ and $\mathbf{u}_Q$, the differences do not depend on the flow rate, but rather on the velocity profile itself.

In Table 4 we present the computed errors between the synthetic solution $\mathbf{u}_d$ and both $\mathbf{u}_Q$, and the controlled solution $\mathbf{u}$.

Comparing the results obtained with the weighting parameters p_1 and p_2, it is clear that a higher weight provides lower error values and, consequently, better optimization results. The results indicate that the relative error can be reduced down to one-fifth, when using the DA procedure.

We have also performed several numerical tests, where the sections in Ω_{part} (see Fig. 5) have been changed, to identify the best location to collect data in order to obtain more accurate results. From Table 5, we observe that the errors obtained in the case where all sections in Ω_{part} are considered are very similar to those obtained for all sections, except sections 5 and 6. When considering all sections except section 4, we can see that the error is significantly higher. From these results, it is possible to conclude that cross sections 5 and 6 could be omitted in the numerical simulations. However, the same does not happen with section 4, for which a larger error indicates that this section is relevant in the simulations. This is an expected result, since the

Table 4 Bypass: relative errors of the controlled solutions $\mathbf{u}$ for both sets of parameters p_1 and p_2, with respect to $\mathbf{u}_d$ and $\mathbf{u}_Q$

Errors	$\mathbf{u}_Q$	Weights	$\mathbf{u}$
RE_Ω	0.122	p_1	0.15241
		p_2	0.03056
$RE_{\Omega_{part}}$	0.099241	p_1	0.151772
		p_2	0.021544
Cost	–	p_1	9.385×10^{-4}
		p_2	1.1×10^{-3}

Table 5 Bypass: relative errors obtained for different Ω_{part} choices, using parameters $p_2 = (10^4, 10^{-3})$

Weight	Errors	All sections	All sect. except 4	All sect. except 5 and 6
p_2	RE_Ω	0.03056	0.03382	0.03079
	$RE_{\Omega_{part}}$	0.021544	0.026353	0.021815
	Cost function	1.1×10^{-3}	1.1×10^{-3}	1.1×10^{-3}

cross sections 5 and 6 belong to a region where the flow rate is lower, and the velocity profile doesn't change significantly.

To verify that the previous conclusions were not a result of the so-called *inverse crime*, we also performed several experiments adding noise to the data observations. The noise was generated by adding to the synthetic solution, in Ω_{part}, a perturbation randomly generated. The noise was therefore taken along the normal distribution with zero mean and standard deviation given by $\bar{\sigma} = V \frac{U_0}{3}$. In this formula, U_0 represents the maximum velocity of $\mathbf{u}_d$ measured at the inlet boundary and $V \in \{0.1, 0.2, 0.4\}$. Such experiments are considered more realistic since, in real situations, data measurements are subject to the inherent noise associated with measurement devices. Table 6 shows the relative errors computed for both cases p_1 and p_2, and for different values of V. The low relative errors confirm what has been suggested in [2, 3], where it was shown that the variational DA approach can be robust to noisy data.

Finally, we present numerical results in the case where, besides the inflow velocity at Γ_{in}, we also control the velocity at one of the outflow boundaries. We chose the outflow boundary adjacent to cross-section 6, represented in Fig. 5. Since the regularization term has now a different weight in the cost function, a new set of parameters was numerically tested. In Table 7, we present the relative errors for the

Table 6 Bypass: relative errors computed in Ω_{part} obtained by including noise in the data, for both sets of parameters p_1 and p_2

Errors	Noise	Weights	u
$RE_{\Omega_{part}}$	10%	p_1	0.151756
		p_2	0.021476
	20%	p_1	0.15176
		p_2	0.021986
	40%	p_1	0.153731
		p_2	0.023047

Table 7 Bypass: control of two artificial boundaries

Errors	$\mathbf{u}_Q$	Weights	u
AE_Ω	3.16013×10^{-5}	p_1	9.96291×10^{-4}
		p_2	8.9355×10^{-6}
		p_3	4.3118×10^{-6}
RE_Ω	0.122	p_1	0.17056
		p_1	0.03056
		p_3	0.01699
$RE_{\Omega_{part}}$	0.099241	p_1	0.17644
		p_2	0.02748
		p_3	0.00419
Cost function	–	p_1	9.96291×10^{-4}
		p_2	1.25×10^{-3}
		p_3	1.3×10^{-3}

Relative errors of the controlled solutions **u** for both sets of parameters p_1 and p_2, with respect to $\mathbf{u}_d$ and $\mathbf{u}_Q$

parameters $p_1 = (10^3, 10^{-3})$, $p_2 = (10^4, 10^{-3})$ and $p_3 = (10^5, 10^{-3})$. It can be observed that, even in the case of two controlled boundaries, the Data Assimilation procedure can substantially reduce the relative errors when compared to the solution $\mathbf{u}_Q$, obtained from a known flow rate at Γ_{in}.

Acknowledgements This work was partially supported by FCT—Fundação para a Ciência e a Tecnologia through the project UID/Multi/04621/2013 of the CEMAT—Center for Computational and Stochastic Mathematics, IST, ULisboa—Portugal, project EXCL/MAT-NAN/0114/2012, and the grant SFRH/BPD/109574/2015.

References

1. B. Wang, X. Zou, J. Zhu, Data assimilation and its applications. Proc. Natl. Acad. Sci. USA **97**(21), 11143–11144 (2000)
2. M. D'Elia, A. Veneziani, Methods for assimilating blood velocity measures in hemodynamics simulations: preliminary results. Procedia Comput. Sci. **1**(1), 1225–1233 (2010)
3. T. Guerra, J. Tiago, A. Sequeira, Optimal control in blood flow simulations. Int. J. Non Linear Mech. **64**, 57–69 (2014)
4. L. Bertagna, A. Veneziani, A model reduction approach for the variational estimation of vascular compliance by solving an inverse fluid-structure interaction problem. Inverse Prob. **30**(5), 055006 (2014)
5. S. Pant, B. Fabreges, J.F. Gerbeau, I.E. Vignon-Clementel, A methodological paradigm for patient-specific multi-scale CFD simulations: from clinical measurements to parameter estimates for individual analysis. Int. J. Numer. Methods Biomed. Eng. **30**(12), 1614–1648 (2014)
6. A. Marsden, Optimization in cardiovascular modeling. Annu. Rev. Fluid Mech. **46**, 519–546 (2014)
7. J. Tiago, A. Gambaruto, A. Sequeira, Patient-specific blood flow simulations: setting Dirichlet boundary conditions for minimal error with respect to measured data. Math. Models Nat. Phenom. **9**(6), 98–116 (2014)
8. J.T. Betts, S.L. Campbell, Discretize then optimize. Technical Document Series, & CT-TECH-03–01. Mathematics and Computing Technology, Phantom Works, Boeing, Seattle, 2003
9. M. Hinze, F. Tröltzsch, Discrete concepts versus error analysis in PDE-constrained optimization. GAMM-Mitt. **33**(2), 148–162 (2010)
10. J. Burkardt, M. Gunzburger, J. Peterson, Insensitive functionals, inconsistent gradients, spurious minima and regularized functionals in flow optimization problems. Int. J. Comput. Fluid Dyn. **16**(3), 171–185 (2002)
11. M. Gunzburger, *Perspectives in Flow Control and Optimization* (SIAM, Philadelphia, 2003)
12. S. Collis, M. Heinkenschloss, Analysis of the streamline upwind/petrov galerkin method applied to the solution of optimal control problems. Tech. Rep. TR02-01. DCAM Rice University, Houston, 2002
13. M. Heinkenschloss, D. Leykekhman, Local error estimates for SUPG solutions of advection-dominated elliptic linear-quadratic optimal control problems. SIAM J. Numer. Anal. **47**(6), 4607–4638 (2010)
14. M. Gunzburger, S. Manservisi, The velocity tracking problem for Navier-Stokes flows with boundary control. SIAM J. Control Optim. **39**, 594–634 (2000)
15. K. Deckelnick, M. Hinze, Semidiscretization and error estimates for distributed control of the instationary Navier-Stokes equations. Numer. Math. **97**(2), 297–320 (2004)

16. J. Tiago, T. Guerra, A. Sequeira, A velocity tracking approach for the data assimilation problem in blood flow simulations. Int. J. Numer. Methods Biomed. Eng. (2017). https://doi.org/10.1002/cnm.2856
17. P. Gill, W. Murray, M.A. Saunders, SNOPT: an SQP algoritm for large-scale constrained optimization. SIAM Rev. **47**, 99–131 (2005)
18. A.M. Robertson, A. Sequeira, M. Kameneva, Hemorheology, in *Hemodynamical Flows: Modeling, Analysis and Simulation*, vol. 37 (Birkhäuser Verlag, Basel, 2008), pp. 63–120
19. M. D'Elia, A. Veneziani, Uncertainty quantification for data assimilation in a steady incompressible Navier-Stokes problem. ESAIM: Math. Model. Numer. Anal. **47**(4), 1037–1057 (2013)
20. N. Arada, Optimal control of shear-thinning fluids. SIAM J. Control Optim. **40**(4), 2515–2542 (2012)
21. T. Guerra, Distributed control for shear-thinning non-Newtonian fluids. J. Math. Fluid Mech. **14**(4), 771–789 (2012)
22. T. Guerra, A. Sequeira, J. Tiago, Existence of optimal boundary control for Navier-Stokes with mixed boundary conditions. Port. Math. **72**(2–3), 267–283 (2015)
23. J. Baranger, K. Najib, Analyse numérique des écoulements quasi-Newtoniens dont la viscosité obéit à la loi puissance ou la loi de Carreau. Numer. Math. **58**, 35–49 (1990)
24. J.W. Barrett, S.W. Liu, Finite element error analysis of a quasi-Newtonian flow obeying the Carreau or power law. Numer. Math. **64**(1), 433–453 (1993)
25. A. Brooks, T.H.J. Hughes, Streamline upwind/petrov-galerkin formulations for a convection dominated flows with a particular emphasis on the incompressible navier-stokes equations. Comput. Methods Appl. Mech. Eng. **32**, 199–259 (1982)
26. Y. Bazilevs, V. Calo, T. Tezduyar, T. Hughes, YZβ discontinuity capturing for advection-dominated processes with application to arterial drug delivery. Int. J. Numer. Meth. Fluids **54**, 593–608 (2007)
27. F. Shakib, T. Hughes, K.J. Zdeně, A new finite element formulation for computational fluid dynamics: X. The compressible Euler and Navier-Stokes equations. Comput. Methods Appl. Mech. Eng. **89**, 141–219 (1991)
28. P. Gill, W. Murray, M.A. Saunders, User's Guide for SNOPT Version 7: Software for Large-Scale Nonlinear Programming (2008)
29. P. Deuflhard, A modified newton method for the solution of ill-conditioned systems of nonlinear equations with application to multiple shooting. Numer. Math. **22**, 289–315 (1974)
30. K. Ito, K. Kunisch, On the choice of the regularization parameter in nonlinear inverse problems. SIAM J. Optim. **2**(3), 376–404 (1992)
31. M. D'Elia, A. Perego, A. Veneziani, A variational data assimilation procedure for the incompressible Navier-Stokes equations in hemodynamics. J. Sci. Comput. **52**(2), 340–359 (2011)
32. A. Gambaruto, J. Janela, A. Moura, A. Sequeira, Sensitivity of hemodynamics in a patient specific cerebral aneurysm to vascular geometry and blood rheology. Math. Biosci. Eng. **8**(2), 409–423 (2011)

Segmentation Techniques for Cardiovascular Modeling

A. A. Danilov, R. A. Pryamonosov, and A. S. Yurova

1 Introduction

In this paper we present methods and algorithms for construction of patient-specific discrete geometric models for cardiovascular biomedical applications. Each application imposes specific restrictions on both the input medical images and the output patient-specific discrete model, and, therefore, requires a specific class of 3D reconstruction methods.

Personalized modeling of cardiac hemodynamics received a great deal of attention, and a vast number of models have been described in the literature. Local hemodynamics modeling requires the patient-specific local reconstruction of coronary and cerebral arteries [1, 2]. Given an imaging dataset, one performs image segmentation, volume reconstruction, and numerical discretization.

Modeling of cardiac electrophysiology may be formalized as the full-scale study of the heart electrical activity from inner-cellular level to the cardiac tissues level [3]. The reconstruction of personalized anatomical model of the pathological heart is one of the crucial steps in electrophysiology modeling. The bidomain model requires an accurate anatomical model of patient heart and myocardium anisotropy structure. Patient-specific segmentation should be focused on the heart tissues as well as surrounding organs in the thorax and abdomen regions.

The cornerstone of medical image processing is the segmentation process that assigns labels to the voxels. Various medical image segmentation techniques have been developed [4–6]. The most promising fully automatic segmentation methods belong to atlas-based techniques. The patient-specific segmentation is obtained from the atlas of presegmented images. This atlas should contain enough different

A. A. Danilov (✉) · R. A. Pryamonosov · A. S. Yurova
Institute of Numerical Mathematics, Russian Academy of Sciences, Moscow, Russia

Moscow Institute of Physics and Technology, Dolgoprudny, Russia

© Springer International Publishing AG, part of Springer Nature 2018
R. P. Mondaini (ed.), *Trends in Biomathematics: Modeling, Optimization and Computational Problems*, https://doi.org/10.1007/978-3-319-91092-5_4

49

cases for accurate mapping of the new patient data. Thus atlas-based approach requires huge amount of segmentation expert work for preparation of atlases and the development of algorithms dealing with big data. Semi-automatic segmentation technologies require interaction with the operator. They are used primarily for precise local segmentation, where only one organ or tissue is processed. In our previous work we used several techniques for adaptation of the once segmented reference human model to different individuals. This technique relies on anthropometric scaling, control points mapping and supervised segmentation [7, 8].

In this work we introduce our previous patient-specific segmentation techniques and present in detail methods for segmentation and mesh generation of heart ventricles using dynamic contrast enhanced Computed Tomography (ceCT) images.

2   Methodology

Vascular segmentation techniques were addressed in detail in our previous works [9, 10]. We will briefly highlight the main steps of our pipeline. Input data are DICOM datasets obtained with contrast enhanced Computed Tomography Angiography (ceCTA). Essential steps of this method consist of aorta segmentation, computation of vesselness values, searching branches of aorta arch or ostia points, and removing segmentation errors near aorta boundary. We use fast variant of the isoperimetric distance trees algorithm [11] for aorta identification. The coronary arteries network is reconstructed by the use of Frangi vesselness filter [12], which is based on Hessian 3D analysis of the ceCTA image and is applicable to all tubular structures in the vascular dataset.

We examined several techniques for automatic segmentation of soft tissues, and developed methods for detailed segmentation of the heart [13], and automatic segmentation of surrounding tissues in the thorax [10]. In this work we will focus on segmentation and mesh generation for dynamic datasets and validation of the soft tissues segmentation.

2.1   Dynamic Cardiac Images Segmentation

We developed the technology for generation of a dynamic mesh for heart ventricles. In this work we focus on the left ventricle. We tested the proposed pipeline on the anonymized dynamic chest ceCT dataset of 100 images with $512 \times 512 \times 480$ voxels and $0.625 \times 0.625 \times 0.25$ mm resolution.

At the first stage we applied 3D non-local means smoothing [14], cropped and resampled the input images. Resulting smoothed images have $96 \times 96 \times 96$ voxels and $1.25 \times 1.25 \times 1$ mm resolution. We selected several images for manual segmentation at different stages of cardiac cycle: the beginning of systole (image #0), the end of systole (image #30), and the middle of rapid inflow during diastole (image #50). We used levelset method from ITK-SNAP package [15] for user-guided segmentation, and segmented four materials: left ventricle, left atrium, aorta, and right ventricle and atrium combined.

At the next stage we applied machine learning techniques to segment all images. We constructed the random forest classifier, trained on the manually segmented images. The result of classification is post-processed using a combination of mathematical operations: dilation, erosion, and construction of connected regions.

At the final stage we reconstruct the position of valve planes by the principal component analysis of the interfaces between left ventricle and left atrium, and between left ventricle and aorta. We compute the mean position of the valve planes across all images. We assume these planes will be fixed during the cardiac cycle for simplicity of mesh generation and numerical modeling.

We construct the unstructured tetrahedral mesh for the first image #0 using Delaunay triangulation from CGAL Mesh library [16]. The left ventricle domain is defined implicitly by segmented image, the valve planes are defined explicitly. We also split each tetrahedron with all four nodes lying on the boundary, enforcing at least one internal node in each tetrahedron. We deform the mesh by node movements for each subsequent image. At the first stage we move only boundary nodes simultaneously propagating and smoothing the surface mesh. We shift each boundary node in the direction of weighted sum of two vectors: vector along the surface normal towards the new position of the boundary surface (weight 0.2), and vector towards the center of surrounding nodes (weight 0.4). We repeat this procedure until the maximum movement distance drops below $\varepsilon = 0.001$ mm, or until the maximum number of 2000 iterations is exceeded. We pay special attention to the nodes on the valve planes: they should always stay on the planes (Fig. 1).

At the second stage we apply simultaneous untangling and smoothing algorithm [17, 18]; the boundary nodes are fixed, and only the internal nodes are shifted. As mentioned above, we enforced all tetrahedra to have at least one internal node, thus we greatly improved the robustness of untangling stage.

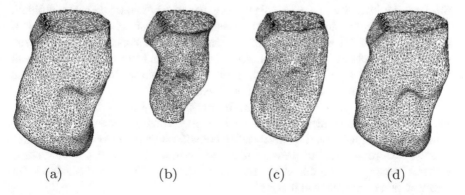

(a) (b) (c) (d)

Fig. 1 Surface triangular mesh of the left ventricle for several ceCT images. (**a**) Image #0. (**b**) Image #30. (**c**) Image #50. (**d**) Image #80

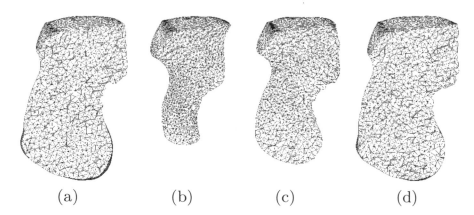

Fig. 2 Volume cross-sections of the tetrahedral mesh of the left ventricle for several ceCT images. (**a**) Image #0. (**b**) Image #30. (**c**) Image #50. (**d**) Image #80

As the final result we constructed the series of topologically invariant dynamic meshes for the left ventricle based on the dynamic ceCT images (Fig. 2) containing 14,033 mesh nodes and 69,257 tetrahedra.

2.2 Segmentation of Abdominal Parenchymal Organs

In this part, we propose a method for segmentation of abdominal parenchymal organs. The main steps of the algorithm are binary mask generation using analysis of CT texture features and further extraction of the 3D organ models.

The fully automatic segmentation of abdominal cavity is a complicated task because of several factors. First, there is a large anatomical variability of patients; second, medical images come from different devices and have different properties; third, similar intensity values for adjacent organs make boundary detection difficult. To partly overcome this problem, we considered contrast-enhanced CT as input data. Multiphase CT-scans are performed in order to enhance contrast between anatomical structures. The main phases of enhancement are as follows: without contrast (non-enhanced CT), arterial, portal and late phase. In our research, the smoothed CT-scans of the portal phase are used.

We propose a method that is robust to inter-patient gray level and anatomical variability. The main idea is based on two properties of parenchymal organs: homogeneous structure and relatively sharp boundaries on the images with contrast. We use texture analysis to measure these properties. In Ref. [19], some texture features were proposed for 2D image classification. We used these ideas for segmentation of 3D medical images.

Textural features are computed using $s \times s \times s$ voxel neighborhood, where s is an odd number greater than 1. In our study, we considered several textural features: contrast, inverse difference moment, second angular moment, etc. From all experimental results, the entropy was chosen as the most informative and easily processed:

$$ENT = -\sum_{i,j} p(i, j) \ln(p(i, j)), \tag{1}$$

where p is the spatial-dependence matrix [19].

This entropy property describes two important anatomical peculiarities used for segmentation process in this study: voxels inside of parenchymal organs have low entropy values because of the homogeneous structure and voxels of the boundary have high values because of diversity of intensities in the neighborhood (Fig. 3).

Figure 4 demonstrates the result of entropy computation for different neighborhood sizes ($s = 3, 5$) and type of adjacency used in computation of p. Experiments

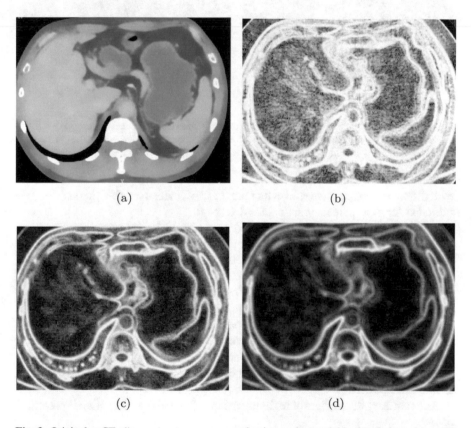

(a)　　　　　　　　　　(b)

(c)　　　　　　　　　　(d)

Fig. 3 Original ceCT slice and entropy computed using various neighborhood sizes. (**a**) ceCT slice. (**b**) $3 \times 3 \times 3$ vox neighborhood. (**c**) $5 \times 5 \times 5$ vox neighborhood. (**d**) $7 \times 7 \times 7$ vox neighborhood

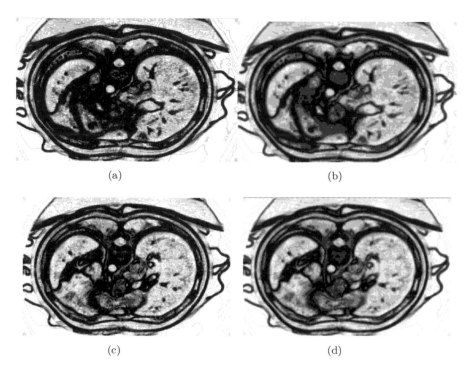

(a)   (b)

(c)   (d)

Fig. 4 Entropy calculation using various neighborhood sizes and adjacency types. (**a**) $3 \times 3 \times 3$ vox neighborhood with 6-adjacency. (**b**) $5 \times 5 \times 5$ vox neighborhood with 6-adjacency. (**c**) $3 \times 3 \times 3$ vox neighborhood with 26-adjacency. (**d**) $5 \times 5 \times 5$ vox neighborhood with 26-adjacency

with different datasets have shown that the $3 \times 3 \times 3$ voxel neighborhood ($s = 3$) is sufficient for most parenchymal organs detection.

In some rare cases the neighborhood size should be increased. In Fig. 5 we present several CT images with high inhomogeneity in the liver. The entropy values calculated with small neighborhood size are not applicable for liver segmentation, and a bigger neighborhood size should be used.

The second step of our algorithm is 3D volume extraction. Binary mask does not represent organs as separated components because of the multiple "leaks". We use active contours method for extraction of parenchymal organs. We compared several implementations of active contours methods [20, 21]. The most convenient one for our purposes is the level set function implemented in Convert3D tool from ITK-SNAP [15]. The coordinates of seed points can be found automatically using prior anatomical knowledge.

The summary of our algorithm for parenchymal organs segmentation is presented below:

1. Pre-process and smooth the input dataset.
2. Compute the spatial-dependence matrix (specified in Ref. [19]) for all voxels of the smoothed dataset.

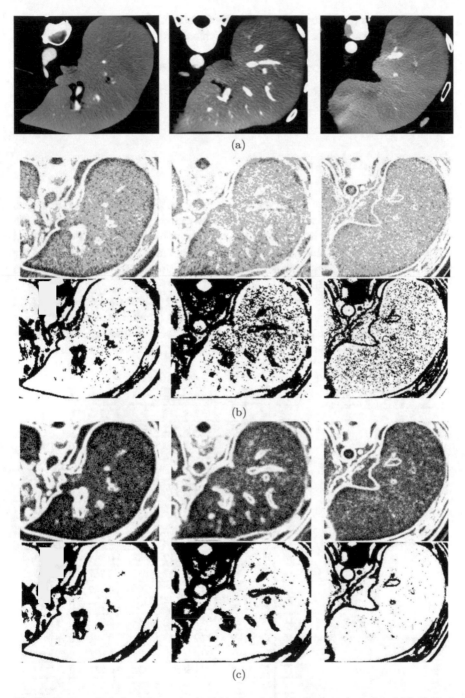

Fig. 5 Entropy calculation using various neighborhood sizes for liver segmentation. (**a**) Individual ceCT slices. (**b**) Entropy and binary mask with $3 \times 3 \times 3$ vox neighborhood. (**c**) Entropy and binary mask with $5 \times 5 \times 5$ vox neighborhood

3. Compute entropy (1) for each voxel using the spatial-dependence matrix.
4. Obtain the binary mask by entropy values thresholding.
5. Set the seed points for organs extraction.
6. Implement active contours method and extract 3D model.

This algorithm is especially useful for segmentation of a liver, which is one of the significant organs during ECG modeling.

3   Validation of Results

We evaluated the results of image segmentation using 20 anonymized ceCT images. Table 1 contains sex and age of each patient as well as general information about ceCT images. For each image four organs were segmented using the proposed texture-based method: spleen, stomach, gallbladder, and liver. We also included the results of automatic liver segmentation produced by proprietary software Aquarius iNtuition Client by TeraRecon company [22].

Each segmented organ was marked independently by three medical experts with one of four possible grades. Each expert counted the number of minor and major errors. Minor errors represent regions of incorrect voxels with insignificant volume compared to the total volume of the object. Major errors represent significant regions

Table 1 Patient and image information for validation test

N	Sex	Age	Dimensions	Resolution (mm)
1	M	34	$512 \times 512 \times 843$	$0.713 \times 0.713 \times 0.4$
2	M	67	$512 \times 512 \times 748$	$0.782 \times 0.782 \times 0.3$
3	F	54	$512 \times 512 \times 578$	$0.781 \times 0.781 \times 0.4$
4	F	47	$512 \times 512 \times 656$	$0.601 \times 0.601 \times 0.3$
5	M	74	$512 \times 512 \times 695$	$0.743 \times 0.743 \times 0.4$
6	M	45	$512 \times 512 \times 747$	$0.740 \times 0.740 \times 0.4$
7	F	78	$512 \times 512 \times 934$	$0.782 \times 0.782 \times 0.4$
8	M	82	$512 \times 512 \times 594$	$0.625 \times 0.625 \times 0.3$
9	M	68	$512 \times 512 \times 932$	$0.885 \times 0.885 \times 0.4$
10	F	75	$512 \times 512 \times 683$	$0.743 \times 0.743 \times 0.4$
11	F	78	$512 \times 512 \times 832$	$0.781 \times 0.781 \times 0.4$
12	M	43	$512 \times 512 \times 1000$	$0.782 \times 0.782 \times 0.4$
13	M	46	$512 \times 512 \times 468$	$0.743 \times 0.743 \times 0.4$
14	M	74	$512 \times 512 \times 638$	$0.751 \times 0.751 \times 0.5$
15	F	89	$512 \times 512 \times 378$	$0.782 \times 0.782 \times 0.5$
16	F	91	$512 \times 512 \times 956$	$0.675 \times 0.675 \times 0.4$
17	F	44	$512 \times 512 \times 832$	$0.781 \times 0.781 \times 0.4$
18	M	52	$512 \times 512 \times 662$	$0.763 \times 0.763 \times 0.3$
19	M	52	$512 \times 512 \times 563$	$0.782 \times 0.782 \times 0.8$
20	M	72	$512 \times 512 \times 763$	$0.907 \times 0.907 \times 0.3$

Table 2 Expert evaluation of segmented images by three medical experts and their average grades

N	Texture-based method				TeraRecon
	Spleen	Stomach	Gallbladder	Liver	Liver
1	1/1/1	2/3/3	1/1/1	3/3/2	4/3/4
2	1/1/1	3/3/3	1/1/1	2/1/1	4/4/3
3	1/1/1	–	1/1/1	2/1/2	2/2/2
4	1/1/1	–	1/1/1	2/1/1	3/3/3
5	1/1/1	3/3/3	1/1/1	1/1/2	3/3/3
6	1/1/1	2/3/2	2/1/1	2/1/2	2/2/2
7	1/1/1	3/3/3	2/1/1	1/1/1	2/2/3
8	1/1/1	1/2/1	1/1/1	1/2/2	2/3/4
9	1/1/1	3/3/3	1/1/1	2/2/2	3/2/3
10	1/1/1	3/2/3	1/1/1	1/1/1	2/2/2
11	1/1/1	1/2/2	1/1/1	1/2/1	3/3/4
12	1/1/1	3/3/3	1/1/1	4/4/4	2/2/3
13	1/2/1	3/3/3	1/1/1	1/2/2	2/2/2
14	1/1/1	3/3/1	1/1/1	3/3/3	2/2/2
15	1/1/1	3/3/2	1/1/1	1/1/1	3/3/3
16	1/1/1	1/1/1	1/1/1	1/1/2	2/2/2
17	1/1/1	3/3/3	1/1/1	2/1/2	3/3/3
18	1/1/1	3/3/2	1/1/1	2/2/2	2/3/3
19	1/1/1	3/3/3	1/1/1	1/1/1	3/3/3
20	1/1/1	3/4/3	1/1/1	2/2/2	3/3/3
Avg_1	1	2.56	1.1	1.75	2.6
Avg_2	1.05	2.78	1	1.65	2.6
Avg_3	1	2.44	1	1.8	2.85

The grades are explained in the text, the smaller the better

of incorrect voxels. Grade 1—*excellent* was used if no errors were observed, grade 2—*good* was used if 1–3 minor errors were observed, grade 3—*satisfactory* was used if 4–6 minor errors were observed or 1 major and 0–3 minor errors were observed, grade 4—*poor* was used in all other cases, e.g. two major errors or more than 6 minor errors were observed.

The evaluation results are presented in Table 2. We observe excellent results for segmentation of spleen and gallbladder due to their high contrast compared with the surrounding tissues. The good-to-satisfactory results of stomach segmentation are limited by inhomogeneity of the contents of the stomach. The liver segmentation by texture-based method provides good results with some minor errors due to similar texture of surrounding tissues. The automatic liver segmentation by TeraRecon software provides good-to-satisfactory results. Further validation tests should be conducted in order to ensure good segmentation quality.

4 Conclusions

We introduced several segmentation techniques for cardiovascular biomedical applications developed in our group. The detailed algorithms and corresponding results are presented in our previous papers [9, 10, 13]. A new technique for segmentation and mesh generation using dynamic ceCT images was proposed. The texture-based method of abdominal parenchymal organs segmentation was presented and validated on several ceCT images. The segmented images and constructed meshes are used in hemodynamics and electrophysiology modeling [7–9].

Acknowledgements This work has been supported by the Russian Science Foundation (RSF) grant 14-31-00024.

References

1. C.K. Zarins, C.A. Taylor, J.K. Min, J. Cardiovasc. Transl. Res. **6**, 708 (2013)
2. P.D. Morris, D. Ryan, A.C. Morton et al., JACC: Cardiovasc. Interv. **6**, 149 (2013)
3. G. Lines, M. Buist, P. Grottum et al., Comput. Visual. Sci. **5**, 215 (2002)
4. M. Holtzman-Gazit, R. Kimmel, N. Peled, D. Goldsher, IEEE Trans. Image Process. **15**, 354 (2006)
5. A.G. Radaelli, J. Peiró, Int. J. Numer. Methods Biomed. Eng. **26**, 3 (2010)
6. S.Y. Yeo, X. Xie, I. Sazonov, P. Nithiarasu, Int. J. Numer. Methods Biomed. Eng. **30**, 232 (2014)
7. A.A. Danilov, D.V. Nikolaev, S.G. Rudnev et al., Russ. J. Numer. Anal. Math. Model. **27**, 431 (2012)
8. Y.V. Vassilevski, A.A. Danilov, T.M. Gamilov et al., Russ. J. Numer. Anal. Math. Model. **30**, 185 (2015)
9. A. Danilov, Y. Ivanov, R. Pryamonosov, Y. Vassilevski, Int. J. Numer. Methods Biomed. Eng. **32**, e02754 (2016)
10. A. Danilov, R. Pryamonosov, A. Yurova, Computation **4**, 35 (2016)
11. L. Grady, *Computer Vision – ECCV 2006* (2006), p. 449
12. A.F. Frangi, W.J. Niessen, K.L. Vincken, M.A. Viergever, *Medical Image Computing and Computer-Assisted Intervention – MICCAI'98* (1998), p. 130
13. A.A. Danilov, R.A. Pryamonosov, A.S. Yurova, *ECCOMAS Congress 2016*, vol. 1 (2016), p. 454
14. A. Buades, B. Coll, J.M. Morel, Multiscale Model. Simul. **4**, 490 (2005)
15. P.A. Yushkevich, J. Piven, H.C. Hazlett et al., Neuroimage **31**, 1116 (2006)
16. L. Rineau, M. Yvinec, Comput. Geom. **38**, 100 (2007)
17. J.M. Escobar, E. Rodríguez, R. Montenegro et al., Comput. Methods Appl. Mech. Eng. **192**, 2775 (2003)
18. J.M. Escobar, E. Rodríguez, R. Montenegro et al., SUS code: simultaneous mesh untangling and smoothing code (2010). http://www.dca.iusiani.ulpgc.es/SUScode
19. R.M. Haralick, K. Shanmugam, I. Dinstein, IEEE Trans. Syst. Man Cybern. **3**, 610 (1973)
20. P. Marquez-Neila, L. Baumela, L. Alvarez, IEEE Trans. Pattern Anal. Mach. Intell. **36**, 2 (2014)
21. J. Sethian, *Level Set Methods and Fast Marching Methods: Evolving Interfaces in Computational Geometry, Fluid Mechanics, Computer Vision, and Materials Science* (Cambridge University Press, Cambridge, 1999)
22. TeraRecon, CT Body Advanced Visualization Package. http://www.terarecon.com/advanced-visualization/ct-body

Dynamics of an Infectious Disease Including Ectoparasites, Rodents and Humans

A. Dénes and G. Röst

1 Introduction

Ectoparasites are parasites that live on or in the skin but not within the body. These parasites, e.g. lice, fleas, mites have long been known as vectors of several infectious diseases including epidemic typhus and plague. It is also commonly known that in several cases, ectoparasites are transmitted to humans from animals, most often by rodents. A well-known example for this is plague, caused by the bacterium *Yersinia pestis*: the fleas transmitting this disease were transmitted to humans by rats [5]. Other notable examples are Omsk haemorrhagic fever, caused by a Flavivirus transmitted by ticks on water voles and muskrats [4]; rickettsialpox, caused by the bacteria *Rickettsia akari* transmitted by mites on mice [6]; murine typhus, caused by the bacteria *Rickettsia typhi*, transmitted by fleas, usually on rats [9]; scrub typhus caused by the parasite *Orientia tsutsugamushi*, transmitted by trombiculid mites, carried by mice. The latter disease is estimated to cause more than a million cases annually in Asia with more than a billion people being at risk, which makes scrub typhus the most medically important rickettsial disease [10] (Figs. 1 and 2).

In this work, we consider an infectious disease caused by a pathogen spread by ectoparasites which are harboured by rodents. We assume that ectoparasites spread by the rodents might be infectious or non-infectious. A given rodent or human can be infested only by one type (either infectious or non-infectious) of the ectoparasite. A human can be infested (and hence possibly infected) through adequate contact with an infested (infected) rodent or another human. We assume that the ectoparasites are not transmitted back from humans to the rodents. Due to infestation and/or treatment, infested and infected humans may become susceptible again.

A. Dénes (✉) · G. Röst
Bolyai Institute, University of Szeged, Szeged, Hungary
e-mail: denesa@math.u-szeged.hu

© Springer International Publishing AG, part of Springer Nature 2018
R. P. Mondaini (ed.), *Trends in Biomathematics: Modeling, Optimization and Computational Problems*, https://doi.org/10.1007/978-3-319-91092-5_5

Fig. 1 The pathogen can
jump from rodents to humans
via ectoparasites. Figure:
courtesy of Júlia Röst

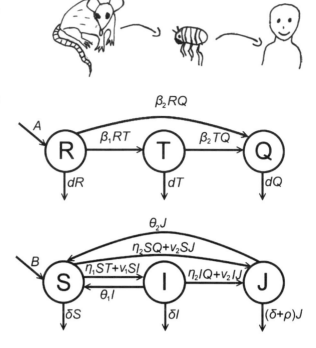

Fig. 2 Transmission diagram
representing transitions
between the rodent and the
human compartments

The structure of the paper is as follows. In Sect. 2, we establish a compartmental model describing the spread of the infestation and the disease. In Sect. 3, we study the subsystem formed by the equations for the rodent compartments, while, using the results of Sect. 3, we study the human subsystem in Sect. 4.

2 The Model

We denote by $R(t)$ the compartment of susceptible rodents, $T(t)$ stands for the rodents infested by non-infectious parasites, while $Q(t)$ denotes the number of rodents infested by infectious parasites. Similarly, we have three compartments for the humans: $S(t)$ denotes susceptibles, $I(t)$ those infested by non-infectious parasites, and $J(t)$ those infested by infectious parasites. A and d stand for the birth, resp. death rates of rodents. The notation β_1 stands for the transmission rate between the compartments R and T, while β_2 is the transmission rate between R and Q, resp. T and Q. B and δ stand for natural birth and death rates for humans, and ρ denotes disease-induced death rate for the infected human compartment J. The parameter ν_1 denotes transmission rate between the compartments S and I, while ν_2 denotes transmission rate from J to S and J to I. The parameter η_1 denotes the transmission rate from rodents infested by non-infectious parasites to susceptible humans, while

η_2 is the transmission rate from rodents infested by infectious parasites to humans. We denote by θ_1, resp. θ_2 the disinfestation, resp. recovery rate from compartments I, resp. J.

Using the above notations, our equations take the following form:

$$
\begin{aligned}
R'(t) &= A - \beta_1 R(t)T(t) - \beta_2 R(t)Q(t) - dR(t), \\
T'(t) &= \beta_1 R(t)T(t) - \beta_2 T(t)Q(t) - dT(t), \\
Q'(t) &= \beta_2 R(t)Q(t) + \beta_2 T(t)Q(t) - dQ(t), \\
S'(t) &= B - \eta_1 S(t)T(t) - \eta_2 S(t)Q(t) - v_1 S(t)I(t) - v_2 S(t)J(t) \\
&\quad - \delta S(t) + \theta_1 I(t) + \theta_2 J(t), \\
I'(t) &= \eta_1 S(t)T(t) + v_1 S(t)I(t) \\
&\quad - \eta_2 I(t)Q(t) - v_2 I(t)J(t) - \delta I(t) - \theta_1 I(t), \\
J'(t) &= \eta_2 S(t)Q(t) + \eta_2 I(t)Q(t) + v_2 S(t)J(t) + v_2 I(t)J(t) \\
&\quad - \delta J(t) - \rho J(t) - \theta_2 J(t),
\end{aligned}
\tag{1}
$$

with positive initial conditions $R(0), T(0), Q(0), S(0), I(0), J(0) \geq 0$. The phase space

$$
\mathbb{R}_+^6 = \{(R, T, Q, S, I, J) \in \mathbb{R}^6 : R, T, Q, S, I, J \geq 0\}
$$

is clearly invariant to system (1).

3 The Rodent Subsystem

3.1 Equilibria, Local Stability

The first three equations of (1) can be decoupled from the remaining ones. The subsystem for the spread among rodents, given by

$$
\begin{aligned}
R'(t) &= A - \beta_1 R(t)T(t) - \beta_2 R(t)Q(t) - dR(t), \\
T'(t) &= \beta_1 R(t)T(t) - \beta_2 T(t)Q(t) - dT(t), \\
Q'(t) &= \beta_2 R(t)Q(t) + \beta_2 T(t)Q(t) - dQ(t)
\end{aligned}
\tag{2}
$$

has a similar structure as the model given by Dénes and Röst [1, 2], though, in the present case, birth and death rates are not equal in contrast to the cited papers. To calculate the equilibria of the full system, we start by calculating those of the rodent subsystem (2), which are easily obtained by solving the algebraic system of equations

$$0 = A - \beta_1 RT - \beta_2 RQ - dR,$$

$$0 = \beta_1 RT - \beta_2 TQ - dT,$$

$$0 = \beta_2 RQ + \beta_2 TQ - dQ,$$

resulting in the four possible equilibria

$$E_R = \left(\tfrac{A}{d}, 0, 0\right), \qquad\qquad E_T = \left(\tfrac{d}{\beta_1}, \tfrac{A}{d} - \tfrac{d}{\beta_1}, 0\right)$$

$$E_Q = \left(\tfrac{d}{\beta_2}, 0, \tfrac{A}{d} - \tfrac{d}{\beta_2}\right), \qquad E_{TQ} = \left(\tfrac{A\beta_2}{d\beta_1}, \tfrac{d}{\beta_2} - \tfrac{A\beta_2}{d\beta_1}, \tfrac{A}{d} - \tfrac{d}{\beta_2}\right). \tag{3}$$

By introducing a single infested/infected individual into one of the equilibria E_R, E_T and E_Q, we obtain three different reproduction numbers. If we introduce a rodent infested by the non-infectious parasites into the disease- and infestation-free equilibrium, we obtain the reproduction number $r_1 = \frac{A\beta_1}{d^2}$.

Introducing a rodent infested by the infectious parasites into the equilibrium E_R, we obtain the reproduction number $r_2 = \frac{A\beta_2}{d^2}$.

If we introduce a rodent infested by the infectious parasites into the equilibrium E_T, we obtain again the same reproduction number r_2. Finally, let us introduce a rodent infested by the non-infectious parasites into the equilibrium E_Q. In this case, the expected sojourn time of an individual infected with the first strain in the T-compartment is $(\beta_2 Q^* + d)^{-1}$, and the number of new infections generated by this individual is $\beta_1 R^*$, where R^* and Q^* stand for the first, resp. third coordinates of the equilibrium E_Q. This way we obtain the reproduction number $r_3 = \frac{\beta_1 d^2}{\beta_2^2 A}$.

It is obvious that the equilibrium E_R always exists, E_T exists if and only if $r_1 > 1$, E_Q exists if and only if $r_2 > 1$, while E_{TQ} exists if and only if $r_2 > 1$ and $r_3 > 1$.

The following proposition on the local stability of the four equilibria can easily be checked, see [3].

Proposition 3.1 *The disease-free equilibrium E_R is locally asymptotically stable if $r_1 < 1$ and $r_2 < 1$ and unstable if $r_1 > 1$ or $r_2 > 1$. The equilibrium E_T is locally asymptotically stable if $r_1 > 1$ and $r_2 < 1$. The equilibrium E_Q is locally asymptotically stable if $r_2 > 1$ and $r_3 < 1$. The equilibrium E_{TQ} is locally asymptotically stable if $r_2 > 1$ and $r_3 > 1$.*

3.2   Persistence

Before we can state our results on the persistence of the three compartments, we will need some notions and theorems from [8].

Definition 3.1 Let X be a nonempty set and $\rho: X \to \mathbb{R}_+$. A semiflow $\Phi: \mathbb{R}_+ \times X \to X$ is called *uniformly weakly ρ-persistent*, if there exists some $\varepsilon > 0$ such that $\limsup_{t \to \infty} \rho(\Phi(t, x)) > \varepsilon$ for all $x \in X$, $\rho(x) > 0$. Φ is called *uniformly* *(strongly) ρ-persistent* if there exists some $\varepsilon > 0$ such that $\lim\inf_{t\to\infty} \rho(\Phi(t,x)) > \varepsilon$ for all $x \in X$, $\rho(x) > 0$. A set $M \subseteq X$ is called *weakly ρ-repelling* if there is no $x \in X$ such that $\rho(x) > 0$ and $\Phi(t,x) \to M$ as $t \to \infty$.

System (2) generates a continuous flow on the phase space

$$X := \left\{ (R, T, Q) \in \mathbb{R}_+^3 \right\}.$$

Theorem 3.1 *$R(t)$ is always uniformly persistent. $T(t)$ is uniformly persistent if $r_1 > 1$ and $r_2 < 1$ as well as if $r_2 > 1$ and $r_3 > 1$. $Q(t)$ is uniformly persistent if $r_2 > 1$.*

Proof To show uniform persistence of the susceptible compartment, we will use the method of fluctuation (see, e.g., [7, Lemma A.1]). We denote by R_∞ the limit inferior of $R(t)$, while T^∞ and Q^∞ denote the limit superior of $T(t)$, resp. $Q(t)$ as $t \to \infty$. Using the fluctuation lemma we know that there exists a time sequence $t_k \to \infty$ such that $R(t_k) \to R_\infty$ and $R'(t_k) \to 0$ as $k \to \infty$. If we apply this to the equation for $R(t)$, we obtain

$$R'(t_k) + \beta_1 R(t_k) T(t_k) + \beta_2 R(t_k) Q(t_k) = A.$$

It is easy to see that for the total rodent population we have $R(t) + T(t) + Q(t) \to \frac{A}{d}$, thus, $0 \le T^\infty \le \frac{A}{d}$ and $0 \le Q^\infty \le \frac{A}{d}$. Using this and letting $k \to \infty$ we get $R_\infty \ge \frac{d}{\beta_1 + \beta_2}$.

To show persistence of the infested compartments, we need some theory from [8]. We use the notation $x = (R, T, Q) \in X$ for the state of the system and the usual notation $\omega(x)$ for the ω-limit set of a point x defined as

$$\omega(x) := \{y \in X : \exists \{t_n\}_{n \ge 1} \text{ s. t. } t_n \to \infty \text{ and } \Phi(t_n, x) \to y \text{ as } n \to \infty\}.$$

We first show the persistence of $T(t)$. Let $\rho(x) = T$. Let us consider the invariant extinction space of T, defined as $X_T := \{x \in X : \rho(x) = 0\}$. We follow [8, Chapter 8] and examine the set $\Omega_{x \in X_T} := \cup_{x \in X_T} \omega(x)$. Applying the Bendixson–Dulac criterion with Dulac function $1/Q$ and the Poincaré–Bendixson theorem, we obtain that all solutions in the extinction space X_T tend to an equilibrium.

Let us first consider the case $r_1 > 1$ and $r_2 \le 1$. Clearly, in this case $\Omega = \{E_R\}$. As a first step, we prove weak ρ-persistence. In order to apply [8, Theorem 8.17], we let $M_1 = \{E_R\}$. Then Ω is a subset of M_1, which is isolated, compact, invariant and acyclic. We have to show that M_1 is weakly ρ-repelling, from which we obtain persistence.

Let us suppose that this does not hold, i.e. there exists a solution such that $\lim_{t\to\infty}(R(t), T(t), Q(t)) = (\frac{A}{d}, 0, 0)$ and $T(t) > 0$. Then for any $\varepsilon > 0$, for sufficiently large t, we have $R(t) > \frac{A}{d} - \varepsilon$ and $Q(t) < \varepsilon$. For such t, we can give the following estimation for $T'(t)$:

$$T'(t) = T(t)(\beta_1 R(t) - \beta_2 Q(t) - d) > T(t)\left(\beta_1 \tfrac{A}{d} - \beta_1 \varepsilon - \beta_2 \varepsilon - d\right),$$

which is positive if ε is sufficiently small as $\tfrac{A\beta_1}{d} > d$ follows from $r_1 > 1$. This contradicts $T(t) \to 0$.

In the second case, when $r_2 > 1$ and $r_3 > 1$, also E_Q exists, so we have $\Omega = \{E_R, E_Q\}$. Now we let $M_1 = \{E_R\}$ and $M_2 = \{E_Q\}$. Clearly, $\Omega \subset M_1 \cup M_2$ and $\{M_1, M_2\}$ is acyclic and M_1 and M_2 are invariant, compact and isolated. We have to show that M_1 and M_2 are weakly ρ-repelling.

Suppose first that M_1 is not weakly ρ-repelling. Then there exists a solution such that $\lim_{t\to\infty}(R(t), T(t), Q(t)) = (\tfrac{A}{d}, 0, 0)$ and $T(t) > 0$. Again, for any $\varepsilon > 0$, for sufficiently large t, we have $R(t) > \tfrac{A}{d}$ and $Q(t) < \varepsilon$ and for such t, we can give the following estimation for $T'(t)$:

$$T'(t) = T(t)(\beta_1 R(t) - \beta_2 Q(t) - d) > T(t)\left(\beta_2 \tfrac{A}{d} - \beta_1 \varepsilon - \beta_2 \varepsilon - d\right),$$

where we used that $\beta_1 > \beta_2$, which follows from $r_2 r_3 > 1$. This expression is positive for ε small enough, which contradicts $T(t) \to 0$.

Now let us suppose that M_2 is not weakly ρ-repelling. Then there exists a solution such that $\lim_{t\to\infty}(R(t), T(t), Q(t)) = \left(\tfrac{d}{\beta_2}, 0, \tfrac{A}{d} - \tfrac{d}{\beta_2}\right)$. Then, for any $\varepsilon > 0$, if t is large enough, then $R(t) > \tfrac{d}{\beta_2} - \varepsilon$ and $Q(t) < \tfrac{A}{d} - \tfrac{d}{\beta_2} + \varepsilon$ and for such t we can give the following estimation for $T'(t)$:

$$\begin{aligned}
T'(t) &= T(t)(\beta_1 R(t) - \beta_2 Q(t) - d) \\
&> T(t)\left(\tfrac{d\beta_1}{\beta_2} - \beta_1\varepsilon - \beta_2\left[\tfrac{A}{d} - \tfrac{d}{\beta_2} + \varepsilon\right] - d\right) \\
&= T(t)\left(\tfrac{d\beta_1}{\beta_2} - \tfrac{A\beta_2}{d} - (\beta_1 + \beta_2)\varepsilon\right),
\end{aligned}$$

which is positive for ε small enough as $r_3 > 1$. This contradicts $T(t) \to 0$.

Let us now turn to the persistence of $Q(t)$ in the case $r_2 > 1$. We set $\rho(x) = Q$. We have the equilibrium E_R if $r_1 \leq 1$ and the two equilibria E_R and E_T if $r_1 > 1$. Similarly to the case of $T(t)$, we define the extinction space of Q as $X_Q := \{x \in X : \rho(x) = 0\} = \{(R, T, 0) \in \mathbb{R}^3_+\}$. In this case we have $\Omega = \cup_{x \in X_Q}\omega(x) = \{E_R\}$ if $r_1 \leq 1$ and $\Omega = \cup_{x \in X_Q}\omega(x) = \{E_R, E_T\}$ if $r_1 > 1$. We define $M_1 = \{E_R\}$ and $M_2 = \{E_T\}$. Just like in the proof of the persistence of $T(t)$, Ω is invariant, and M_1 and M_2 are isolated and acyclic.

To show that M_1 is weakly ρ-repelling, we can proceed in an analogous way as in the case of $T(t)$.

In the case $r_1 > 1$, we have to show that M_2 is weakly ρ-repelling. Suppose this does not hold. Then there exists a solution such that $\lim_{t\to\infty}(R(t), T(t), Q(t)) = (\tfrac{d}{\beta_1}, \tfrac{A}{d} - \tfrac{d}{\beta_1}, 0)$ and $Q(t) > 0$. Then, for any $\varepsilon > 0$, if t is sufficiently large, then $R(t) > \tfrac{d}{\beta_1} - \varepsilon$ and $T(t) > \tfrac{A}{d} - \tfrac{d}{\beta_1} - \varepsilon$ and for such t we can give the following estimation for $Q(t)$:

$$Q'(t) = Q(t)(\beta_2 R(t) + \beta_2 T(t) - d)$$

$$> Q(t)\left(\beta_2\left(\tfrac{d}{\beta_1} - \varepsilon\right) + \beta_2\left(\tfrac{A}{d} - \tfrac{d}{\beta_1} - \varepsilon\right) - d\right)$$

$$= Q(t)\left(\tfrac{A\beta_2}{d} - d - 2\beta_2\varepsilon\right),$$

which is positive if ε is small enough, as $r_2 > 1$, which contradicts $Q(t) \to 0$.

We have shown uniform weak persistence in all cases; to show uniform (strong) persistence, we apply Theorem 4.5 from [8]. Our flow is clearly continuous, the subspaces $X_T, X_Q, X \setminus X_T$ and $X \setminus X_Q$ are invariant. The existence of a compact attractor is also clear, as all solutions enter a compact region after some time. This means that all conditions of [8, Theorem 4.5] hold and thus we obtain uniform strong persistence. □

3.3   Global Stability

Theorem 3.2

(1) Equilibrium E_R is globally asymptotically stable if $r_1 < 1$ and $r_2 < 1$.

(2) Equilibrium E_T is globally asymptotically stable on $X \setminus X_T$ if $r_1 > 1$ and $r_2 < 1$. E_R is globally asymptotically stable on X_T.

(3) Equilibrium E_Q is globally asymptotically stable on $X \setminus X_Q$ if $r_2 > 1$ and $r_3 < 1$. E_R is globally asymptotically stable on X_Q if $r_1 < 1$ and E_T is globally asymptotically stable on X_Q if $r_1 > 1$.

(4) Equilibrium E_{TQ} is globally asymptotically stable on $X \setminus (X_T \cup X_Q)$ if $r_2 > 1$ and $r_3 > 1$. E_T is globally asymptotically stable on X_Q and E_Q is globally asymptotically stable on X_T.

Proof First we note that the rodent subsystem (2) can be reduced to two dimensions by introducing the notation $F(t) := R(t) + T(t)$. We obtain the system

$$F'(t) = A - \beta_2 F(t)Q(t) - dF(t),$$
$$Q'(t) = \beta_2 F(t)Q(t) - dQ(t). \tag{4}$$

This system has two equilibria, $\left(\tfrac{A}{d}, 0\right)$ and $\left(\tfrac{d}{\beta_2}, \tfrac{A}{d} - \tfrac{d}{\beta_2}\right)$, with the latter one only existing if $r_2 > 1$. We use the Dulac function $1/Q$ to show that there is no periodic solution of (4):

$$\frac{\partial}{\partial F}\frac{A - \beta_2 FQ - dF}{Q} + \frac{\partial}{\partial Q}\frac{\beta_2 FQ - dQ}{Q} = -\beta_2 - \frac{d}{Q} < 0.$$

Thus, applying the Bendixson–Dulac criterion, we obtain that there is no periodic solution of (4), and by the Poincaré–Bendixson theorem we get that all solutions tend to an equilibrium.

In the first two cases, when $r_2 < 1$, only the first equilibrium exists. Thus, in this case $Q(t) \to 0$ and $F(t) \to \frac{A}{d}$ as $t \to \infty$, and therefore, the second equation of (2) takes the following form on the limit set:

$$T'(t) = \beta_1 \left(\frac{A}{d} - T(t)\right) T(t) - dT(t) = \gamma T(t) - \beta_1 T^2(t)$$

with $\gamma = \left(\frac{A\beta_1}{d} - d\right)$.

The solution started from $T(t) = 0$ is the constant solution $T(t) \equiv 0$, while the nontrivial solutions take the form

$$\frac{Ce^{\gamma t}}{1 + \frac{\beta_1}{\gamma} Ce^{\gamma t}} \tag{5}$$

for $C \in \mathbb{R}_+$. Clearly, for $r_1 < 1$ (which is equivalent to $\gamma \leq 0$), the solutions tend to zero on the limit set, therefore, for all solutions, $T(t) \to 0$ as $t \to \infty$.

If $r_1 > 0$ (i.e. $\gamma > 0$), we have $\lim_{t \to \infty} T(t) = \frac{A}{d} - \frac{d}{\beta_1}$ on the limit set; using the persistence of $T(t)$ we obtain that for all solutions, $T(t) \to \frac{A}{d} - \frac{d}{\beta_1}$ as $t \to \infty$.

In the case $r_2 > 1$, also the second equilibrium exists. However, we know from the previous subsection that for $r_2 > 1$, the compartment $Q(t)$ is uniformly persistent, so no solution with positive initial value in $Q(t)$ can tend to the first equilibrium. Thus, the limit of all such solutions is the second equilibrium and $Q(t) \to (\frac{A}{d} - \frac{d}{\beta_2})$ as $t \to \infty$. We can proceed in a similar way as in the case $r_2 < 1$: on the limit set, we can transform the second equation of (2) to

$$T'(t) = \beta_1 \left(\frac{d}{\beta_2} - T(t)\right) T(t) - \beta_2 \left(\frac{A}{d} - \frac{d}{\beta_2}\right) T(t) - dT(t)$$

$$= \gamma T(t) - \beta_1 T^2(t)$$

with $\gamma = \left(\frac{d\beta_1}{\beta_2} - \frac{A\beta_2}{d}\right)$. Similarly as above, we can see that the solution started from $T(t) = 0$ is the constant solution $T(t) \equiv 0$, while the nontrivial solutions take the form (5). In the case $r_3 < 1$ (which is equivalent to $\gamma \leq 0$), the solutions tend to 0, while if $r_3 > 1$ (which is equivalent to $\gamma > 0$), we have $\lim_{t \to \infty} T(t) = \frac{d}{\beta_2} - \frac{A\beta_2}{d\beta_1}$, and this is what we wanted to show.   □

Remark 3.1 We note that changing global asymptotic stability to attractivity, the results of Theorem 3.2 also hold when the given reproduction numbers are equal to 1, instead of being smaller than 1.

4 The Human Subsystem

Let us now turn to the human subsystem of (1) consisting of the last three equations.

In the sequel, we assume that the rodent subsystem is in a steady state, and substitute any of the equilibria of the rodent subsystem into these equations to obtain the system

$$
\begin{aligned}
S'(t) = {} & B - \eta_1 T^* S(t) - \eta_2 Q^* S(t) - v_1 S(t) I(t) - v_2 S(t) J(t) \\
& - \delta S(t) + \theta_1 I(t) + \theta_2 J(t), \\
I'(t) = {} & \eta_1 T^* S(t) + v_1 S(t) I(t) \\
& - \eta_2 Q^* I(t) - v_2 J(t) I(t) - \delta I(t) - \theta_1 I(t), \\
J'(t) = {} & \eta_2 Q^* S(t) + \eta_2 Q^* I(t) + v_2 S(t) J(t) + v_2 I(t) J(t) \\
& - \delta J(t) - \rho J(t) - \theta_2 J(t),
\end{aligned}
\tag{6}
$$

where T^* and Q^* are the second, resp. third coordinates in any of the four equilibria (3).

To find all possible equilibria of (6), first we introduce the notation $G(t) := S(t) + I(t)$ to obtain the system

$$
\begin{aligned}
G'(t) &= B - \eta_2 Q^* G(t) - v_2 G(t) J(t) - \delta G(t) + \theta_2 J(t), \\
J'(t) &= \eta_2 Q^* G(t) + v_2 G(t) J(t) - \delta J(t) - \rho J(t) - \theta_2 J(t).
\end{aligned}
\tag{7}
$$

We will apply the Bendixson–Dulac criterion with Dulac function $1/J$ and the Poincaré–Bendixson theorem to obtain that in this case, all solutions of system (7) tend to one of the equilibria. Indeed, we have

$$
\begin{aligned}
& \frac{\partial}{\partial G} \frac{B - \eta_2 Q^* G - v_2 G J - \delta G + \theta_2 J}{J} \\
& + \frac{\partial}{\partial J} \frac{\eta_2 Q^* G + v_2 G J - \delta J - \rho J - \theta_2 J}{J} \\
& = -\frac{\eta_2 Q^*}{J} - v_2 - \frac{\delta}{J} - \frac{\eta_2 G}{J^2} < 0,
\end{aligned}
$$

from which we obtain the assertion above.

This equation may have two equilibria:

$$
\left(\frac{D + B v_2 - \sqrt{(D - B v_2)^2 + 4 B \eta_2 Q^* v_2 (\delta + \rho)}}{2 \delta v_2}, \frac{-D + B v_2 + \sqrt{(D - B v_2)^2 + 4 B \eta_2 Q^* v_2 (\delta + \rho)}}{2 (\delta + \rho) v_2} \right)
$$

and

$$\left(\frac{D+Bv_2+\sqrt{(D-Bv_2)^2+4B\eta_2 Q^* v_2(\delta+\rho)}}{2\delta v_2} , \frac{-D+Bv_2-\sqrt{(D-Bv_2)^2+4B\eta_2 Q^* v_2(\delta+\rho)}}{2(\delta+\rho)v_2} \right)$$

denoted by E_1 and E_2, respectively, with $D = \delta^2 + Q^*\eta_2\rho + \delta(Q^*\eta_2 + \theta_2 + \rho)$. The first coordinate of E_1 is always positive, since this coordinate may be rewritten as

$$\frac{D + Bv_2 - \sqrt{(D + Bv_2)^2 - 4B\delta v_2(\delta + \theta_2 + \rho)}}{2\delta v_2}.$$

It can easily be seen that the first coordinate of E_2 is always positive.

Let us first consider the case $Q^* > 0$, (i.e. when the rodent subsystem tends to the equilibrium E_Q or E_{TQ}, which is equivalent to $r_2 > 1$). In this case, the second coordinate of E_1 is always positive, while second coordinate of E_2 is negative if $Q^* > 0$. Hence, in the case $Q^* > 0$, there is only one equilibrium and using the Poincaré–Bendixson theorem, we obtain that all solutions tend to E_1.

In the case $r_2 \le 1$, i.e. when $Q^* = 0$, the system takes the simpler form

$$\begin{aligned} G'(t) &= B - v_2 G(t) J(t) - \delta G(t) + \theta_2 J(t), \\ J'(t) &= v_2 G(t) J(t) - \delta J(t) - \rho J(t) - \theta_2 J(t). \end{aligned} \tag{8}$$

This system has the two equilibria

$$e_1 := \left(\frac{B}{\delta}, 0 \right) \quad \text{and} \quad e_2 := \left(\frac{\delta + \theta_2 + \rho}{v_2}, \frac{Bv_2 - \delta(\delta + \theta_2 + \rho)}{v_2(\delta + \rho)} \right).$$

Now, it is easy to see that the first of these equilibria always exists, while the second one only exists if

$$\mathcal{R}_0^J := \frac{Bv_2}{\delta(\delta + \theta_2 + \rho)} > 1.$$

Just as above, we obtain that all solutions of (8) tend to one of these equilibria. In the case $\mathcal{R}_0^J \le 1$, this equilibrium is clearly e_1. Using similar methods as for the rodent subsystem, we will show that $J(t)$ is always uniformly strongly persistent if $\mathcal{R}_0^J > 1$. To show this, we choose $\rho(x) = J$. Consider the extinction space X_J defined as $X_J := \{x \in \mathbb{R}_+^2 : \rho(x) = 0\}$; now $\Omega = M_1 := e_1$, which is obviously invariant, isolated and acyclic. Let us suppose that M_1 is not weakly ρ-repelling, i.e. there is a solution which tends to e_1 such that $J(t) > 0$. Then, given any $\varepsilon > 0$, for t sufficiently large, we can give the following estimate for $J'(t)$:

$$\begin{aligned} J'(t) &= v_2 G(t) J(t) - \delta J(t) - \rho J(t) - \theta_2 J(t) \\ &> J(t) \left(v_2 \frac{B}{\delta} - v_2\varepsilon - \delta - \rho - \theta_2 \right), \end{aligned}$$

which is positive as $\mathcal{R}_0^J > 1$. Hence, in the case $\mathcal{R}_0^J > 1$, all solutions of (8) started with positive initial value $J(0)$ tend to the equilibrium e_2.

We have now finished the analysis of (7) and showed that in each case, depending on the reproduction numbers r_2 and $\mathcal{R}_0^J$, all solutions of this equation tend to an equilibrium. Let us denote by J^* the second coordinate of this equilibrium and substitute this value into the first two equations of (6) to obtain

$$
\begin{aligned}
S'(t) = \ & B - \eta_1 T^* S(t) - \eta_2 Q^* S(t) - v_1 S(t) I(t) - v_2 J^* S(t) \\
& - \delta S(t) + \theta_1 I(t) + \theta_2 J^*, \\
I'(t) = \ & \eta_1 T^* S(t) + v_1 S(t) I(t) - \eta_2 Q^* I(t) \\
& - v_2 J^* I(t) - \delta I(t) - \theta_1 I(t).
\end{aligned}
\tag{9}
$$

This equation has a similar structure as (7). The two possible equilibria of system (9) are

$$
\left(\frac{v_1(B+\theta_2 J^*)+P-\sqrt{(v_1(B+\theta_2 J^*)-P)^2+H}}{2v_1 K}, \ \frac{v_1(B+\theta_2 J^*)-P+\sqrt{(v_1(B+\theta_2 J^*)-P)^2+H}}{2v_1 K} \right)
$$

and

$$
\left(\frac{v_1(B+\theta_2 J^*)+P+\sqrt{(v_1(B+\theta_2 J^*)-P)^2+H}}{2v_1 K}, \ \frac{v_1(B+\theta_2 J^*)-P-\sqrt{(v_1(B+\theta_2 J^*)-P)^2+H}}{2v_1 K} \right)
$$

denoted by $\mathcal{E}_1$ and $\mathcal{E}_2$, respectively, where the notations K, P and H are defined as $K = (\delta + \eta_2 Q^* + v_2 J^*)$, $P = K(\delta + \theta_1 + \eta_1 T^* + \eta_2 Q^* + v_2 J^*)$ and $H = 4\eta_1 v_1 T^*(B + \theta_2 J^*)K$. Again, we can apply the Bendixson–Dulac criterion, in this case with the Dulac function $1/I$, to show that all solutions tend to an equilibrium:

$$
\begin{aligned}
& \frac{\partial}{\partial S} \frac{B - \eta_1 T^* S - \eta_2 Q^* S - v_1 SI - v_2 J^* S - \delta S + \theta_1 I + \theta_2 J^*}{I} \\
& + \frac{\partial}{\partial I} \frac{\eta_1 T^* S + v_1 SI - \eta_2 Q^* I - v_2 J^* I - \delta I - \theta_1 I}{I} \\
& = \frac{-\eta_1 T^*}{I} - \frac{\eta_2 Q^*}{I} - v_1 - \frac{v_2 J^*}{I} - \frac{\delta}{I} - \frac{\eta_1 T^* S}{I^2},
\end{aligned}
$$

which is negative for all $I, S > 0$.

Similarly as in the case of the equilibria of system (7), it is easy to see that the first coordinates of $\mathcal{E}_1$ and $\mathcal{E}_2$ are always positive, while the second coordinate of $\mathcal{E}_2$ is negative if $T^* > 0$ (i.e. when $r_1 > 1$ and $r_2 < 1$, meaning that E_T is globally asymptotically stable or $r_2 > 1$ and $r_3 > 1$ meaning that E_{TQ} is globally asymptotically stable). Hence, in this case there is only one equilibrium, and by the Poincaré–Bendixson theorem, all solutions tend to this equilibrium.

From the above, we obtain that if $T^* > 0$ and $Q^* > 0$, i.e. when $r_2 > 1$ and $r_3 > 1$, then all solutions tend to the equilibrium $(\mathcal{E}_1^1, \mathcal{E}_1^2, E_1^2)$, where upper index i denotes the ith coordinate of a given equilibrium.

In the case $T^* > 0$ and $Q^* = 0$, i.e. when $r_1 > 1$ and $r_2 \leq 1$, $\mathcal{E}_1$ is the only equilibrium of (9), and the reproduction number $\mathcal{R}_0^J$ determines which equilibrium is the limit of the solutions of (8). Hence, if $r_1 > 1$, $r_2 \leq 1$ and $\mathcal{R}_0^J \leq 1$ then all solutions of (6) tend to the equilibrium

$$\left(\mathcal{E}_1^1, \mathcal{E}_1^2, 0\right),$$

while if $r_1 > 1$, $r_2 \leq 1$ and $\mathcal{R}_0^J > 1$ then all solutions of (6) tend to the equilibrium

$$\left(\mathcal{E}_1^1, \mathcal{E}_1^2, \frac{B v_2 - \delta(\delta + \theta_2 + \rho)}{v_2(\delta + \rho)}\right).$$

In the case $T^* = 0$ (i.e. when $r_1 < 1$ and $r_2 < 1$, meaning that E_R is globally asymptotically stable or $r_2 > 1$ and $r_3 < 1$, meaning that E_Q is globally asymptotically stable), system (9) reduces to

$$S'(t) = B - \eta_2 Q^* S(t) - v_1 S(t) I(t) - v_2 S(t) J^*$$
$$\quad - \delta S(t) + \theta_1 I(t) + \theta_2 J^*, \tag{10}$$
$$I'(t) = v_1 S(t) I(t) - \eta_2 Q^* I(t) - v_2 I(t) J^* - \delta I(t) - \theta_1 I(t),$$

which has two equilibria

$$\mathcal{E}_1 = \left(\frac{\eta_2 Q^* + v_2 J^* + \delta + \theta_1}{v_1}, \frac{v_1(B + \theta_2 J^*) - (\eta_2 Q^* + v_2 J^* + \delta)(\eta_2 Q^* + v_2 J^* + \delta + \theta_1)}{v_1(\eta_2 Q^* + v_2 J^* + \delta)}\right),$$

resp.

$$\mathcal{E}_2 = \left(\frac{B + \theta_2 J^*}{\eta_2 Q^* + v_2 J^* + \delta}, 0\right).$$

One may easily observe that the second equilibrium always exists, while the sign of the second coordinate of the first equilibrium depends on the parameters and the limits Q^* and J^*: the first equilibrium exists if and only if

$$\mathcal{R}_0^I := \frac{v_1(B + \theta_2 J^*)}{(\eta_2 Q^* + v_2 J^* + \delta)(\eta_2 Q^* + v_2 J^* + \delta + \theta_1)} > 1.$$

In the case $\mathcal{R}_0^I \leq 1$, there is only one equilibrium, $\mathcal{E}_2$, so it is clear from the above that all solutions of (6) tend to the equilibrium

$$\left(\frac{B + \theta_2 J^*}{\eta_2 Q^* + \nu_2 J^* + \delta}, 0, J^* \right).$$

In the case $\mathcal{R}_0^I > 1$, we will again use persistence theory to show that all solutions of (10) tend to the equilibrium $\mathcal{E}_1$. We now choose $\rho(x) = I$ and consider the extinction space $X_I := \{x \in \mathbb{R}_+^2 : \rho(x) = 0\}$. It is clear that now $\Omega = M_1 := \{\mathcal{E}_2\}$, which is invariant, acylic and isolated. Let us suppose that M_1 is not weakly ρ-repelling, i.e. there exists a solution which tends to $\mathcal{E}_2$ such that $I(t) > 0$. Then, for any $\varepsilon > 0$, for large enough t, we can estimate $I'(t)$ as

$$I'(t) = I(t)(\nu_1 S(t) - \eta_2 Q^* - \nu_2 J^* - \delta - \theta_1)$$

$$> I(t) \left(\nu_1 \left(\frac{B + \theta_2 J^*}{\eta_2 Q^* + \nu_2 J^* + \delta} + \varepsilon \right) - \eta_2 Q^* - \nu_2 J^* - \delta - \theta_1 \right),$$

which is positive as $\mathcal{R}_0^I > 1$. From this we obtain that in the case $\mathcal{R}_0^I > 1$, all solutions of (10) started with positive initial value $I(0)$ tend to $\mathcal{E}_1$.

On the ω-limit set of solutions of (6), Eq. (10) holds, which has at most two equilibria. Hence, the global attractor of (10) consists either of a single equilibrium or two equilibria and connecting orbits between them. When there is only one equilibrium, then the solutions of (6) tend to this equilibrium. When two equilibria exist, then $J(t)$ is uniformly persistent, hence, the ω-limit set of positive solutions of (6) can only be the equilibrium with the positive J coordinate.

Now we go through all possibilities regarding the value of Q^* and J^* to give a precise characterization. In the case $Q^* > 0$ (i.e. $r_2 > 1$), there is only one equilibrium of (7), hence $J(t)$ tends to E_1^2. This means that in the case $r_2 > 1$ and $\mathcal{R}_0^I \leq 1$ all solutions of (6) tend to the equilibrium

$$\left(\frac{B + \theta_2 E_1^2}{\eta_2 \left(\frac{A}{d} - \frac{d}{\beta_2} \right) + \nu_2 E_1^2 + \delta}, 0, E_1^2 \right),$$

while in the case $r_2 > 1$ and $\mathcal{R}_0^I > 1$, all solutions of (6) tend to the equilibrium

$$\left(\frac{\eta_2 Q^* + \nu_2 E_1^2 + \delta + \theta_1}{\nu_1}, \frac{\nu_1(B + \theta_2 E_1^2) - (\eta_2 Q^* + \nu_2 E_1^2 + \delta)(\eta_2 Q^* + \nu_2 E_1^2 + \delta + \theta_1)}{\nu_1(\eta_2 Q^* + \nu_2 E_1^2 + \delta)}, E_1^2 \right)$$

with $Q^* = \left(\frac{A}{d} - \frac{d}{\beta_2} \right)$.

In the case $r_2 \leq 1$ (i.e. $Q^* = 0$), the reproduction number $\mathcal{R}_0^J$ determines the limit of $J(t)$. In the case $r_1 \leq 1, r_2 \leq 1, \mathcal{R}_0^J \leq 1, \mathcal{R}_0^I \leq 1$, all solutions of (6) tend to the equilibrium

$$\left(\frac{B}{\delta}, 0, 0 \right).$$

In the case $r_1 \leq 1$, $r_2 \leq 1$, $\mathcal{R}_0^J \leq 1$, $\mathcal{R}_0^I > 1$, all solutions of (6) tend to the equilibrium

$$\left(\frac{\delta + \theta_1}{\nu_1}, \frac{B}{\delta} - \frac{\delta + \theta_1}{\nu_1}, 0 \right).$$

In the case $r_1 \leq 1$, $r_2 \leq 1$, $\mathcal{R}_0^J > 1$, $\mathcal{R}_0^I \leq 1$, all solutions of (6) tend to the equilibrium

$$\left(\frac{\delta + \theta_2 + \rho}{\nu_2}, 0, \frac{\nu_2 B - \delta(\delta + \theta_2 + \rho)}{\nu_2(\delta + \rho)} \right).$$

In the case $r_1 \leq 1$, $r_2 \leq 1$, $\mathcal{R}_0^J > 1$, $\mathcal{R}_0^I > 1$, all solutions of (6) tend to the equilibrium

$$\left(\frac{\nu_2 B + \theta_1 \rho + \delta(\theta_1 - \theta_2)}{\nu_1(\delta + \rho)}, \frac{\delta\theta_2 - \nu_2 B}{\nu_1(\delta + \rho)} + \frac{\delta + \theta_2 + \rho}{\nu_2} - \frac{\theta_1}{\nu_1}, \frac{\nu_2 B - \delta(\delta + \theta_2 + \rho)}{\nu_2(\delta + \rho)} \right).$$

5 Discussion

We have established a six-compartment model to describe the spread of an infectious disease spread by ectoparasites which are transmitted to humans by rodents. We have identified three reproduction numbers for the rodent subsystem. These threshold numbers determine which of the four possible equilibria of the rodent subsystem is globally attractive. Assuming that the rodent subsystem is already in a steady state, we studied the human subsystem and calculated the possible equilibria of this subsystem depending on which of the rodent equilibria is globally attractive. We also determined which equilibrium of the human subsystem is globally attractive. Our results show that in each case, depending on the different reproduction numbers, one equilibrium is globally attractive. Our results show that if one type of the parasite (infectious or noninfectious) is present in the rodent population, then the same type will also be present in the human population. Using our results, we may study the possibilities of eradicating the disease. There are three main ways to control the disease: we may decrease the transmission rates $\eta_{1,2}$ between humans and rodents, increase the disinfestation rates $\theta_{1,2}$ of humans to shorten the duration of infestation of humans and we may reduce d which means culling of the rodents.

Controlling only the human population (increasing the disinfestation rates $\theta_{1,2}$) only results in a mitigation not sufficient to eradicate the disease. The same holds for decreasing the transmission rates from rodents to humans, except the extreme case of decreasing the transmission rates $\eta_{1,2}$ to zero. In this latter case, one may decrease the human reproduction numbers to be less than 1 by increasing the disinfestation rates $\theta_{1,2}$ and thus eliminate the infestation.

By controlling the rodent population (increasing the death rate d), one can reduce the reproduction numbers r_1 and r_2 to be both less than 1 and this way one may eliminate the infestation among the rodents. Also in this case, infestation from rodents to humans can be eliminated and this way the human reproduction numbers determine which equilibrium of the human subsystem will be globally attractive. Hence, also in such a case, by increasing the disinfestation rate among humans may result in the elimination of the parasites and of the disease.

Acknowledgements A. Dénes was supported by Hungarian Scientific Research Fund OTKA PD and National Research, Development and Innovation Office NKFIH KH 125628 and the János Bolyai Research Scholarship of the Hungarian Academy of Sciences. G. Röst was supported by the EU-funded Hungarian grant EFOP-3.6.1-16-2016-00008 and Marie Sklodowska-Curie Grant No. 748193.

A. Dénes thanks to the International Union of Biological Sciences (IUBS) for partial support of living expenses in Moscow, during the 17th BIOMAT International Symposium, October 29–November 04, 2017.

References

1. A. Dénes, G. Röst, Biomathematics **1**, 1209256, 1–5 (2012)
2. A. Dénes, G. Röst, Nonlinear Anal. Real World Appl. **18**, 100–107 (2014)
3. A. Dénes, G. Röst, Impact of excess mortality on the dynamics of diseases spread by ectoparasites, in *Interdisciplinary Topics in Applied Mathematics, Modeling and Computational Science*, ed. by M. Cojocaru, I.S. Kotsireas, R.N. Makarov, R. Melnik, H. Shodiev. Springer Proceedings in Mathematics & Statistics, vol. 117 (Springer, Cham, 2015), pp. 177–182
4. M.R. Holbrook, J.F. Aronson, G.A. Campbell, S. Jones, H. Feldmann, A.D.T. Barrett, J. Infect. Dis. **191**, 100–108 (2005)
5. M.J. Keeling, C.A. Gilligan, Proc. R. Soc. B **267**, 2219–2230 (2000)
6. Ch.D. Paddock, M.E. Eremeeva, Rickettsialpox, in *Rickettsial Diseases*, ed. by D. Raoult, Ph. Parola (CRC Press, New York, 2007)
7. H.L. Smith, *An Introduction to Delay Differential Equations with Applications to the Life Sciences* (Springer, New York, 2011)
8. H.L. Smith, H.R. Thieme, *Dynamical Systems and Population Persistence* (AMS, Providence, 2011)
9. Y. Tselentis, A. Gikas, Murine typhus, in *Rickettsial Diseases*, ed. by D. Raoult, Ph. Parola (CRC Press, New York, 2007)
10. G. Watt, P. Kantipong, Orientia tsutsugamushi and scrub typhus, in *Rickettsial Diseases*, ed. by D. Raoult, Ph. Parola (CRC Press, New York, 2007)

p-Adic Side of the Genetic Code and the Genome

Branko Dragovich and Nataša Ž. Mišić

1 Introduction

The living organisms are the most complex systems on our planet Earth and probably in the whole universe. A central role in the life functioning at the level of living cells plays the genetic code (*GC*). The *GC* is a rule how 64 codons (building blocks of the genes) code 20 amino acids (building blocks of the proteins) and one stop signal. Mathematically, the genetic code is a map from a set of 64 elements onto a set of 21 elements. There are more than three times codons with respect to protein amino acids, so every amino acid could be coded by three codons. However amino acids are not coded by equal number of codons. In the standard *GC* two amino acids are coded by one codon, while other amino acids are coded by either two, three, four or six codons (stop signal is coded by three codons), see Table 1. This property that some amino acids are coded by more than one codon is called the *degeneration of the GC*. By an estimation [1] there is more than 10^{84} possibilities to map set of 64 elements onto set of 21 elements, while in the living organisms there is practically one basic genetic code with few dozen slight modifications. Then a question arises: What are the essential properties of the basic genetic code and how to describe them in the simple mathematical form. A satisfactory answer to this question belongs to the adequate mathematical modeling.

The standard *GC* was deciphered in the mid-1960s. Usually it is presented by the genetic code table (see Table 1), where an explicit relation between codons

B. Dragovich (✉)
Institute of Physics, University of Belgrade, Belgrade, Serbia
Mathematical Institute, Serbian Academy of Sciences and Arts, Belgrade, Serbia
e-mail: dragovich@ipb.ac.rs

N. Ž. Mišić
Research and Development Institute Lola Ltd, Belgrade, Serbia

© Springer International Publishing AG, part of Springer Nature 2018
R. P. Mondaini (ed.), *Trends in Biomathematics: Modeling, Optimization and Computational Problems*, https://doi.org/10.1007/978-3-319-91092-5_6

Table 1 Table of the standard genetic code

UUU	Phe	(F)	UCU	Ser	(S)	UAU	Tyr	(Y)	UGU	Cys	(C)
UUC	Phe	(F)	UCC	Ser	(S)	UAC	Tyr	(Y)	UGC	Cys	(C)
UUA	Leu	(L)	UCA	Ser	(S)	UAA	Ter	(*)	UGA	Ter	(*)
UUG	Leu	(L)	UCG	Ser	(S)	UAG	Ter	(*)	UGG	Trp	(W)
CUU	Leu	(L)	CCU	Pro	(P)	CAU	His	(H)	CGU	Arg	(R)
CUC	Leu	(L)	CCC	Pro	(P)	CAC	His	(H)	CGC	Arg	(R)
CUA	Leu	(L)	CCA	Pro	(P)	CAA	Gln	(Q)	CGA	Arg	(R)
CUG	Leu	(L)	CCG	Pro	(P)	CAG	Gln	(Q)	CGG	Arg	(R)
AUU	Ile	(I)	ACU	Thr	(T)	AAU	Asn	(N)	AGU	Ser	(S)
AUC	Ile	(I)	ACC	Thr	(T)	AAC	Asn	(N)	AGC	Ser	(S)
AUA	Ile	(I)	ACA	Thr	(T)	AAA	Lys	(K)	AGA	Arg	(R)
AUG	Met	(M)	ACG	Thr	(T)	AAG	Lys	(K)	AGG	Arg	(R)
GUU	Val	(V)	GCU	Ala	(A)	GAU	Asp	(D)	GGU	Gly	(G)
GUC	Val	(V)	GCC	Ala	(A)	GAC	Asp	(D)	GGC	Gly	(G)
GUA	Val	(V)	GCA	Ala	(A)	GAA	Glu	(E)	GGA	Gly	(G)
GUG	Val	(V)	GCG	Ala	(A)	GAG	Glu	(E)	GGG	Gly	(G)

The 20 amino acids found in proteins are listed with three-letter and single-letter notation. The mRNA codons representing each amino acid are also listed. All 64 possible 3-letter combinations of nucleotides C, A, U and G are used either to encode one of these amino acids or as one of the three stop codons that signal the end of a codon sequence. While mRNA can be decoded unambiguously, it is not possible to exactly predict the corresponding mRNA sequence from its amino acids sequence, because most amino acids have multiple codons

and amino acids is given. Already at the first sight, one can note that the relations between codons and amino acids are not random but well arranged.

First attempts to model the *GC* were undertaken soon after discovery of the double helix structure of DNA in 1953, i.e. before the standard genetic code was deciphered. For a review of these, theoretically interesting and biologically wrong, early models (of Gamow and Crick), see, e.g., Hayes [2]. Soon after deciphering of the genetic code, Rumer [3] correctly noted that the first two bases (nucleotides) in the codon are more important in coding than the third one.

There are many papers devoted to mathematical investigation of the genetic code, and we refer [4–7] here only to a few of them which illustrate algebraic modeling. They mainly connect irreducible representations of some algebras with structure of the 64 codons. Although these algebraic approaches contain some interesting results, they are rather complex mathematical constructions. In this context, it is worth recalling the emergence of special theory of relativity and quantum mechanics. In both cases there were experimental results without satisfactory interpretation by known classical theories, which occurred to be complex along efforts to adapt them to experimental data. However, invention of new theoretical concepts and application of the corresponding mathematical methods provided correct description and appearance of these two theories (relativistic and quantum) which make the fundamental of modern theoretical physics.

In 2006, a quite new and simple approach to model the genetic code was proposed [8]. The basic idea is in the following. Two codons which code the same amino acid are closer in the information sense than those codons which code different amino acids. This closeness is in an ultrametric space which elements are codons. The natural way to realize such ultrametric space is by introducing adequate p-adic space of codons. In the Dragovich [8] approach was constructed related 5-adic space and p-adic distance between codons was considered for $p = 5$ and $p = 2$. This model very well describes the structure of the codon space which appears in coding of amino acids. In a series of papers [8–13] was shown that the concept of p-adic distance is adequate for description of the genetic code structure and promising approach to further investigation of similarity in the genomes.

This article is mainly devoted to developments of p-adic modeling of the genetic code and the genome initiated by B. Dragovich, where $p = 5$ and 2. In the case, the vertebrate mitochondrial code (Table 4) is considered as the basic code. To the nucleotides are assigned digits in three digit 5-adic numbers which correspond to the codons. With respect to the smallest 5-adic distance, 64 codons form 16 codon quadruplets. Each of these quadruplets is composed of two doublets according to 2-adic distance. In this way, one obtains 32 codon doublets—within every doublet the first two nucleotides are the same, while the third nucleotides are either purines or pyrimidines. Every doublet is assigned either to an amino acid or to the stop signal. In the vertebrate mitochondrial code there are 12 amino acids (aa) coded by single doublets, 6 aa and stop signal are coded by two doublets, and 2 aa are coded by three doublets. Note that the standard GC can be obtained from this mitochondrial one by the following formal replacements in codon assignments: AUA: Met $\rightarrow$ Ile, AGA and AGG: Ter $\rightarrow$ Arg, UGA: Trp $\rightarrow$ Ter, where Ter denotes stop (terminal) codon.

This paper is organized in the following way. In Sect. 2 we give basic information on the genome and the genetic code. Section 3 contains a short and basic introduction to ultrametric spaces and, in particular, to ultrametricity with p-adic distance. Various aspects of the p-adic genetic code are considered in Sect. 4. A p-adic approach to ultrametric side of the genome is presented in Sect. 5. Some concluding remarks are stated in the last section.

2 On the Genome and the Genetic Code

We recall here some knowledge from molecular biology related to DNA, RNA, proteins, their constituents, structures and functions.

The genome of an organism is its whole genetic material which consists of the coding and non-coding parts of DNA. In some viruses, genetic material consists of RNA. Investigation of the genome is the subject of genomics.

DNA (desoxyribonucleic acid) is a macromolecule which contains two polynucleotide chains with double-helical structure discovered by Crick and Watson in 1953. Nucleotides consist of a base, a sugar and a phosphate group. The sugar and phosphate groups provide a helical backbone of DNA. There are four nitrogenous bases: cytosine (C), adenine (A), thymine (T) and guanine (G). Cytosine and thymine belong to pyrimidines, while adenine and guanine are purines. From information point of view, bases and their corresponding nucleotides have the same meaning. Two helical chains share complementary base pairing—to base C in one chain corresponds G in the other one, and the same rule is valid for pairing of A and T. Hence in DNA there is an equal number of C and G, and also an equal number of A and T. DNA are packed in the chromosomes and their main role is to store genetic information. For example, DNA of *Homo sapiens* contains over 3×10^9 base pairs. Only about 1.5% of human DNA code proteins, the rest is non-coding DNA related to regulatory processes and unknown functions.

RNA (ribonucleic acid) is a single-strand polynucleotide molecule, where thymine is substituted by uracil (U) which is also pyrimidine. As a result of gene transcription from DNA one obtains messenger RNA (mRNA). In such process nucleotides C, A, T, G from a gene are transcribed into their complements G, U, A, C. After splicing mRNA comes to ribosomes where performs synthesis of proteins on the basis of information contained in mRNA. In this gene expression some non-coding RNAs (ncRNA) also participate, like microRNA (miRNA), transfer RNA (tRNA) and ribosomal RNA (rRNA).

Codons are ordered triplets composed of the four nucleotides C, A, T(U), G. There are $4 \times 4 \times 4 = 64$ codons. Each of the codons contains information which determines an amino acid or stop signal in the process of the protein synthesis.

Proteins [14] are polypeptide macromolecules composed of amino acids primarily arranged in a linear chain. They are substantial ingredients of all living cells and determine the phenotype of any organism. Proteins have primary, secondary and tertiary structure which is closely related to their function. Proteomics is a study of the proteome which is a set of all proteins of an organism.

Amino acids are molecules that consist of amino, carboxyl and side chain group (R). There are 20 standard protein amino acids (alanine, cytosine, aspartate, glutamate, phenylalanine, glycine, histidine, isoleucine, lysine, leucine, methionine, asparagine, proline, glutamine, arginine, serine, threonine, valine, tryptophan, tyrosine) which differ with respect to R group. The sequence of codons in mRNA determines sequence of amino acids in a protein.

As it was stated in Introduction, the genetic code is a concrete connection between 64 codons and 20 amino acids with stop signal. It is not direct chemical connection between codons and amino acids, but there are some intermediate steps. After transcription of a gene from DNA, the corresponding RNA passes a splicing process in which introns (some sequences of nucleotides) being removed and one gets a maturated mRNA which is a definite sequence of codons. Such mRNA serves as a matrix which strongly defines sequence of amino acids in proteins. The reading of mRNA and synthesis of proteins performs in the ribosomes which are very complex molecular systems composed of the rRNA and ribosomal proteins. An amino acid incoming to protein synthesis is attached to its tRNA, while another side of tRNA contains the corresponding anticodon. An anticodon consists of nucleotides which are mainly complementary to nucleotides in the related codon. According to Crick's wobble hypothesis [15] there is a smaller number of tRNA than codons which code amino acids. Namely, the first to nucleotides of a codon are recognized by tRNA according to complementary rule, but the third nucleotide can be recognized by the first position in the tRNA anticodon as follows: (1) C recognizes G; (2) A recognizes U; (3) U can recognize A and G; (4) G can recognize C and U; and (5) I (inosine) can recognize C, A and U.

3 Ultrametric and *p*-Adic Distances

To measure distances between two objects with positions x and y at the straight line one uses standard absolute value, i.e. $d(x, y) = |x - y|$. As distance is smaller one says that objects are closer. M. Fréchet (1878–1973) in 1906 generalized notion of distance and introduced the metric space (M, d), where M is a set and d is a distance function. Distance d is a real-valued function of any two elements $x, y \in M$ which satisfy the following properties:

$$(i) \ d(x, y) = 0 \Leftrightarrow x = y,$$

$$(ii) \ d(x, y) = d(y, x),$$

$$(iii) \ d(x, y) \leq d(x, z) + d(z, y),$$

where last property is called triangle inequality. An ultrametric space is a metric space which also satisfies ultrametric (also known as strong triangle or non-Archimedean) inequality

$$d(x, y) \leq \max\{d(x, z), d(z, y)\}. \tag{1}$$

Ultrametric space was introduced in 1944 by M. Krasner (1912–1985), although examples of ultrametric spaces have been used earlier in taxonomy, which started 1735 by C. Linné (1707–1778) as biological classification with hierarchical structure. Living beings with more common ancestors are ultrametrically closer than those with less common. A very important class of ultrametric spaces contains *p*-adic numbers which were introduced in 1897 by K. Hensel (1861–1941).

The ultrametric spaces have many unusual properties which are a consequence of the ultrametric inequality (1). For example, by suitable notation of points x, y, z, inequality (1) can be rewritten in the form $d(x, y) \leq d(x, z) = d(y, z)$. This means that all ultrametric triangles are isosceles. One can note also the following properties: (1) There is no partial intersection of the balls; (2) Any point of a ball can be its center; (3) Each ball is both open and closed—clopen ball. For a proof of these properties of ultrametric balls, see, e.g., Schikhof's book [16].

To illustrate ultrametric spaces, it is worth considering an alphabet of k letters with the corresponding words of the length n (n-letter words). Then there are k^n words. Let the related set of words be denoted by $W_{k,n}(k^n)$. For simplicity and what we will have in the sequel, we can consider ultrametricity of the case $W_{4,3}(64)$.

As ultrametric distances, we shall consider: ordinary ultrametric distance, the Baire distance and p-adic distance.

3.1 Ordinary Ultrametric Distance

Ordinary ultrametric distance between any two different words x and y is $d(x, y) = n - (m - 1)$, where $m (m = 1, 2, \ldots, n)$ is the first position at which letters differ counting from the beginning. It takes n values, i.e. $d(x, y) = 1, 2, \ldots, n$. Note that one can redefine this distance by scaling it as $d_s(x, y) = \frac{n-m+1}{n}$ and then the scaled distances are between $1, \frac{n-1}{n}, \ldots, \frac{2}{n}, \frac{1}{n}$.

In the concrete case $W_{4,3}(64)$ there are 64 three-letter words (see Table 2). Possible distances $d(x, y)$ are 1, 2, 3 and the corresponding scaling ones are

$$
d_s(x, y) = \frac{4 - m}{3} =
\begin{cases}
1, & m = 1 \\
\frac{2}{3}, & m = 2 \\
\frac{1}{3}, & m = 3.
\end{cases}
\tag{2}
$$

For example, $d_s(abc, bac) = 1$, $d_s(abc, acb) = \frac{2}{3}$, $d_s(abc, abb) = \frac{1}{3}$.

3.2 The Baire Distance

This distance is usually defined as $d_B(x, y) = 2^{-(m-1)}$, where m is as defined in the above, i.e. it is the first position in words x and y at which letters differ, and it can be $m = 1, 2, \ldots, n$. Thus the Baire distance takes values $1, \frac{1}{2}, \frac{1}{2^2}, \ldots, \frac{1}{2^{n-1}}$. Note that instead of the base 2 one can take any natural number larger than 2.

In the case $W_{4,3}(64)$ the Baire distance is

$$
d_B(x, y) = 2^{-(m-1)} =
\begin{cases}
1, & m = 1 \\
\frac{1}{2}, & m = 2 \\
\frac{1}{4}, & m = 3.
\end{cases}
\tag{3}
$$

For example, $d_B(abc, bac) = 1$, $d_B(abc, acb) = \frac{1}{2}$, $d_B(abc, abb) = \frac{1}{4}$.

3.3 p-Adic Distance

p-Adic absolute value (p-adic norm) of a non-zero integer x is $|x|_p = p^{-k}$, where k is degree of a prime number p in x, and $|0|_p = 0$. Since $k \in \mathbb{N}$, p-adic absolute value of any integer x is $|x|_p \leq 1$. p-Adic distance between two integers x and y is $d_p(x, y) = |x - y|_p$. This distance is related to divisibility of $x - y$ by prime p (more divisible—lesser distance). With respect to a fixed prime p as a base, any positive integer has its unique expansion $x = x_0 + x_1 p + x_2 p^2 + \cdots + x_n p^n$, where $x_i \in \{0, 1, \ldots, p - 1\}$ are digits. If in this expansion x_k is the first digit different from zero, then p-adic norm of x is $|x|_p = p^{-k}$.

To have connection with the above alphabet and words it is natural to make a correspondence between letters $\{a, b, c, d\}$ and some four digits. In this way the role of letters play digits (see Table 2). The smallest prime number base which contains four digits is $p = 5$ and we use digits $\{1, 2, 3, 4\}$ without digit 0. One can construct some sets of 5-adic integers in the form

$$x = x_0 + x_1 5 + \cdots + x_k 5^k \quad \text{or} \quad x \equiv x_0 x_1 \ldots x_k, \quad x_i \in \{1, 2, 3, 4\}. \tag{4}$$

Table 2 Table of words constructed of four letters and arranged in the ultrametric form

111 aaa	211 baa	311 caa	411 daa
112 aab	212 bab	312 cab	412 dab
113 aac	213 bac	313 cac	413 dac
114 aad	214 bad	314 cad	414 dad
121 aba	221 bba	321 cba	421 dba
122 abb	222 bbb	322 cbb	422 dbb
123 abc	223 bbc	323 cbc	423 dbc
124 abd	224 bbd	324 cbd	424 dbd
131 aca	231 bca	331 cca	431 dca
132 acb	232 bcb	332 ccb	432 dcb
133 acc	233 bcc	333 ccc	433 dcc
134 acd	234 bcd	334 ccd	434 dcd
141 ada	241 bda	341 cda	441 dda
142 adb	242 bdb	342 cdb	442 ddb
143 adc	243 bdc	343 cdc	443 ddc
144 add	244 bdd	344 cdd	444 ddd

The same has done for three-digit 5-adic numbers, where four digits are identified as $a = 1$, $b = 2$, $c = 3$, $d = 4$. These cases illustrate ultrametric spaces of $W_{4,3}(64)$. Here 64 three-digit 5-adic numbers (three-letter words) are presented so that within boxes 5-adic distance is the smallest, i.e. $d_5(x, y) = \frac{1}{25}$, while 5-adic distance between any two boxes in vertical line is $\frac{1}{5}$ and otherwise is equal to 1. Ultrametric tree illustration of these cases is in Fig. 1

Note that our notation of natural numbers by position of digits is opposite with respect to the usual decimal one. The four letters $\{a, b, c, d\}$ can be identified with the four digits $\{1, 2, 3, 4\}$ as follows: $a = 1$, $b = 2$, $c = 3$, $d = 4$. Then there are 64 words presented in two different ways—by three letters and three digits, see Table 2.

In the case $W_{4,3}(64)$ there are three-letter words represented now by three-digit 5-adic numbers (see Table 2). The corresponding 5-adic distance of a pair of words (numbers) $x = x_0 + x_1 5 + x_2 5^2 \equiv x_0 x_1 x_2$ and $y = y_0 + y_1 5 + y_2 5^2 \equiv y_0 y_1 y_2$ is

$$d_5(x, y) = |x_0 x_1 x_2 - y_0 y_1 y_2|_5 = \begin{cases} 1, & x_0 \neq y_0 \\ \frac{1}{5}, & x_0 = y_0, x_1 \neq y_1 \\ \frac{1}{25}, & x_0 = y_0, x_1 = y_1, x_2 \neq y_2 . \end{cases} \tag{5}$$

For example, $d_5(123, 213) = 1$, $d_5(123, 132) = \frac{1}{5}$, $d_5(123, 122) = \frac{1}{25}$.

We shall see later that p-adic distance between words is finer and more informative than the ordinary and the Baire distances. Namely, for the same set of natural numbers one can also employ p-adic distance with $p \neq 5$.

The most advanced examples of the ultrametric spaces are the fields of p-adic numbers $\mathbb{Q}_p$, where index p denotes any prime number. There are infinitely many fields $\mathbb{Q}_p$ which are not mutually isomorphic—for every prime number p there is its own $\mathbb{Q}_p$. There are also infinitely many algebraic extensions, which are some analogs of the classical field $\mathbb{C}$ of complex numbers. The field $\mathbb{Q}_p$ can be constructed by completion of the field $\mathbb{Q}$ of rational numbers in the same procedure as it is usually done for the field $\mathbb{Q}_\infty \equiv \mathbb{R}$ of real numbers, just one has to take $|\cdot|_p$ instead of the usual absolute value $|\cdot|_\infty \equiv |\cdot|$. p-Adic numbers and their functions are rather well-developed part of modern mathematics, see, e.g., books [16, 17]. Many applications from Planck scale physics via complex systems to the universe as a whole, known as p-adic mathematical physics, have been considered, e.g. see recent review articles [18, 19]. p-Adic and standard models (over real and complex numbers) are connected within adelic framework, see adelic quantum mechanics [20, 21]. In this modeling of the genetic code only p-adic distance is used.

The above examples illustrate how ultrametric distance measures dissimilarity between two words, or in other words, dissimilarity between two elements of an ultrametric space. Also these ultrametric examples can be represented by trees. Namely, instead of the four letters $\{a, b, c, d\}$ or digits $\{1, 2, 3, 4\}$ in the three-letter words one can take four line segments to draw edges of the related tree (see Fig. 1).

4 The p-Adic Genetic Code

Now we want to apply the above presented ultrametric and, in particular, p-adic distance to the investigation of the genetic code properties. To this end we have to establish connection between set of nucleotides $\{C, A, U, G\}$ and set of

Fig. 1 Ultrametric tree related to Table 2. Tree is related to $W_{4,3}(64)$ case and also to the vertebrate mitochondrial code presented at Table 4. One can easily calculate ordinary ultrametric distance and see that distance between any three tree end points satisfies the strong triangle (ultrametric) inequality

Table 3 Eight possible connections between the nucleotides {C, G, U, A} and the digits {1, 2, 3, 4} which take care that 2-adic distance between two pyrimidines (C, U), as well as between two purines (A, G), is $\frac{1}{2}$

C = 1	A = 2	U = 3	G = 4
U = 1	G = 2	C = 3	A = 4
C = 1	G = 2	U = 3	A = 4
U = 1	A = 2	C = 3	G = 4
A = 1	C = 2	G = 3	U = 4
G = 1	U = 2	A = 3	C = 4
A = 1	U = 2	G = 3	C = 4
G = 1	C = 2	A = 3	U = 4

digits {1, 2, 3, 4}. There are 4! possible connections. However, taking into account chemical properties of nucleotides and coded amino acids, 4! possibilities can be reduced to 8 options presented in Table 3. Namely, there are two pyrimidines which have similar structure (one ring) and two purines which also have similar structure (two rings). Note that there is strong connection in living organisms between structure and function. This similarity within two pyrimidines, as well as similarity within two purines, can be described by 2-adic distance. Fortunately, by 2-adic distance one can also express dissimilarity between purines and pyrimidines. Since $d_2(3, 1) = d_2(4, 2) = |2|_2 = \frac{1}{2}$ one has to connect nucleotides and digits so that $d_2(\text{U,C}) = d_2(\text{G,A}) = \frac{1}{2}$ and $d_2(purine, pyrimidine) = 1$. As the most suitable we take the following identification: C = 1, A = 2, U = 3, G = 4.

Since the vertebrate mitochondrial ((*VM*)) code is simpler than the standard genetic code, we apply this approach first of all to the *VM* code, see Table 4 with symmetry in distribution of codon doublets and quadruplets with respect to the middle vertical line. It is worth noting that an amino acid in this *VM* code is coded either by one, two or three pairs of codons. Every such pair of codons has the same first two nucleotides and at third position are two pyrimidines or two purines. In fact there are 16 codon quadruplets within which 5-adic distance is the smallest, i.e. $\frac{1}{25}$. A pair of two codons which has simultaneously 5-adic distance $\frac{1}{25}$ and 2-adic distance $\frac{1}{2}$ is called doublet. In the case of *VM* code, 64 codons can be viewed as 32 codon doublets, which are distributed as follows: 12 amino acids (His, Gln, Asn,

Table 4 The vertebrate mitochondrial code with p-adic ultrametric structure

111 CCC Pro	211 ACC Thr	311 UCC Ser	411 GCC Ala
112 CCA Pro	212 ACA Thr	312 UCA Ser	412 GCA Ala
113 CCU Pro	213 ACU Thr	313 UCU Ser	413 GCU Ala
114 CCG Pro	214 ACG Thr	314 UCG Ser	414 GCG Ala
121 CAC His	221 AAC Asn	321 UAC Tyr	421 GAC Asp
122 CAA Gln	222 AAA Lys	322 UAA Ter	422 GAA Glu
123 CAU His	223 AAU Asn	323 UAU Tyr	423 GAU Asp
124 CAG Gln	224 AAG Lys	324 UAG Ter	424 GAG Glu
131 CUC Leu	231 AUC Ile	331 UUC Phe	431 GUC Val
132 CUA Leu	232 AUA Met	332 UUA Leu	432 GUA Val
133 CUU Leu	233 AUU Ile	333 UUU Phe	433 GUU Val
134 CUG Leu	234 AUG Met	334 UUG Leu	434 GUG Val
141 CGC Arg	241 AGC Ser	341 UGC Cys	441 GGC Gly
142 CGA Arg	242 AGA Ter	342 UGA Trp	442 GGA Gly
143 CGU Arg	243 AGU Ser	343 UGU Cys	443 GGU Gly
144 CGG Arg	244 AGG Ter	344 UGG Trp	444 GGG Gly

Digits are related to nucleotides as follows: C = 1, A = 2, U = 3, G = 4. 5-Adic distance between codons: $\frac{1}{25}$ inside quadruplets, $\frac{1}{5}$ between different quadruplets in the same column, 1 otherwise. Each quadruplet can be viewed as two doublets, where every doublet code one amino acid or termination signal (Ter). 2-Adic distance between codons in doublets is $\frac{1}{2}$. Two doublets which code the same aa belong to the same quadruplet. Amino acids leucine (Leu) and serine (Ser) are coded by three doublets—the third doublet is at $\frac{1}{2}$ 2-adic distance with respect to the corresponding doublet in quadruplet, which contains the first two doublets

Lys, Tyr, Asp, Glu, Ile, Met, Phe, Cys and Trp) are coded by single doublets, 6 *aa* (Pro, Thr, Ala, Val, Arg and Gly) and stop signal are related to two doublets, and 2 *aa* (Ser and Leu) are coded by three doublets.

We can note that each of amino acids serine (Ser) and leucine (Leu) has one doublet which is separated from their quadruplet codons at 5-adic distance equal 1, but codons in the doublet are still at $\frac{1}{2}$ 2-adic distance with respect to codons in the corresponding doublet within quadruplet.

Skipping digit 0 is suitable in p-adic modeling of the genetic code. Namely, to use the digit 0 for a nucleotide is inadequate, because it may lead to non-uniqueness in the representation of the sequence of codons in DNA and RNA by natural numbers.

4.1   On Evolution of the Genetic Code

The origin and early evolution of the genetic code are among the most intriguing and important problems of the origin and evolution of the life. Since there are no fossils from that time, it gives rise to some speculations. However, it might be possible that some of these hypotheses could be tested as related traces in the modern genomes.

Biological evolution is an adaptive development of simpler living systems to more complex ones. Living organisms are open systems in permanent interaction with environment, guided by some internal rules and environmental factors.

One can conjecture [9] on the evolution of the genetic code using *p*-adic approach to the genetic code and the genomic space, assuming that simpler codons coded older amino acids.

According to this hypothesis standard codons evolved in three steps.

(i) *Single nucleotide codons.*

In this primitive code, single four nucleotides $\{C, A, U, G\}$ play role of codons and code four amino acids $\{Gly, Ala, Asp, Val\}$, which are the oldest according to their temporal appearance, see Table 5. These amino acids are in the last column of Table 4 and it is natural to assign mid-nucleotides to these amino acids. Then the primitive code reads: $C = 1 = Ala$, $A = 2 = Asp$, $U = 3 = Val$, $G = 4 = Gly$.

(ii) *Dinucleotide codons.*

Dinucleotides are codons made of two nucleotides. According to this hypothesis they appeared by adding four nucleotides in front of every four primitive codons: $C \rightarrow CC, AC, UC, GC$; $A \rightarrow CA, AA, UA, GA$; $U \rightarrow CU, AU, UU, GU$; $G \rightarrow CG, AG, UG, GG$. Addition of nucleotide G to the primitive code does not change its meaning, i.e. $GC = 41 = Ala$, $GA = 42 = Asp$, $GU = 43 = Val$, $GG = 44 = Gly$, see Tables 4 and 6. Other 12 dinucleotides code 11 new amino acids so that serine (Ser) is coded twice, see Table 5.

Table 5 Temporal appearance of the 20 standard amino acids

(1)	Gly	(2)	Ala	(3)	Asp	(4)	Val
(5)	Pro	(6)	Ser	(7)	Glu	(8)	Leu
(9)	Thr	(10)	Arg	(11)	Ile	(12)	Gln
(13)	Asn	(14)	His	(15)	Lys	(16)	Cys
(17)	Phe	(18)	Tyr	(19)	Met	(20)	Trp

For details of this temporal appearance of the 20 standard amino acids, see Trifonov's paper [22]

Table 6 Table of amino acids coded by the codons which have pyrimidine at the third position

11(11) CC Pro	21(12) AC Thr	31(13) UC Ser	41(14) GC Ala
12(21) CA His	22(22) AA Asn	32(23) UA Tyr	42(24) GA Asp
13(31) CU Leu	23(32) AU Ile	33(33) UU Phe	43(34) GU Val
14(41) CG Arg	24(42) AG Ser	34(43) UG Cys	44(44) GG Gly

Only serine (Ser) appears twice. By this way, there is a formal connection between the amino acids and the root (dinucleotide) of codons coding them. Identifying these amino acids with related codon roots (i.e. first two digits of 5-adic numbers) one gets some ultrametricity between above amino acids (on importance of 16 codon roots, see Rumer's paper). Since the amino acids which are coded by codons having the same nucleotide at the second position have the similar chemical properties, it is better to use ultrametric distance assigning digits to amino acids in opposite way, as it is done in the brackets. This interchange of digits could be related to evolution of the genetic code

(iii) *Trinucleotide codons.*

Trinucleotide codons exist in all living cells. They were formed by adding all four nucleotides C, A, U(T), G at the end of each 16 dinucleotides, i.e. at the third position. These 64 codons provided possibility to code all 20 amino acids and stop signal by more than 1 codon. Those dinucleotide codons that got pyrimidine at the third position did not change their meaning, cf. Tables 4 and 6.

According to this genetic code evolution, formation of a trinucleotide codon started by nucleotide which is now at the second position, then it was added nucleotide which is now at the first position, and finally appeared nucleotide at the third position. This evolution of codons is in agreement with evidence that codons with the same nucleotide at the second position code amino acids with similar chemical properties. Also this evolution hypothesis in three steps gives an explanation why set of 64 codons is an ultrametric space, see Table 4 for the vertebrate mitochondrial code.

Further evolution of the genetic code is related to variations in assignment of some amino acids to some trinucleotide codons. These variations can be viewed as slight changes in the *VM* code. For example, the standard code obtains by the following changes:

- 232 (AUA): Met → Ile,
- 242 (AGA) and 244 (AGG): Ter → Arg,
- 342 (UGA): Trp → Ter.

It seems that all versions of the genetic code are optimally arranged with respect to possible mutations and evolution of other cell ingredients (coevolution [23]).

4.2   On the Genetic Code as an Ultrametric Network

Many systems can be considered as networks, which are sets of nodes (vertices) joined together by edges (links). There are many examples in biological and social systems. Let us see how the genetic code can be viewed as a *p*-adic ultrametric network.

One can start from two separate systems of biomolecules—one related to 4 nucleotides and another based on 20 standard amino acids. Four types of nucleotides are chemically linked to a large number of various sequences known as DNA and RNA. Standard amino acids can also be chemically linked and form various peptides and proteins. These sequences of DNA and RNA, as well as peptides and proteins, are well-known examples of biological networks.

By the genetic code amino acids are linked to codons, which are elements of an ultrametric space. Since standard amino acids can also be formally regarded as elements of an ultrametric space (see Table 7), one can say that the genetic code links two ultrametric networks to one larger ultrametric network of 85 elements (64 codons + 20 aa + 1 stop signal). Note that one can also consider ultrametric distance between codons and amino acids with stop signal.

Table 7 The rewritten and extended Table 6, where the first two digits are replaced

11 Pro	12 Thr	13 Ser	14 Ala			
21 His	22 Asn	23 Tyr	24 Asp	212 Gln	222 Lys	242 Glu
31 Leu	32 Ile	33 Phe	34 Val	322 Met		
41 Arg		43 Cys	44 Gly	432 Trp		

Third digits are added to the amino acids which are coded by one doublet with purine at the third position. Table contains ultrametrics between amino acids, which corresponds to some of their physicochemical properties. 5-Adic distance between amino acids in rows is either $\frac{1}{5}$ or $\frac{1}{25}$, otherwise it is equal to 1

If we look at codons as an ultrametric network with information content, then they are nodes mutually linked by p-adic distance. Recall that there are three possibilities of 5-adic distance between codons: $\frac{1}{25}$, $\frac{1}{5}$ and 1. With respect to these distances, we can, respectively, call the corresponding subsets of codons as small, intermediate and large community. Thus, any codon has 3 neighbors at 5-adic distance $\frac{1}{25}$ and makes a small community. Any codon is also linked to 12 and 48 other codons to make an intermediate and large community, respectively. Consequently, any codon belongs simultaneously to a small, intermediate and large community.

Amino acids in Table 7 have the following physicochemical similarities.

- First row: small size and moderate in hydropathy.
- Second row: average size and hydrophilic.
- Third row: average size and hydrophobic.
- Fourth row: special case of diversity.

5 The *p*-Adic Genome

In the previous section we demonstrated that codons and amino acids are elements of some p-adic ultrametric spaces. Ultrametric approach should be useful also in investigation of similarity (dissimilarity) between definite sequences of DNA, RNA and proteins. These sequences can be genes, microRNA, peptides, or some other polymers. Since elements of genes (proteins) are codons (amino acids), which have ultrametric properties, it is natural to use their ultrametric similarity in determination of similarity between genes (proteins). It means that one can consider not only ultrametric similarity between two sequences (strings) but also ultrametrically improved Hamming distance.

5.1 p-Adic Modification of the Hamming Distance

Let $a = a_1 a_2 \cdots a_n$ and $b = b_1 b_2 \cdots b_n$ be two strings of equal length. Hamming distance between these two strings is $d_H(a, b) = \sum_{i=1}^{n} d(a_i, b_i)$, where $d(a_i, b_i) = 0$ if $a_i = b_i$, and $d(a_i, b_i) = 1$ if $a_i \neq b_i$. In other words, $d_H(a, b) = n - v$, where v is the number of positions at which elements of both strings are equal. We introduce p-adic Hamming distance in the following way: $d_{pH}(a, b) = \sum_{i=1}^{n} d_p(a_i, b_i)$, where $d_p(a_i, b_i) = |a_i - b_i|_p$ is p-adic distance between numbers a_i and b_i. When $a_i, b_i \in \mathbb{N}$ then $d_p(a_i, b_i) \leq 1$. If also $a_i - b_i \neq 0$ is divisible by p, then $d_p(a_i, b_i) < 1$. There is the following relation: $d_{pH}(a, b) \leq d_H(a, b) \leq d(a, b)$, where $d(a, b)$ is the ordinary ultrametric distance. In the case of strings as parts of DNA, RNA and proteins, this modified distance is finer and should be more appropriate than Hamming distance itself. For example, elements a_i and b_i can be nucleotides, codons and amino acids with above assigned natural numbers, and primes $p = 2$ and $p = 5$.

6 Concluding Remarks

In this paper we presented some simple examples of ultrametric spaces. We applied p-adic distances for modeling of 64 codons and 20 standard amino acids. Ultrametric space of codons is illustrated by the corresponding tree. We emphasized that degeneracy of the vertebrate mitochondrial code has strong ultrametric structure. A p-adic approach to evolution of codons and the genetic code is presented. It is shown that codons and amino acids can be viewed as ultrametric networks which are connected by the genetic code. Investigation of similarity (dissimilarity) between genes, microRNA, proteins and some other polymers by p-adic ultrametric approach, in particular by the p-adic Hamming distance, is proposed.

We plan to employ this ultrametric approach to investigation of concrete DNA, RNA and protein sequences. This approach can also be applied to analyze similarity of words in some human languages and some systems of hierarchical structure.

Acknowledgements This work was supported in part by Ministry of Education, Science and Technological Development of the Republic of Serbia, projects: OI 173052, OI 174012, TR 32040 and TR 35023.

References

1. E.V. Koonin, A.S. Novozhilov, Origin and evolution of the genetic code: the universal enigma. Life **61**(2), 99–111 (2009)
2. B. Hayes, The invention of the genetic code. Am. Sci. **86**(1), 8–14 (1998)
3. Y.B. Rumer, On systematization of codons in the genetic code. Dokl. Acad. Nauk USSR **167**(6), 1393–1394 (1966) [in Russian]

4. J.E.M. Hornos, Y.M.M. Hornos, Algebraic model for the evolution of the genetic code. Phys. Rev. Lett. **71**, 4401–4404 (1993)
5. M. Forger, S. Sachse, Lie superalgebras and the multiplet structure of the genetic code I: codon representations. J. Math. Phys. **41**(8), 5407–5422 (2000)
6. J.D. Bashford, I. Tsohantjis, P.D. Jarvis, Codon and nucleotide assignments in a supersymmetric model of the genetic code. Phys. Lett. A **233**, 481–488 (1997)
7. L. Frappat, A. Sciarrino, P. Sorba, Crystalizing the genetic code. J. Biol. Phys. **27**, 1–38 (2001). arXiv:physics/0003037
8. B. Dragovich, A. Dragovich, A *p*-adic model of DNA sequence and genetic code. p-Adic Numbers Ultrametric Anal. Appl. **1**(1), 34–41 (2009). arXiv:q-bio.GN/0607018v1
9. B. Dragovich, A. Dragovich, *p*-Adic modelling of the genome and the genetic code. Comput. J. **53**(4), 432–442 (2010). arXiv:0707.3043v1 [q-bio.OT]
10. B. Dragovich, *p*-Adic structure of the genetic code (2012). arXiv:1202.2353 [q-bio.OT]
11. B. Dragovich, Genetic code and number theory. Facta Universitatis: Phys. Chem. Techn. **14**(3), 225–241 (2016). arXiv:0911.4014 [q-bio.OT]
12. B. Dragovich, A.Y. Khrennokov, N.Ž. Mišić, Ultrametrics in the genetic code and the genome. Appl. Math. Comput. **309**, 350–358 (2017). arXiv:1704.04194 [q-bio.OT]
13. A. Khrennikov, S. Kozyrev, Genetic code on a diadic plane. Phys. A: Stat. Mech. Appl. **381**, 265–272 (2007). arXiv:q-bio/0701007
14. A.V. Finkelshtein, O.B. Ptitsyn, *Physics of Proteins* (Academic Press, London, 2002)
15. F.H.C. Crick, Codon–anticodon pairing: the wobble hypothesis. J. Mol. Biol. **19**, 548–555 (1966)
16. W.H. Schikhof, *Ultrametric Calculus: An Introduction to p-Adic Calculus* (Cambridge University Press, Cambridge, 1984)
17. I.M. Gel'fand, M.I. Graev, I.I. Pyatetski-Shapiro, *Representation Theory and Automorphic Functions* (Saunders, Philadelphia, 1969)
18. B. Dragovich, A.Y. Khrennikov, S.V. Kozyrev, I.V. Volovich, On *p*-adic mathematical physics. p-Adic Numbers Ultrametric Anal. Appl. **1**(1), 1–17 (2009). arXiv:0904.4205v1 [math-ph]
19. B. Dragovich, A.Y. Khrennikov, S.V. Kozyrev, I.V. Volovich, E.I. Zelenov, *p*-Adic mathematical physics: the first 30 years. p-Adic Numbers Ultrametric Anal. Appl. **9**(2), 87–121 (2017). arXiv:1705.04758[math-ph]
20. B. Dragovich, Adelic model of harmonic oscillator. Theor. Math. Phys. **101**, 1404–1415 (1994). arXiv:hep-th/0402193
21. B. Dragovich, Adelic harmonic oscillator. Int. J. Mod. Phys. A **10**, 2349–2365 (1995). arXiv:hep-th/0404160
22. E.N. Trifonov, The triplet code from first principles. J. Biomol. Struc. Dyn. **22**(1), 1–11 (2004)
23. J.T.F. Wong, A co-evolution theory of the genetic code. Proc. Nat. Acad. Sci. USA **72**, 1909–1912 (1975)

Stochastic Assessment of Protein Databases by Generalized Entropy Measures

R. P. Mondaini and S. C. de Albuquerque Neto

1 Introduction

In this work we continue the development of the topic of identifying domain regions in protein databases via functions of random variables like generalized entropy measures [1–4]. We take as granted the familiarity with protein as sequences of amino acids and with protein domains as characterized by expert biologists [5–8]. From $\underline{m}$ protein family domains, we isolate n_l amino acids on each domain, $l = 1, 2, \ldots, m$. We then discard all domains such that $n_l < n$ with n specified a priori and we also delete $(n_l - n)$ amino acids on all remaining domains with $n_l > n$. According to Fig. 1, we then form blocks of $m \times n$ amino acids each. The sample space of the subsequent statistical analysis to be undertaken is then formed from the association of at least one of these blocks to each protein family domain.

Our aim is to identify the Evolution of proteins on protein databases via the distribution of random variables functions like entropy measures.

By considering t ordered values of the n columns of the $m \times n$ block above, $j_1 < j_2 < \cdots < j_{t-1} < j_t$ with $1 \le t \le n$ such that

R. P. Mondaini (✉) · S. C. de Albuquerque Neto
Federal University of Rio de Janeiro, Centre of Technology, COPPE, Rio de Janeiro, RJ, Brazil
e-mail: Rubem.Mondaini@ufrj.br

© Springer International Publishing AG, part of Springer Nature 2018
R. P. Mondaini (ed.), *Trends in Biomathematics: Modeling, Optimization and Computational Problems*, https://doi.org/10.1007/978-3-319-91092-5_7

```
LKNIQANIPKEVLTVITGVAGSGKSTLIHSVFLKEYPDAIVID...SNPATYTGIMDPIRKAFGKENDVSPSLF
LKNVTVDIPEGVLTVVTGVAGSGKSSLIHHAFLPEHPEAIVID...SNPATYTGIMDDLRKRFAKANGQSPSLF
LHGIDARFPLGAFTAVTGVSGSGKSSLVSQALVELVGEQLGQE...LPQAGALPRTT
IQDLSVKFPLQNLVAITGVSGSGKSSLILQTLLPFAQEELNRA...QIEGLEKLDKVIYLDQSP
LHDVTVDFPLGKFVCVTGVSGGGKSTLTIETLYKNAAMKLNGA...IKGFEHLDKVIDIDQRAIGRTPRS
LKKINVNIPLGKFVVVTGVSGSGKSTLVNQIIVNAIAKNLGTT...IRGLFNIDKLIAINQSPIGRTPRSNPA
LKDVTLTLPVGLFSCITGVSGSGKSTLINDTLYSIAQRQLNGA...EIQGLEHFDKVIDIDQSPIGRTPRSNPA
LQNVTARFPLGKFIAVTGVSGSGKSTLINSILKKAIAQKLNRN...ITGIEHVDR
LKNVNVKIPLGVFSAVTGVSGSGKSTLVNEILYKTLARDLNRA...IRGLEHIEKVVEIDQSPIGRTPRSNPA
LNGLNFTVGAGERVGIVGRTGAGKSTLAAVLFRLLENVEGTIL...KLSQLRSRLAIIPQDPFLFSGTLR
IQDINFQCSLSSRIAVIGPNGAGKSTLINVLTGELLPTSGEVY...IKQHAFAHIDNHLDKTPSEYIQWRFQTG
LVDAKLRLKAGQRYALVGRNGSGKSTLLKAIAEKLIPGISETT...LTDVNSDLRPSHIPTDIGESGQGRGVLQ
LKQIDVDFPLGEFVVVTGVSGSGKSTLVNDVLKRVLAQKLNRN...VSGIKNIERLVNIDQSPIGRTPRSNPA
                    :                                      :
LQDIDVKFPLGKFVAVTGVSGSGKSTLVNSILKKAIAQKLNRN...VEGIENIER
LKNIDVEFPLRVLTSVTGVSGSGKSTLVNEILYKGLNKKINKS...IIGEENIDKIINIDQSP
LKNVSVEFPLGEFVAVTGVSGSGKSTLVNQILKKALAQKLNRN...ITGYEAIEKIVDIDQSPIGRTPRSNPA
LKHINLEIPLEKFTCVTGCSGCGKSSLVYDTIYAESQRGFLEG...LMDKPKVGKIENLRPALNISQNYY
VKDISFNIYQGEVLGLVGESGSGKSTTGSALIGLARHSFGDII...GEKVTKELTDFMVNNVQMIFQDPSNSLN
LQNVTATFPLGTLTAVTGVSGSGKSSLVNDILYRVLANELNGA...VTGLENLDKVVHVDQNPIGRTPRSNPA
LKNINIRIPIDVVTVLTGVAGSGKSSLVKELKNSLDVPYIDLA...STPATYLDILDPIRKLFAQ
```

Fig. 1 A $m \times n$ block of amino acids—a representative of a protein family

$$j_1 = 1, 2, \ldots, (n - t + 1)$$

$$j_2 = (j_1 + 1), \ldots, (n - t + 2)$$

$$\vdots \tag{1}$$

$$j_{t-1} = (j_1 + t - 2), \ldots, (n - 1)$$

$$j_t = (j_1 + t - 1), \ldots, (n)$$

and the random variables to be given by probabilities of occurrence:

$$p_{j_1 j_2 \ldots j_t}(a_1, a_2, \ldots, a_t) = \frac{n_{j_1 j_2 \ldots j_t}(a_1, a_2, \ldots, a_t)}{m}, \qquad 1 \le t \le n, \tag{2}$$

where

$$a_1, a_2, \ldots, a_t = \text{A, C, D, E, F, G, H, I, K, L, M, N, P, Q, R, S, T, V, W, Y} \tag{3}$$

are the amino acids in the one-letter code, and $n_{j_1 j_2 \ldots j_t}(a_1, a_2, \ldots, a_t)$ is the number of occurrence of amino acids on each row of the ordered set of t columns $j_1, j_2, \ldots, j_t$.

We also have

$$\sum_{a_1,a_2,\ldots,a_t} n_{j_1\ j_2\ldots j_t}(a_1, a_2, \ldots, a_t) = m \implies \sum_{a_1,a_2,\ldots,a_t} p_{j_1\ j_2\ldots j_t}(a_1, a_2, \ldots, a_t) = 1 \qquad (4)$$

There are then $\binom{n}{t}$ objects $p_{j_1\ j_2\ldots j_t}$ and each of those has $(20)^t = 10^{\left(1+\frac{\log 2}{\log 10}\right)t}$ $\approx 10^{1.3t}$ components which are given by Eq. (2).

2   From a Binomial Distribution to a Poisson Process

"Poissonization" is a current process in Mathematical Statistics. A discrete binomial probability distribution with discrete time is embedded in a Poisson process of continuous time [9]. This approach will be described in the present section. Let $p(n_j(t), t)$ be the probability of finding n_j amino acids of the same kind on the jth column. Let σ be the probability per unit time that a transition will occur of an amino acid from this column to a subsequent column. Then $\sigma \Delta t$ is the probability that this transition will occur on the interval of time Δt. Then $(1 - \sigma \Delta t)$ is the probability that no transition to another column will occur during the same interval of time Δt. The master equation of this process can be written:

$$p(n_j(t + \Delta t), t + \Delta t) = \sigma \Delta t p(n_j(t) - 1, t) + (1 - \sigma \Delta t) p(n_j(t), t). \qquad (5)$$

We can then write for $\Delta t \to 0$,

$$\frac{\partial p(n_j(t), t)}{\partial t} = \sigma \left(p(n_j(t) - 1, t) - p(n_j(t), t) \right) \qquad (6)$$

$$\frac{\partial p(n_0(t), t)}{\partial t} = -\sigma p(n_0(t), t) \qquad (7)$$

At this point we introduce a pictorial note which aims to substantiate some reasonings of the present section. Let us imagine that $\underline{m}$ icosahedron dice are being tossed simultaneously by a "master devil player"—the owner of the "Ribosome factory". Let us also imagine that there is an adjoint library to the factory where the results of all throws (books titled "How to build the a-amino acid", a = A, C, D, E, F, G, H, I, K, L, M, N, P, Q, R, S, T, V, W, Y) are allocated. After the first throw of dice by the master devil, his assistant will deliver m books to the 1st librarian. During the 2nd throw of dice, the 1st librarian will deliver his books to the 2nd librarian, and after this 2nd throw, the 1st librarian will receive the new set of $\underline{m}$ books from the assistant. During the 3rd throw, the 2nd librarian will deliver his books to the 3rd librarian and the 1st will deliver his new books to the 2nd librarian. The assistant will then deliver another set of m books to the 1st librarian and so on up to nth throw of the master devil and the delivery of books to the nth librarian. Dices are assumed to be fair, but there is a previous game played both with tetrahedra (faces A, C, G, U) and icosahedra to follow the guidelines of the Genetic

Code and actually, a successful game for organizing protein families should take into consideration unfair icosahedron dice.

Initial conditions can be written for Eqs. (6), (7), by taking into consideration that all amino acids should be stored at the "Ribosome factory" before the start of the process. This means:

$$p(n_0, 0) = 1; \qquad p(n_{j \neq 0}, 0) = 0 \tag{8}$$

Equations (6)–(8) after integration from 0 to t will lead to a Poisson process:

$$p(n_0(t), t) = e^{-\sigma t}$$

$$p(n_1(t), t) = e^{-\sigma t} \sigma t$$

$$p(n_2(t), t) = e^{-\sigma t} \frac{(\sigma t)^2}{2} \tag{9}$$

$$\vdots$$

$$p(n_j(t), t) = e^{-\sigma t} \frac{(\sigma t)^{n_j}}{(n_j)!}$$

The moments of this Poisson distributions will be given by

$$\langle (n_j)^k \rangle = \sum_{n_j=1}^{\infty} (n_j)^k \, e^{-\sigma t} \frac{(\sigma t)^{n_j}}{(n_j)!} = e^{-\sigma t} \left(\sigma t \frac{\partial}{\partial (\sigma t)} \right)^{k-1} \left(\sigma t \, e^{\sigma t} \right) \tag{10}$$

We then have:

$$\langle n_j \rangle = \sigma t \tag{11}$$

$$\langle (n_j)^2 \rangle = \sigma t + \sigma^2 t^2 \tag{12}$$

$$\langle (n_j)^3 \rangle = \sigma t + 3\sigma^2 t^2 + \sigma^3 t^3 \tag{13}$$

$$\langle (n_j)^4 \rangle = \sigma t + 7\sigma^2 t^2 + 6\sigma^3 t^3 + \sigma^4 t^4 \tag{14}$$

$$\langle (n_j)^5 \rangle = \sigma t + 15\sigma^2 t^2 + 25\sigma^3 t^3 + 10\sigma^4 t^4 + \sigma^5 t^5 \tag{15}$$

3   From a Multinomial Distribution to the Gibbs–Shannon Entropy Measure

We start this section by describing in more detail the tasks of the "librarians" of Sect. 2. We now derive a multinomial distribution from the information of allocation of amino acids on the slots of each of the n columns. Let $n_j(a)$ be the number of a-amino acids to be arranged in $n_j(a)$ slots among the m slots of the jth column.

The number of possibilities for arranging, e.g., A-amino acid is

$$N_j(A) = \binom{m}{n_j(A)} = \frac{m!}{n_j(a)!(m - n_j(a))!} \tag{16}$$

In order to arrange $n_j(C)$ amino acids on the slots of the same jth column, we have $(m - n(A))$ available slots. The number of possibilities of arranging these $n_j(C)$ amino acids is $\binom{m-n_j(A)}{n_j(C)}$ and the number of possibilities of arranging $n_j(A) + n_j(C)$ amino acids on the slots of the jth column is

$$N_j(A, C) = \binom{m}{N_j(A)}\binom{m - n_j(A)}{n_j(C)} = \frac{m!}{n_j(A)!\, n_j(C)!(m - n_j(A) - n_j(C))!} \tag{17}$$

There are then $(m - n_j(A) - n_j(C))$ available slots in the jth. The number of possibilities of arranging $n_j(D)$ amino acids is $\binom{m-n_j(A)-n_j(C)}{n_j(D)}$. The number of possibilities of arranging $(n_j(A) + n_j(C) + n_j(D))$ amino acids is

$$N_j(A, C, D) = \binom{m}{N_j(A)}\binom{m - n_j(A)}{n_j(C)}\binom{m - n_j(A) - n_j(C)}{n_j(D)}$$

$$= \frac{m!}{(n_j(A))!(n_j(C))!(n_j(D))!(m - n_j(A) - n_j(C) - n_j(D))!} \tag{18}$$

We then write for the number of possibilities of arranging $n_j(A) + n_j(C) + \cdots + n_j(W) + n_j(Y) = m$ amino acids on the jth slot:

$$N_j(A, C, \ldots, W, Y) = \frac{m!}{\prod_a (n_j(a))!} \tag{19}$$

Each amino acid has a probability of occurrence in the jth column of $p_j(a) = 1/20, \forall j, \forall(a)$.

The probability of finding $n_j(a)$ amino acids in the jth column is $(1/20)^{n_j(a)}$ and the probability of $\sum_a n_j(a) = m$ amino acids is given by

$$\prod_a \left(\frac{1}{20}\right)^{n_j(a)} = \left(\frac{1}{20}\right)^{\sum_a n_j(a)} = \left(\frac{1}{20}\right)^m \tag{20}$$

The probability of observing all possible configurations of m amino acids in the jth column is:

$$p\big(n_j(A), n_j(C), \ldots, n_j(W), n_j(Y)\big) = \frac{m!}{(20)^m \prod_a n_j(a)} \tag{21}$$

This corresponds to a special case of a multinomial distribution which is obtained from

$$\big(x_j(A)+\cdots+x_j(Y)\big)^m = \sum_{n_j(A),\ldots,n_j(Y)=0}^{m} \frac{m!\big(x_j(A)\big)^{n_j(A)} \cdot \ldots \cdot \big(x_j(Y)\big)^{n_j(Y)}}{n_j(A)! \cdot \ldots \cdot n_j(Y)!} \tag{22}$$

We have here $\big(x_j(a) \equiv p_j(a)\big)$, where $p_j(a) = \frac{n_j(a)}{m}$ is the probability of occurrence of the a-amino acid in the jth column as obtained from Eq. (2) for $t = 1$, and (22) will turn into

$$1 = \left(\sum_a p_j(a)\right)^m = \sum_{n_j(A),\ldots,n_j(Y)=0}^{m} \frac{m!\big(p_j(A)\big)^{n_j(A)} \cdot \ldots \cdot \big(p_j(Y)\big)^{n_j(Y)}}{n_j(A)! \cdot \ldots \cdot n_j(Y)!} \tag{23}$$

The maximum number $N_j(A, C, \ldots, W, Y)$ is obtained when $n_j(A) = n_j(C) = \cdots = n_j(W) = n_j(Y)$ which means $p_j(A) = p_j(C) = \cdots = p_j(W) = p_j(Y) = 1/20$, and we have

$$\big(p_j(A)\big)^{n_j(A)} \cdot \ldots \cdot \big(p_j(Y)\big)^{n_j(Y)} = \left(\frac{1}{20}\right)^{n_j(A)+\cdots+n_j(Y)} = \left(\frac{1}{20}\right)^m \tag{24}$$

We then have

$$1 = \sum_{n_j(a)=0}^{m} \frac{m!}{(20)^m \big((n_j(a))!\big)^{20}}, \quad \forall a \tag{25}$$

which is a particular case of Eq. (23).

In order to introduce the Gibbs–Shannon entropy associated to the distribution of amino acids on each column of the $m \times n$ block, [10] we write from Eq. (21)

$$\log p\big(n_j(A), n_j(C), \ldots, n_j(W), n_j(Y)\big) = \log m! - m \log 20 - \sum_a \log \big(n_j(a)\big)! \tag{26}$$

We now consider that $m \gg 1$, $n_j(a) \gg 1$, $\forall a$ and we apply the Stirling approximation to Eq. (26),

$$\log p\big(n_j(a)\big) \approx m \log m - m - m \log 20 - \sum_a \big(n_j(a) \log n_j(a) - n_j(a)\big) \tag{27}$$

Equation (27) can be also written as

$$\frac{1}{m} \log p(n_j(a)) \approx - \log 20 - \sum_a p_j(a) \log p_j(a) \tag{28}$$

where

$$S_j = - \sum_a p_j(a) \log p_j(a) \tag{29}$$

is the Gibbs–Shannon entropy measure [10] associated to the distribution of amino acids of Eq. (21).

This development is able to stress the possibility of working with entropy measures which have the Gibbs–Shannon entropy as a special case. These entropy measures, like the Sharma–Mittal two-parameter set

$$SM_j(r, s) = - \frac{1}{1-r} \left(1 - \left(\sum_a (p_j(a))^s \right)^{\frac{1-r}{1-s}} \right) \tag{30}$$

The limit forms corresponding to Gibbs–Shannon's could be written as

$$\lim_{s \to 1} \lim_{r \to s} SM_j(r, s) = \lim_{s \to 1} \lim_{r \to 2-s} SM_j(r, s) = \lim_{s \to 1} \lim_{r \to 1} SM_j(r, s)$$

$$= S_j = - \sum_a p_j(a) \log p_j(a) \tag{31}$$

These entropy measures should be proficient for analyzing the results associated to intermediate values of the number of amino acids on each jth slot when the Stirling approximation is not valid anymore. This will be studied on a forthcoming publication.

We now check the result for the maximum of the Gibbs–Shannon entropy measure on the $m \times n$ block of amino acids. Since $n_j(A) = \cdots = n_j(Y)$ do correspond to the maximum value of $N_j(A, C, \ldots, W, Y)$, this also corresponds to the maximum value of $p(n_j(A), \ldots, n_j(Y))$ which is given by

$$p_{\max}(n_j(a)) = \frac{m!}{(20)^m \left((n_j(a))! \right)^{20}} \tag{32}$$

and from the Stirling approximation,

$$p_{\max}(n_j(a)) \approx \frac{e^{-m}(m)^m}{(20)^m \left(e^{-n_j(a)} (n_j(a))^{n_j(a)} \right)^{20}}$$

$$= \frac{e^{-m}(m)^m}{(20)^m e^{-m}\left(n_j(a)\right)^m} \equiv 1 \tag{33}$$

We then have from Eq. (28)

$$0 = \frac{1}{m}\log p_{\max}\left(n_j(a)\right) \approx -\log 20 + (S_j)_{\max}$$

or

$$(S_j)_{\max} = \log 20 \tag{34}$$

This is a trivial result and can be taken as a proof of the convenience of the Gibbs–Shannon entropy for describing the distribution of amino acids in the slots of the $m \times n$ blocks and on the limit of a very large number of amino acids.

4 The Fokker–Planck Equation

A Poisson process has been derived at the beginning of Sect. 2 from a master equation, Eq. (5). We now take into consideration the explicit dependence of the probability with the number of amino acids [9]. The probability of transfer of an amino acid from the jth column will lead to

$$p\left(n_j(t) - 1, t\right) = p\left(n_j(t), t\right) - \frac{\partial p\left(n_j(t), t\right)}{\partial n_j(t)} + \frac{1}{2}\frac{\partial^2 p\left(n_j(t), t\right)}{\partial n_j^2(t)} \tag{35}$$

We can then write from Eq. (5) after taking the limit $\Delta t \to 0$,

$$\frac{\partial p\left(n_j(t)\right)}{\partial t} + \sigma \frac{\partial p\left(n_j(t), t\right)}{\partial n_j(t)} - \frac{1}{2}\sigma \frac{\partial^2 p\left(n_j(t), t\right)}{\partial n_j^2(t)} = 0 \tag{36}$$

The solution of this equation will provide a probability distribution function of each kind of amino acid. This is a simple Fokker–Planck equation [11–13] and it can be also written from the specification of the moments according to [13]. We should have:

$$\lim_{t \to 0}\frac{\langle \Delta n_j(t)\rangle}{t} = \lim_{t \to 0}\frac{\langle\left(n_j(t) - 0\right)\rangle}{t - 0} = \lim_{t \to 0}\frac{\sigma t}{t} = \sigma \tag{37}$$

$$\lim_{t \to 0}\frac{\langle\left(\Delta n_j(t)\right)^2\rangle}{t} = \lim_{t \to 0}\frac{\langle\left(n_j(t) - 0\right)^2\rangle}{t - 0} = \lim_{t \to 0}\frac{\sigma t + \sigma^2 t^2}{t} = \sigma \tag{38}$$

$$\lim_{t \to 0} \frac{\langle (\Delta n_j(t))^k \rangle}{t} = \lim_{t \to 0} \frac{\langle (n_j(t) - 0)^k \rangle}{t - 0} = 0, \quad k \geq 3 \tag{39}$$

Equations (37)–(39) means that we should restrict the kinetic analysis to be built from Eq. (36) to the two first moments $\langle n_j(t) \rangle$, $\langle (n_j(t))^2 \rangle$, since Eqs. (13)–(15) are inconsistent with Eq. (39).

From the definition of the moments

$$1 = \int_0^\infty p(n_j(t), t) \mathrm{d}n_j(t) \tag{40}$$

$$\langle n_j(t) \rangle = \int_0^\infty n_j(t) p(n_j(t), t) \mathrm{d}n_j(t) \tag{41}$$

$$\vdots$$

$$\langle (n_j(t))^k \rangle = \int_0^\infty (n_j(t))^k p(n_j(t), t) \mathrm{d}n_j(t) \tag{42}$$

where we have assumed that for large $m \times n$ blocks, $m \gg 1$, $n \gg 1$, all the sums could be replaced by an integration and the boundary conditions could be given by

$$\lim_{n_j \to 0} p(n_j(t), t) = 0 \tag{43}$$

$$\lim_{n_j \to \infty} (n_j(t))^k p(n_j(t), t) = 0 \tag{44}$$

which means that all transferred amino acids will remain inside a region of large size. After using Eqs. (40)–(42), (43), (44), we can write

$$\frac{\partial \langle n_j(t) \rangle}{\partial t} = \sigma \tag{45}$$

$$\frac{\partial \langle (n_j(t))^2 \rangle}{\partial t} = 2\sigma \langle n_j(t) \rangle + \sigma \tag{46}$$

$$\frac{\partial \langle (n_j(t))^3 \rangle}{\partial t} = 3\sigma \langle (n_j(t))^2 \rangle + 3\sigma \langle n_j(t) \rangle \tag{47}$$

$$\frac{\partial \langle (n_j(t))^4 \rangle}{\partial t} = 4\sigma \langle (n_j(t))^3 \rangle + 6\sigma \langle (n_j(t))^2 \rangle \tag{48}$$

$$\vdots$$

$$\frac{\partial \langle (n_j(t))^k \rangle}{\partial t} = k\sigma \langle (n_j(t))^{k-1} \rangle + \frac{k(k-1)}{2}\sigma \langle (nj(t))^{(k-2)} \rangle \tag{49}$$

and we have after integration from 0 to t:

$$\langle n_j(t) \rangle = \sigma t \tag{50}$$

$$\langle \left(n_j(t) \right)^2 \rangle = \sigma t + \sigma^2 t^2 \tag{51}$$

$$\langle \left(n_j(t) \right)^3 \rangle = 3\sigma^2 t^2 + \sigma^3 t^3 \tag{52}$$

$$\langle \left(n_j(t) \right)^4 \rangle = 3\sigma^2 t^2 + 6\sigma^3 t^3 + \sigma^4 t^4 \tag{53}$$

$$\langle \left(n_j(t) \right)^5 \rangle = 15\sigma^3 t^3 + 10\sigma^4 t^4 + \sigma^5 t^5 \tag{54}$$

All equations of this set will obviously satisfy Eqs. (37)–(39) and we notice the equality of Eqs. (50)–(51) with Eqs. (11), (12) respectively.

5   The Maximization of Gibbs–Shannon Entropy Measure

We should stress that the present modelling does not apply to the evolution of an "orphan" protein. Orphan proteins are characterized by $m = 1$ and we have from Eq. (2), with $t = 1$,

$$p_j(a) = \frac{n_j(a)}{1} = 0, 1, \quad \forall j, \forall a \tag{55}$$

From Eqs. (55), (29), we can write

$$S_j = -(0 \cdot \log 0 + 0 \cdot \log 0 + \cdots + 1 \cdot \log 1 + \cdots + 0 \cdot \log 0) = 0 \tag{56}$$

If a successful modelling is supposed to improve the amount of information to be obtained, we should start from a configuration of maximum entropy [10] and not from a minimum one like that of an "orphan" protein. No model will be successful if we start from a minimum of entropy.

The constrained maximization of the Gibbs–Shannon entropy which should be restricted to the moments $\langle n_j(t) \rangle$ and $\langle \left(n_j(t) \right)^2 \rangle$ and has been explained on the last section can be written as

$$0 = \delta \int_0^\infty \Big[-p\big(n_j(t), t\big) \log p\big(n_j(t), t\big) - \lambda_0(t) p\big(n_j(t), t\big)$$

$$- \lambda_1(t) n_j(t) p\big(n_j(t), t\big) - \lambda_2(t) \big(n_j(t)\big)^2 p\big(n_j(t), t\big) \Big] \delta p\big(n_j(t), t\big) \mathrm{d}n_j(t)$$

$$\tag{57}$$

where $\lambda_1(t)$, $\lambda_2(t)$ are Lagrangian multipliers. We then have

$$0 = \int_0^\infty \left[-\log p(n_j(t), t) - 1 - \lambda_0(t) - \lambda_1(t)n_j(t) - \lambda_2(t)n_j^2(t) \right]$$
$$\cdot \delta p(n_j(t), t) dn_j(t) \tag{58}$$

where $\delta p(n_j(t), t)$ are the arbitrary variations of $p(n_j(t), t)$. The probability distribution corresponding to the maximal entropy is

$$p(n_j(t), t) = e^{-(1+\lambda_0(t))} e^{-\lambda_1(t)n_j(t) - \lambda_2(t)n_j^2(t)} \tag{59}$$

The Lagrangian multipliers will be derived from Eqs. (40)–(42). The multiplier $\lambda_0(t)$ is given by:

$$e^{1+\lambda_0(t)} \equiv Z(t) = \int_0^\infty e^{-\lambda_1(t)n_j(t) - \lambda_2(t)n_j^2(t)} dn_j(t) \tag{60}$$

where $Z(t)$ is the Partition Function:

$$Z(t) = \frac{1}{2\sqrt{\lambda_2(t)} \cdot M(y(t))} \tag{61}$$

where

$$y(t) = \frac{\lambda_1(t)}{2\sqrt{\lambda_2(t)}} \tag{62}$$

and

$$M(y(t)) = \frac{1}{\sqrt{\pi}} \cdot \frac{e^{-y^2(t)}}{1 - \text{erf}(y(t))} \tag{63}$$

with erf(y) as the Error Function [11]

$$\text{erf}(y) = \frac{2}{\sqrt{\pi}} \int_0^\infty e^{-z^2} dz \tag{64}$$

Some of the moments $\langle (n_j(t))^k \rangle$ could be given from Eq. (61) as

$$\langle n_j(t) \rangle = -\frac{\partial Z/\partial \lambda_1}{Z} = \frac{1}{\sqrt{\lambda_2}} \frac{1}{N(y)}(1 - y N(y)) \tag{65}$$

$$\langle (n_j(t))^2 \rangle = \frac{\partial^2 Z/\partial \lambda_1^2}{Z} = \frac{1}{\lambda_2} \frac{1}{N(y)}(-2y + (1 + 2y^2)N(y)) \tag{66}$$

$$\langle(n_j(t))^3\rangle = -\frac{\partial^3 Z/\partial\lambda_1^3}{Z} = \frac{1}{\lambda_2\sqrt{\lambda_2}}\frac{1}{N(y)}\left(1 + y^2 - \left(\frac{3}{2}y + y^3\right)N(y)\right) \quad (67)$$

$$\langle(n_j(t))^4\rangle = \frac{\partial^4 Z/\partial\lambda_1^4}{Z} = \frac{1}{2\lambda_2^2}\frac{1}{N(y)}\left(-5y - 2y^3 + \left(\frac{3}{2} + 2y^4\right)N(y)\right) \quad (68)$$

where

$$N(y) \equiv \frac{1}{M(y)} \quad (69)$$

We then have from Eq. (59),

$$p(n_j(t), t) = 2\sqrt{\lambda_2(t)}M(y(t))e^{-\lambda_1 n_j(t) - \lambda_2 n_j^2(t)} \quad (70)$$

Some useful formulae:

$$M'(y) = -2yM(y) + M^2(y) \quad (71)$$

$$M''(y) = -2(1 - 2y^2)M(y) - 6yM^2(y) + 2M^3(y) \quad (72)$$

Asymptotic expansions:

$$0 < y \ll 1, \ M(y) \approx \frac{1}{\sqrt{\pi}}\left(1 + \frac{2}{\sqrt{\pi}}\right)y; \quad y \gg 1, \ M(y) \approx y \quad (73)$$

To each point $(\bar{y}, M(\bar{y}))$ of the function will correspond to a solution of the equations and a distribution function:

$$\frac{1}{\sqrt{\lambda_2}}\frac{1}{N(y)}(1 - y N(y)) = \sigma t \quad (74)$$

$$\frac{1}{\lambda_2}\frac{1}{N(y)}\left[-2y + (1 + 2y^2)N(y)\right] = \sigma t \quad (75)$$

A class of solutions will be obtained by looking for a point $y = \bar{y}$ in this parameter space such that $M^2(\bar{y}) \approx 0$. We then consider the Taylor expansion:

$$M(y) \approx M(\bar{y}) + M'(\bar{y})(y - \bar{y}) + \mathcal{O}(y - \bar{y})^2$$
$$= (1 - 2\bar{y}(y - \bar{y}))M(\bar{y}) + \mathcal{O}(y - \bar{y})^2 \quad (76)$$

and we have:

$$\frac{1}{\lambda_2(t)}\left[\frac{1}{2} + 2\bar{y}^3 M(\bar{y}) + (1 - 2\bar{y}^2)M(\bar{y})y\right] = \sigma t \quad (77)$$

$$\frac{1}{\sqrt{\lambda_2(t)}}\left[y - M(\bar{y})\left(1 - 2\bar{y}(y - \bar{y})\right)\right] = \sigma t \qquad (78)$$

and $y = \lambda_1(t)/2\sqrt{\lambda_2(t)}$.

From Eqs. (77), (78), we obtain the Lagrangian Multipliers:

$$\lambda_1(\bar{y}) \approx \frac{1 + 2\left(1 - 2\bar{y}^2 + 2\bar{y}^3\right)M(\bar{y})}{1 + 2\bar{y}M(\bar{y})} \qquad (79)$$

$$\lambda_2(\bar{y}) \approx \frac{1 + 2\left(1 - 2\bar{y}^2 + 2\bar{y}^3\right)M(\bar{y})}{2\sigma t} \qquad (80)$$

In Fig. 2, we depict the graphs of functions $M(y)$ and $\lambda_1(t)$ as well as the function $\lambda_2(t)$ for three values of the non-dimensional parameter $\sigma t = 1, 2, 3$.

The family of distribution functions corresponding to Eqs. (74), (75) and (77), (78) is given by

$$p\left(n_j(t), t\right) = 2\sqrt{\frac{\lambda_2(\bar{y})}{\pi}} \cdot \frac{e^{-\lambda_2(\bar{y})}\left(n_j(t) - \dfrac{\sigma t}{1 + 2\bar{y}M(\bar{y})}\right)^2}{1 + \mathrm{erf}\left(\dfrac{\sigma t\sqrt{\lambda_2(\bar{y})}}{1 + 2\bar{y}M(\bar{y})}\right)} \qquad (81)$$

In Fig. 3, we show the graphs of three examples of probability distributions of the family above (Table 1).

Fig. 2 ($M(\bar{y})$, solid line), ($\lambda_1(\bar{y})$, red solid line), ($\lambda_2(\bar{y})$, $\sigma t = 1$, blue solid line), ($\lambda_2(\bar{y})$, $\sigma t = 2$, green solid line), ($\lambda_2(\bar{y})$, $\sigma t = 3$, orange solid line)

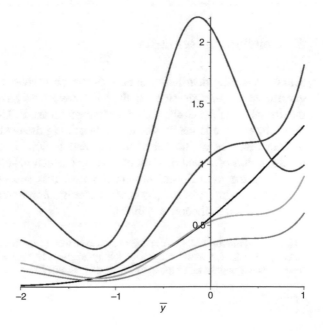

Fig. 3 $(p_1(n_j(t), t)$, red
solid line), $(p_2(n_j(t), t)$, blue
solid line), $(p_3(n_j(t), t)$,
green solid line)

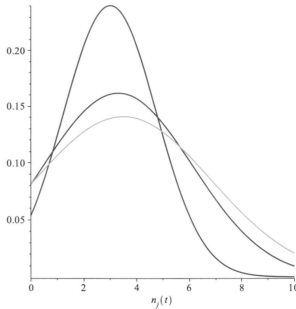

Table 1 Probability distributions (Eq. (81)) corresponding to three sets of values of the parameters

$p(n_j(t), t)$	$M(\bar{y})$	$M^2(\bar{y})$	$\lambda_2(\bar{y})$	$\bar{y}$	σt
$p_1(n_j(t), t)$	≈ 0	0	1	-3	3
$p_2(n_j(t), t)$	≈ 0.030245	≈ 0	0.06333	-1.5	3
$p_3(n_j(t), t)$	≈ 0.053828	≈ 0	0.045123	-1.3	3

6 Concluding Remarks

Many probability distribution functions can be derived from the maximization of
generalized entropy measures. In the present work we have chosen to work with the
maximization of the Gibbs–Shannon entropy measure. The essential message to be
remembered here is the feasibility of describing the evolution of the probabilities
of occurrence of amino acids on specified protein families as driven from the
maximization of a generic entropy measure which is supposed to be effective for
describing the formation of protein families. It is supposed that this will help to
clarify the evolution of the association of protein families into Clans. This will be
the subject of a forthcoming publication.

Acknowledgements S.C. de Albuquerque Neto thanks to the International Union of Biological
Sciences (IUBS) for partial support of living expenses in Moscow, during the 17th BIOMAT
International Symposium, October 29–November 04, 2017.

References

1. R.P. Mondaini, S.C. de Albuquerque Neto, Entropy measures and the statistical analysis of protein family classification, in *BIOMAT 2015* (2016), pp. 193–210
2. R.P. Mondaini, S.C. de Albuquerque Neto, The pattern recognition of probability distributions of amino acids in protein families, in *BIOMAT 2016* (2017), pp. 29–50
3. R.P. Mondaini, A survey of geometric techniques for pattern recognition of probability of occurrence of amino acids in protein families, in *BIOMAT 2016* (2017), pp. 304–326
4. R.P. Mondaini, Entropy measures based method for the classification of protein domains into families and clans, in *BIOMAT 2013* (2014), pp. 209–218
5. R.D. Finn et al., Pfam: clans, web tools and services. Nucleic Acids Res. **34**, D247–D251 (2006)
6. M. Punta et al, The Pfam protein families database. Nucleic Acids Res. **40**, D290–D301 (2012)
7. R.D. Finn et al., The Pfam protein families database. Nucleic Acids Res. **42**, D222–D230 (2015)
8. R.D. Finn et al., The Pfam protein families database. Nucleic Acids Res. **44** D279–D285 (2016)
9. W. Bialek, *Biophysics – Searching for Principles* (Princeton University Press, Princeton, 2012)
10. E.T. Jaynes, *Probability Theory - The Logic of Science* (Cambridge University Press, Cambridge, 2003)
11. Yu.B. Rumer, M.Sh. Ryvkin, *Thermodynamics, Statistical Physics and Kinetics* (Mir Publishers, 1980), pp. 576
12. H. Risken, *The Fokker-Planck Methods of Solutions and Applications*, 2nd edn. Springer Series in Synergetics (Springer, Berlin, 2008)
13. N.G. Van Kampen, *Stochastic Process in Physics and Chemistry*, 3rd edn. (North-Holland, Amsterdam, 2007)

Why Is Evolution Important in Cancer and What Mathematics Should Be Used to Treat Cancer? Focus on Drug Resistance

Luís Almeida, Rebecca H. Chisholm, Jean Clairambault, Tommaso Lorenzi, Alexander Lorz, Camille Pouchol, and Emmanuel Trélat

1 Introduction to Mathematical Modelling in Cancer

Mathematical models of cancer growth and therapy have already known numerous developments and publications in the past 20 years or so. They belong to two general classes: agent-based models, ruled by stochastic rules of growth (for division, death,

L. Almeida
CNRS, Paris, France

Team Mamba, INRIA, Paris, France

Sorbonne Universités, Laboratoire Jacques-Louis Lions, UPMC, Paris, France
e-mail: almeida@ljll.math.upmc.fr

R. H. Chisholm
Melbourne School of Population and Global Health, University of Melbourne, Parkville, VIC, Australia
e-mail: chisholm.r@unimelb.edu.au

J. Clairambault (✉) · C. Pouchol
Team Mamba, INRIA, Paris, France

Sorbonne Universités, Laboratoire Jacques-Louis Lions, UPMC, Paris, France
e-mail: jean.clairambault@inria.fr; pouchol@ljll.math.upmc.fr

T. Lorenzi
Department of Mathematics and Statistics, University of St Andrews, Fife, Scotland
e-mail: tl47@st-andrews.ac.uk

A. Lorz
CEMSE Division, King Abdullah University of Science and Technology, Thuwal, Saudi Arabia

Sorbonne Universités, Laboratoire Jacques-Louis Lions, UPMC, Paris, France

Team Mamba, INRIA, Paris, France
e-mail: alexander.lorz.1@kaust.edu.sa

E. Trélat
Sorbonne Universités, Laboratoire Jacques-Louis Lions, UPMC, Paris, France

Team Cage, INRIA, Paris, France
e-mail: emmanuel.trelat@upmc.fr

© Springer International Publishing AG, part of Springer Nature 2018
R. P. Mondaini (ed.), *Trends in Biomathematics: Modeling, Optimization and Computational Problems*, https://doi.org/10.1007/978-3-319-91092-5_8

motion, interactions with the environment) in which the individual agents are cancer cells, and continuous models that rely on ordinary or partial differential equations, sometimes delay differential equations, whose solutions are densities of cancer cell populations. The benefits and limitations of these two respective classes of models, with examples, are discussed, e.g., in [6]. As regards anticancer treatments, the continuous version allows to take advantage of mathematical optimisation and optimal control algorithms that have been designed in this framework, originally in engineering settings. A short review of models designed with this therapeutic control vision is presented, for instance, in [2]. It is sometimes possible to obtain a continuous model starting from an agent-based one by averaging methods; alternatively, one can also develop in parallel the two types of models applied to the same biological problem and compare the predicted behaviour of the modelled cell populations, as e.g., in [5], or in a general setting, in [3].

The goal of such models of cancer growth may be to merely understand the biological phenomenon of cancer growth, by designing accurate models that are all the more relevant to describe a biological reality as they are identified and validated on biological measurements in vitro in culture dishes, in vivo on laboratory animals, or from observations (e.g., radiological images) on humans, to be confronted to theoretical growth curves depending on a priori unknown parameters (the physicist's viewpoint). But it may also be of a different nature, to represent the effects of treatments on tumours, with the aim to optimise them. In the latter case, these effects may be described either by their molecular effects on known drug targets (keeping in mind that precision targeting is often alluring, since drugs may have unpredictable effects on non recognised targets) or by their functional effects on the possible fates of cell populations, namely proliferation, extinction, differentiation or senescence. The respective advantages of these two points of view are also discussed, with examples, in [6]. Whatever the chosen point of view, molecular or functional, the goal of these models is here clearly established as understanding and improving the efficacy of anticancer treatments (the physician's viewpoint).

2   Drug Resistance in Cancer

2.1   The Two Main Pitfalls of Cancer Therapeutics

Unwanted toxic side effects on healthy cell populations and emergence of resistance to treatments in cancer cell populations are the two main pitfalls of cancer therapeutics in the clinic. Toxicity is always a concern for the clinician, as it limits the tolerable doses of drugs delivered to the patient, who otherwise might see his tumours eradicated, but at the expense of deadly insults to essential organs or functions (haematopoiesis, digestion, skin covering, liver function, heart function). It has lately been proposed that instead of delivering for short periods of time the maximum tolerated dose (MTD), it might be as efficient to deliver small drug doses (the so-called metronomic strategy), thus minimising toxicity, with as good results on the cancer cell population. To what extent is the immune system involved in the efficacy of this new way of designing delivery schedules is not completely clear and might depend on the anticancer drug in use [36, 37], however metronomic therapies certainly challenge the MTD strategy in both limitation of toxicity and improvement of efficacy. Note that an initial interpretation of the success of metronomic therapies was more mechanic, postulating that too high amounts of drugs destroy the blood vessels that bring the drugs to the tumour [1, 24]. It is not exclusive of the immunogenic explanation [36, 37], which proposes that giving small drug doses may reveal hidden (internalised) cancer antigens by shattering a small number of cancer cells, enough to trigger an efficient immune response towards the whole cancer cell population. Both explanations still remain to be more biologically documented—especially with respect to the immune response—mathematically modelled and tried in clinical settings; however, they address the question of toxicity in an apparently paradoxical way ("more is not necessarily better") that is a challenge for modellers.

Drug resistance, the other major pitfall of cancer therapeutics, is a treatment efficacy limitation of another nature; it may be defined as adaptation of the target cancer cell populations to the hostile environment created by the drug. Resistance to treatments in cancer cell populations, insofar as it is not constitutive of organisms therapies apply to, but secondary, i.e., induced by treatments as a stress response. In many cases (in fact, in most cases), treatments that show remarkable initial efficacy by drastically shrinking tumours see their response decrease with time, until they become totally inefficient as tumours regrow. Furthermore, the newly growing cancer cell populations, that have become resistant to the drug in use, are out of reach for this therapy, and often for others that have not been employed (multi-drug resistance). At the molecular scale, different mechanisms have been identified, such as overexpression of drug efflux pumps (ABC transporters [14], such as the P-glycoprotein, also known as MDR1, or ABCB1), of intracellular drug processing enzymes or of DNA repair enzymes, and it has been proposed to combine cytotoxic drugs with inhibitors of these mechanisms, unfortunately eventually to no avail. As mentioned above, the molecular point of view in pharmacological treatments in principle offers a satisfying framework to perform cancer treatment optimisation, but so-called targeted therapies (i.e., that target intracellular molecular pathways), with a few exceptions, result in disappointing outcomes (see, e.g., [11, 12]).

2.2 From the Single Cancer Cell to Cancer Cell Populations

Indeed, these treatments share the same flaw, which is that they focus on a given molecular target (or on several molecular targets), considering cancer as the disease of the same single cell extended to large quantities, instead of taking into account the population of cancer cells in its diversity, which might offer a key explanation of their failure [13]. Such population diversity (or heterogeneity) is not necessarily of genetic nature but linked to epigenetic changes in the chromatin, thus reversible [31], at least on the initiation of drug resistance (mutations can come later to irreversibly establish resistance in a subclone of the cell population), and may result in differently expressed phenotypes in different cells, potentially inducing different resistance mechanisms as responses to cytotoxic stress in a population of cells bearing all the same genotype.

Introducing the *population* of cancer cells (indeed, the actual target of anticancer treatments) naturally sets the scenery for Darwinian evolution of cells exposed to anticancer drugs seen as an environmental selection pressure, as will be developed in the next section. This viewpoint, introduced in theoretical ecology for quite a long time already, is rather new in biology and medicine (where it has given rise to the new field of *Darwinian medicine*), however, does not allow to decide whether the selection is of pure Darwinian nature (selection of the fittest, cells that were already present in the population before exposure to the drug) or may involve a part of Lamarckian adaptation (no resistant cells initially present, but stochastic triggering of resistance mechanisms in a few cells for which the response to stress happens to be well adapted to resist the cytotoxic effects of the drug in use). This alternative, discussed in a mathematical setting in [5] was already the object of the biological experiment by Luria and Delbrück [23], concluding to sheer Darwinian selection. However, Luria and Delbrück's experiment was performed not on cancer cells exposed to drugs, but on bacteria exposed to bacterium-eating viruses (phages), while human and animal cancer cells bear a genome—and epigenome, i.e., chromatin (histones)—that is by far richer than the bacterial genome, which in our case does not allow to conclude. Nevertheless, the cell population point of view clearly opens new ways to understand and overcome drug resistance in cancer.

3 Cancer as Evolutionary Disease

3.1 Evolution of Multicellularity and Cancer

Darwinian evolution (together with possible Lamarckian adaptation) of cancer—and healthy—cell populations (but healthy cell populations are in principle well controlled as regards their possibilities of phenotype evolution) must of course be considered on the short-time level of a human life or disease, but the much larger time of evolution in the course of billions of years, from unicellular organisms towards the organised and coherent forms of multicellularity represented by present animals and plants, may also shed light on cancer as evolutionary disease. Cancer is a disease of multicellular organisms, that may be defined as loss of coherence between tissues due to loss of coherence control by those genes that have been essential in the evolution towards multicellularity. In [9], it is advocated that the genes that are altered in cancer are precisely the ones that have been employed by evolution to design multicellular organisms. Indeed, evolution proceeds, as stated by

Nobel prize laureate François Jacob in [18], by *tinkering*, i.e., it proceeds by trials and errors taking advantage of any existing material, and, as regards multicellularity, such tinkering may result in localised (in organs and physiological functions elicited by corresponding genes) fragilities, that secondarily, under environmental pressure, may be caught off guard and result in localised cancers. Such loss of coherence control, unmasking a pluripotent phenotype that is also named *plasticity*, may in particular be seen in the process of de-differentiation of cancer cells, i.e., adoption of a pluripotential phenotype (eventually yielding the so-called 'cancer stem cell', whose existence is likely to be transient [21]) making the cells that bear it, as endowed with a rich panel of non-repressed genes, able to develop a wide variety of responses to cytotoxic stress. The involvement of such failed multicellularity (i.e., unpreserved normal differentiation) control genes in revealing an ancient 'toolkit' of preexisting adaptations [9] still remains to be documented, but it certainly offers new ways of considering cancer as an evolutionary disease and drug resistance in cancer as an evolutionary—and adaptive—mechanism.

3.2 Heterogeneity and Plasticity in Cancer Cell Populations

Heterogeneity in cancer cell populations has been documented in advanced solid tumours as of genetic nature, with evidence of multiple branched mutations [10], but, as mentioned above, it may also consist of sheer epigenetic and reversible modifications [31] linked to enzymatic activities located on the chromatin, i.e., without mutations in the genome. However, as recently shown in [30], such fast epigenetic, non genetic, reprogramming of a sparse subpopulation of cancer cells may eventually result in a stably resistant state.

Another look at heterogeneity induced in cancer cell populations by exposure to cytotoxic drugs is presented in [35]. In this article, it is proposed that so-called *cold genes* that have been identified as expressed in the genome of cancer cells (multiple myeloma cells) have a very ancient origin, being conserved without changes throughout evolution from unicellularity, and may be responsible for stress response in extremely hostile and unpredictable situations (resulting from events comparable, *mutatis mutandis*, to the impact on animal life of the meteorite that 66 million years ago hit the Earth—creating the Chicxulub crater in Yucatán— subsequently putting an end to the dominance of dinosaurs), by possibly launching secondary expression of various resistance mechanisms. In this respect, these very ancient 'cold genes', elaborated in a remote past of our planet, when conditions of life were different from the present (UV radiation, acidity, low oxygen concentration in the oceans and in the atmosphere), might be the genetic toolkit of preexisting adaptations mentioned above, or part of it.

The variety of resistance mechanisms developed by cancer cell populations exposed to lethal doses of cytotoxic drugs—an extremely hostile and unpredictable situation for any cell population—has been related to what is called *bet hedging* in theoretical ecology. The term 'bet hedging' is used to qualify behaviour relying on an ensemble of traits that make a population of living individuals adaptable to an unpredictable environment, using a so-called 'risk-spreading strategy', that at the scale of the population may result in keeping safe only a small part of it, but a part that will be able to reconstitute the whole population, with preservation of its common genome, after such adaptation to the new environment [26]. Bet hedging in tumours has been proposed as a stochastic 'cancer strategy'. It is also presented as "an ultimate explanation of intra-tumour heterogeneity" in chapter XVII of the book [33].

Plasticity, mentioned above about de-differentiation of cancer cells and the transient state of cancer stem cell, may be evidenced at the level of the single cell (derepression of genes that must be epigenetically repressed in physiology to produce the differentiation that yields about 200 different functional cell types in the human organism), but also at the level of the cell population, since the spreading of such pluripotent cells makes the population adaptable to environmental changes (*plastic*), possibly by using expression of cold genes in a tiny subpopulation and stochastic (or distributed) bet hedging of resistance phenotypes.

The plasticity—physiologically normal in highly undifferentiated cell states, close to stem cells, but totally pathological in cell populations for which a defined terminal physiological function exists—of the *epigenetic landscape* of a given human genome, as metaphorically proposed in [34], recently revised from a systems biology viewpoint by Sui Huang, see, e.g., [15, 16], provides another approach to plasticity and evolution of cancer, that has been, for instance, exploited to study lineage commitment in haematopoiesis by using bifurcation analysis of an ordinary differential equation model [17].

4   Continuous Mathematical Models

4.1   *Phenotype-Structured Mathematical Models*

The modelling framework of adaptive dynamics we present here is more likely to correspond biologically to epigenetic modifications rather than to genetic mutations, as the evolution in phenotype is in this mathematical setting always reversible (not to mention that, in the case of the application to drug resistance in cancer that we have in mind, eventual induction of emergent resistant cell clones due to mutations under drug pressure is never to be excluded in the long run). From the biologist's point of view, we study phenotypically heterogeneous, but genetically homogeneous, cancer cell populations under stress (in particular by cytotoxic drugs).

The models considered here are all based on the so-called *logistic* ODE model, which we recall here. It is given by the equation

$$\frac{dN}{dt} = (r - dN)\, N,$$

which describes the time-evolution of the number of cells $N(t)$, starting from a prescribed initial condition N_0.

Coefficient r denotes the net selection rate of the cells, namely the difference between their proliferation and death rates, while the logistic term dN stands for an additional death rate proportional to the number of cells.

The underlying assumption is that competition for nutrients and space inside the population does not allow for exponential unconstrained proliferation. Mathematically, it is indeed true that if $N_0 \leq \frac{r}{d}$, then $N(t)$ converges increasingly toward the *carrying capacity* $\frac{r}{d}$.

Let us now introduce a basic phenotype-structured model, where the quantity of interest is a number of cells $n(t, x)$ at time $t > 0$, and of *phenotype expression level* (hereafter simply designed as *phenotype*) $x \in [0, 1]$ standing for the resistance to a given drug. We stress that this phenotype is taken to be continuous, because, as already mentioned, it can be correlated to biological characteristics which themselves are continuous. Here, $[0, 1]$ is taken for simplicity but multi-dimensional phenotypes can of course be considered.

The model reads

$$\frac{\partial n}{\partial t}(t, x) = (r(x) - d(x)\rho(t))\, n(t, x),$$

where $\rho(t) := \int_0^1 n(t, x)\, dx$ is the total number of cells at time t, starting from some initial condition $n^0(\cdot)$. As before, $r(x)$ is the net proliferation rate of cells of cells of phenotype x, while $d(x)\rho(t)$ is the natural extension of the previous logistic term. Note that more general logistic death terms through a Kernel K can be considered, in the form $\int_0^1 K(x, y)n(t, y)\, dy$.

The basic model described above is characterised by two main phenomena: *convergence* and *concentration*. The first one means convergence of $\rho(t)$ towards $\max\left(\frac{r}{d}\right)$, and concentration of the density n on the set of phenotypes where $\frac{r}{d}$ reaches its maximum, namely $\arg\max\left(\frac{r}{d}\right)$. This is why this class of models is extensively used in adaptive dynamics to model selection: only the cells in the fittest phenotypic states can survive, which corresponds mathematically to the convergence of $n(t, \cdot)$ to a sum of Dirac masses located on the set $\arg\max\left(\frac{r}{d}\right)$.

This modelling framework also extends to several populations, in which case the competition between the populations is modelled through Lotka–Volterra-like terms. Let us introduce a model of two interacting populations, which will be further developed in the next section with the modelling of chemotherapy as control terms. It is concerned with two densities of healthy and cancer cells $n_H(t, x)$ and $n_C(t, x)$ respectively, where x is again a continuous phenotype, in the application we have in mind describing the level of resistance to a given drug. The model is given by

$$\frac{\partial n_H}{\partial t} = \left[r_H(x) - d_H(x)\left(a_{HH}\rho_H(t) + a_{HC}\rho_C(t)\right) \right] n_H(t, x), \tag{1}$$

$$\frac{\partial n_C}{\partial t} = \left[r_C(x) - d_C(x)\left(a_{CC}\rho_C(t) + a_{CH}\rho_H(t)\right) \right] n_C(t, x), \tag{2}$$

where, as before, $\rho_H(t) = \int_0^1 n_H(t,x)\,dx$, $\rho_C(t) = \int_0^1 n_C(t,x)\,dx$. The logistic terms now incorporate an intraspecific competition term weighted by coefficients a_{HC} and a_{CH}. Because healthy and cancer cells compete harder within their own population than with the other population (in other words, cells belong to different ecological niches, e.g., for metabolic and energetic reasons linked in particular to respiratory oxidative phosphorylation in one case and glycolysis in the other), it is quite natural to assume

$$a_{HC} < a_{HH}, \quad a_{CH} < a_{CC}.$$

Under hypothesis (4.1), it is proved in [29] that the behaviour of (1) is again convergence and concentration, where the asymptotic values of ρ_H, ρ_C and the sets on which n_H, n_C concentrate can also be explicitly computed.

4.2 Optimal Control for Anticancer Therapeutics

Optimal control methods (reviewed in [32]) applied to models of cancer therapeutics using systems of ordinary differential equations [4, 19, 20] or of partial differential equations [29] are the appropriate tool to theoretically optimise cancer therapeutics, in particular by taking into account the inevitable emergence of drug resistance in cancer cell populations.

The built-in targets for theoretical therapeutic control that are present in the phenotype-structured PDE models we advocate here are not supposed to represent well-defined molecular effects of the drugs in use, but rather functional effects, i.e., related to cell death (cytotoxic drugs), or to proliferation in the sense of slowing down the cell division cycle without killing cells (cytostatic drugs). We propose that cell life-threatening drugs (cytotoxics) induce by far more resistance in the highly plastic cancer cell populations than drugs that only limit their growth (cytostatics), and that a rational combination of the two classes of drugs—and possibly others, adding relevant targets to the model—may be optimised to propose therapeutic control strategies to avoid the emergence of drug resistance in tumours.

We address this optimal control problem in the context of two populations, healthy and cancer as in the model is given by (1), now complemented with two types of drugs of infusion rates u_1 and u_2 for cytotoxic and cytostatic drugs, respectively. The resistance phenotype x they are endowed with is defined with respect to the cytotoxic drug pressure, and is taken to range from sensitiveness ($x = 0$) to resistance ($x = 1$). The controlled model thus reads

$$\begin{cases} \frac{\partial n_H}{\partial t}(t,x) = \left[\frac{r_H(x)}{1+\alpha_H u_2(t)} - d_H(x)I_H(t) - \mu_C(x)u_1(t) \right] n_H(t,x), \\ \frac{\partial n_C}{\partial t}(t,x) = \left[\frac{r_C(x)}{1+\alpha_C u_2(t)} - d_C(x)I_C(t) - \mu_H(x)u_1(t) \right] n_C(t,x), \end{cases} \quad (3)$$

On a fixed therapeutic time-window $[0, T]$, the optimal control problem is to choose the controls u_1 and u_2 so as to minimise the number of cancer cells $\rho_C(T)$, while satisfying the three following constraints.

- *remaining under maximum tolerated doses:* $0 \leq u_1(t) \leq u_1^{\max}$, $0 \leq u_2(t) \leq u_2^{\max}$,
- *avoiding the emergence of too big a tumour:* $\frac{\rho_H(t)}{\rho_H(t)+\rho_C(t)} \geq \theta_{HC}$,
- *limiting unwanted adverse effects to the healthy cell population:* $\rho_H(t) \geq \theta_H \rho_H^0$.

This optimal control problem is motivated by the inefficacy of using constant high doses of drugs, a strategy which on the long run violates the last two constraints. This is indeed what is observed in the simulation presented in Fig. 1: although the tumour size first decreases, it is at the expense of the cancer cell density concentrating on a resistant phenotype. The treatment becomes inefficient and relapse occurs.

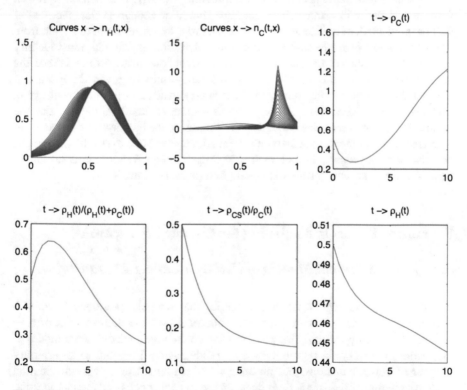

Fig. 1 What should never be done in the clinic! Simulation with $u_1(t) = \mathrm{Cst} = 3.5$ and $u_2(t) = \mathrm{Cst} = 2$, in time $T = 10$. Here $\rho_{CS}(t) = \int_0^1 (1 - x)n_C(t, x)\,dx$ is a measure of the number of sensitive cells in the cancer cell population. *Constant* high doses of the cytotoxic drug yield concentration of cancer cells around a resistant phenotype. The cancer cell population increases steadily while healthy cells decrease

In [29], the previously defined optimal control problem is analysed both numerically and theoretically. As the time T increases, it is found that the optimal control strategy becomes increasingly close to a two-phase strategy.

- The first phase is long, and only constant low doses as given, so as to saturate the second constraint. At the end of this first phase, the drug pressure has been low enough to ensure that the cancer cell density has concentrated on a sensitive phenotype.
- The second phase is short and starts with maximum tolerated doses for both drugs, leading to a quick decrease of both cell numbers because they are efficient on a sensitive cancer cell population. Once the third constraint (on the heathy cell density) has been reached, cytostatic drugs switch to some intermediate value (which can be computed in feedback form) which allows for a further decrease of the tumour size while keeping the healthy cell number at its lower bound.

A numerical simulation of the optimal strategy is presented in Fig. 2.

For a practical implementation of the previous strategy, it is natural to repeat it in a quasi-periodic manner. One can hope that after enough cycles, the tumour will be eradicated, or at least made chronic. In order to decide when to switch from the second short phase to another cycle with a long first phase, one must identify markers for resistance. Indeed, as long as constant low doses do not violate the second constraint on the relative tumour size, they must be given to ensure that the (assumed to be plastic) tumour is becoming sensitive to the treatment again. The switch to the second phase can be led as soon as the markers indicate that the tumour has become sensitive enough again. Finally, if the healthy cells tissue is too damaged (namely the third constraint saturates), one can hope to still let the tumour decrease with a properly chosen cytotoxic drug infusion. When this is no longer possible, one must switch back to the long first phase (no infusion).

5 Future Tracks in Modelling for Cancer Therapeutics

5.1 Beyond Present Models to Optimise Cancer Therapeutics

The models of adaptive population dynamics that we have presented here, with their built-in targets for control, rely on a nonlocal Lotka–Volterra vision of cell-cell population competition. This point of view could be extended to other modes of interaction, which could be mutualistic or predator-prey like, and to an arbitrary number of cell populations. For an analysis of a mutualistic integro-differential 2×2 system modelling interactions between breast cancer cells and their supporting stroma (adipocytes), we refer to [27], whereas a more general Lotka–Volterra-like model for N populations is analysed in [28]. The inferred asymptotic behaviour is again convergence and concentration.

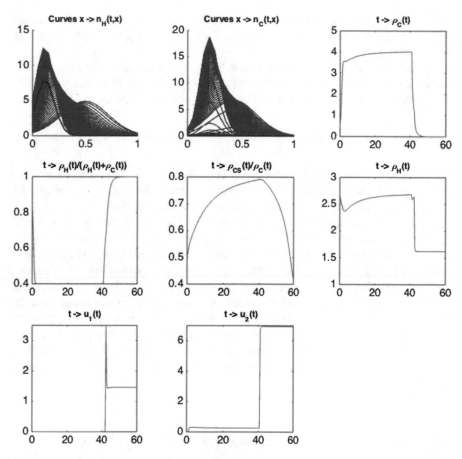

Fig. 2 Simulation of the optimal control problem for $T = 60$. Here, the phenotype is concentrated on a sensitive value *at the end of the first delivery phase*, and eventually likely more resistant—but for a very rare surviving fraction—in the cancer cell population, if one compares the curves showing $\rho_C(t)$ and $\rho_{CS}(t)/\rho_C(t)$ towards the end of the treatment course. Delivered at a high dose for a brief duration only, followed by a medium dose for the remaining time, the cytotoxic drug $u_1(t)$ impinges a drastic decrease, if not total eradication, in the cancer cell population, while preserving healthy cell numbers over a predefined threshold

Let us now come to extensions of the integro-differential setting by considering the basis 1-dimensional model that we first mentioned. Recall first that more general logistic interaction terms can be considered. The other natural extension is to model epimutations (occurring on the relevant time-scale, which is here that of a tumour). They can either be modelled by a Laplacian, leading to

$$\frac{\partial n}{\partial t}(t, x) = (r(x) - d(x)\rho(t))\, n(t, x) + \beta \Delta n(t, x),$$

with Neumann boundary conditions, or more generally through a mutation kernel. Note that both modelling are linked at the limit through a proper rescaling of the kernel, as explained in [25].

A complementary advection term can be added, accounting for cells actively adapting to their environment, seeking for phenotype changes that make them fitter. These can be seen as stress-induced epimutations and with them the model becomes

$$\frac{\partial n}{\partial t}(t, x) + \frac{\partial}{\partial x}(v(x)n(t, x)) = (r(x) - d(x)\rho(t)) \, n(t, x) + \beta \Delta n(t, x).$$

Note that in [5], the advection happens to be compulsory to observe quick enough dynamics to fit those obtained in the experiments presented in [31]. Besides, in [7], the effects of an additional advection term are rigorously studied.

A further advantage of these more general PDE models is that they are able to represent possible asymptotic coexistence of phenotypes, which is not the typical output of the integro-differential models. Moreover, whereas in [5] the agent-based model and the PDE model were treated concurrently in parallel, in [8], a general method describing the passage from the former to the latter is rigorously presented.

A final possible extension worth-mentioning is the addition of a space variable, since it is believed that the heterogeneity of a tumour varies from its periphery to its centre. This is also particularly relevant in view of optimal control through chemotherapy since drugs will efficiently access the outer rim of the tumour but less its core. For possible cancer models taking both phenotype and space into account, we refer to [22].

5.2   Need for Models with a Larger Evolutionary Perspective

From the biological part of this article, it clearly appears that the abovementioned models, sophisticated though they may be, are not enough to study in mathematical settings the evolution of multicellularity and its intrinsic failure, namely cancer, nor are they presently enough to design optimised therapeutic strategies that can overcome drug resistance in cancer. Open questions to biologists remain if one wants to make available a framework within which mathematical modelling may be designed. What are the genes that must be silenced in physiology and are re-expressed in de-differentiated cancer cells? What are the observable links between genes that are known to be essential for multicellularity and what are the genes that are altered in cancer (the same, following [9])? What models to study physiological coherence between tissues in the same organism, i.e., what sticks together in harmony the 200 different cell types of a human organism? What part of the genome bears the so-called cold genes, what part the individual signature of an organism that is transmitted throughout differentiation (the 'self', certainly to be related with the major histocompatibility complex, MHC), and what part the genes that are normally sequentially silenced in the history of differentiations? As regards mechanisms of drug resistance, what part of launching in a cancer cell population is deterministic (triggering cold genes) and what part is stochastic? And such list of open questions is not intended to be comprehensive.

6   Conclusion: A Challenging New Field for Mathematicians

In this short description of cancer as evolutionary disease, focusing on the question of drug resistance and its possible overcoming by optimised strategies in the clinic, we have presented what has been recently developed in the framework of mathematical modelling, that is, adaptive dynamics of cell populations represented by phenotype-structured models relying on partial differential equations, together with optimal control methods to guide their asymptotic behaviour. We have also proposed immediate tracks for future extensions of these existing models, and only sketched the scenery for future mathematical models that still lack biological answers to guide their design. We are nevertheless confident in the fast progress of cancer biology to help mathematicians design models that can be helpful in prevention, prediction and control of cancer in the clinic, provided that the right questions are posed, mathematically challenged and experimentally tackled.

References

1. S. Benzekry, Ph. Hahnfeldt, J. Theor. Biol. **335**, 235 (2013)
2. F. Billy, J. Clairambault, Discr. Cont. Dyn. Syst. Ser. B **18**, 865 (2013)
3. H.M. Byrne, D. Drasdo, J. Math. Biol. **58**, 657 (2009)
4. C. Carrère, J. Theor. Biol. **413**, 24 (2017)
5. R.H. Chisholm, T. Lorenzi, A. Lorz, A.K. Larsen, L. Almeida, A. Escargueil, J. Clairambault, Cancer Res. **75**, 930 (2015)
6. R.H. Chisholm, T. Lorenzi, J. Clairambault, Biochem. Biophys. Acta Gen. Subj. **1860**, 2627 (2016)
7. R.H. Chisholm, T. Lorenzi, A. Lorz, Commun. Math. Sci. **14**, 1181 (2016)
8. R.H. Chisholm, T. Lorenzi, L. Desvillettes, B.D. Hughes, Z. Angew. Math. Phys. **67:100**, 1 (2016)
9. P.C.W. Davies, C.H. Lineweaver, Phys. Biol. **7**, 1 (2011)
10. M. Gerlinger et al., N. Engl. J. Med. **336**, 883 (2012)
11. R.J. Gillies, D. Verduzco, R.A. Gatenby, Nat. Rev. Cancer **12**, 487 (2012)
12. A. Goldman, M. Kohandel, J. Clairambault, Curr. Stem Cell Rep. **3**, 253 (2017)
13. A. Goldman, M. Kohandel, J. Clairambault, Curr. Stem Cell Rep. **3**, 260 (2017)
14. M.M. Gottesman, T. Fojo, S.E. Bates. Nat. Rev. Cancer **2**, 48 (2002)
15. S. Huang, Semin. Cancer Biol. **21**, 183 (2011)
16. S. Huang, Cancer Metastasis Rev. **32**, 423 (2013)
17. S. Huang, Y.P. Guo, G. May, T. Enver, Dev. Biol. **305**, 695 (2007)
18. F. Jacob, Science **196**, 1161 (1977)
19. U. Łędżewicz, H. Schättler, Discr. Cont. Dyn. Syst. Ser. B **6**, 129 (2006)
20. U. Łędżewicz, H. Schättler, *Mathematical Models of Tumor-Immune Dynamics* (Springer, New York, 2013), p. 157

21. Y. Li, J. Laterra, Cancer Res. **72**, 576 (2012)
22. A. Lorz, T. Lorenzi, J. Clairambault, A. Escargueil, B. Perthame, Bull. Math. Biol. **77**, 1 (2015)
23. S.E. Luria, M. Delbrück, Genetics **28**, 491 (1943)
24. E. Pasquier, M. Kavallaris, N. André, Nat. Rev. Clin. Oncol. **7**, 455 (2010)
25. B. Perthame, *Transport Equations in Biology* (Birkhäuser, Basel, 2007)
26. T. Philippi, J. Seger, Tends Ecol. Evol. **4**, 41 (1989)
27. C. Pouchol, Modelling interactions between tumour cells and supporting adipocytes in breast cancer. https://hal.inria.fr/hal-01252122 (2015)
28. C. Pouchol, E. Trélat, arXiv 1702.06187. https://hal.inria.fr/hal-01618357 (2017)
29. C. Pouchol, J. Clairambault, A. Lorz, E. Trélat, arXiv 1612.04698 (2016); J. Maths Pures Appl. (2017, to appear)
30. S.M. Shaffer et al., Nature **546**, 431 (2017)
31. S.V. Sharma et al., Cell **141**, 69 (2010)
32. E. Trélat, *Contrôle Optimal* (Vuibert, Paris, 2005), 246 pp. Reviewed in Mathscinet MR2224013, 2007f:49001
33. B. Ujvari, B. Roche, F. Thomas (eds.), *Ecology and Evolution of Cancer* (Academic, London, 2017)
34. C.H. Waddington, *The Strategy of Genes* (George Allen & Unwin, London, 1957)
35. A. Wu et al., Proc. Natl. Acad. Sci. USA **112**, 10467 (2015)
36. L. Zitvogel, L. Apetoh, F. Ghiringhelli, G. Kroemer, Nat. Rev. Immunol. **8**, 59 (2008)
37. L. Zitvogel, O. Kepp, G. Kroemer, Nat. Rev. Clin. Oncol. **8**, 151 (2011)

Optimal Resource Allocation for HIV Prevention and Control

Dmitry Gromov, Ingo Bulla, and Ethan O. Romero-Severson

1 Introduction

When dealing with economically and socially significant infectious diseases, in particular AIDS and tuberculosis, the central problem is to optimally distribute the limited resources among different treatment and prophylaxis programs. The main difficulty in doing so is that while the individual-level effect of these interventions can be determined using controlled trials, their effectiveness as public health interventions cannot be ascertained with certainty. This is due to the fact that affected populations are different both in terms of the disease transmission dynamics, but also in the efficacy of available instruments given a specific population structure. Identifying the optimal strategy of resource allocation must be based on a (dynamic) model of the underlying medical, biological, and social processes that captures the relevant features of the population.

Two standard approaches to modeling the dynamics of a disease are using either an agent-based [2] or a population balance (compartmental) model [6] while ODE-based compartmental models are generally preferred due to their computational tractability. However, when using dynamic models a number of challenges arise. These are related to the fact that certain parameters of the model cannot be assessed while the others show a high variability. There are two possible ways to approach

D. Gromov (✉)
Faculty of Applied Mathematics and Control Processes, St. Petersburg State University,
Saint Petersburg, Russia

I. Bulla
Institut für Mathematik und Informatik, Universität Greifswald, Greifswald, Germany

E. O. Romero-Severson
Theoretical Biology and Biophysics Group, Los Alamos National Laboratory,
Los Alamos, NM, USA
e-mail: eoromero@lanl.gov

© Springer International Publishing AG, part of Springer Nature 2018
R. P. Mondaini (ed.), *Trends in Biomathematics: Modeling, Optimization and Computational Problems*, https://doi.org/10.1007/978-3-319-91092-5_9

this problem. The first one consists in estimating the parameters using the available data (in particular the data that can be obtained when testing individuals in the framework of the intervention program) while another one concentrates on deriving somewhat universal rules that can be applied to a wide range of situations.

In this paper we aim at presenting a number of problems and their solutions with a particular emphasis on using optimal control based methods. As an illustration, we consider the HIV epidemic in the men-having-sex-with-men (MSM) population. By the introduction of highly active antiretroviral therapy (HAART) [12], HIV has morphed from a lethal threat into a controllable chronic disease, at least for the vast majority of HIV patients in the developed world. Nevertheless, little progress has been made in driving HIV back in MSM, who are at much higher risk of HIV infection than heterosexuals and are thus the key population for HIV interventions in the industrialized world.

When on HAART, the infected patient becomes practically non-infectious [7]. This is referred to as the Treatment as Prophylaxis (TaP) modality in this paper. An alternative consists in using a preventive drug in uninfected persons called pre-exposure prophylaxis (PrEP), which has recently become commercially available. Opposed to TaP, PrEP does not target individuals infected with HIV but rather blocks infection in uninfected people for a short period of time (1–2 days) and when taken continuously protects against infection even given frequent exposure [3]. The target group of PrEP thus comprises persons running a high-risk of infection with HIV, that is injecting drug users, sex workers, and MSM.

To evaluate the cost effectiveness of TaP or PrEP, until now the population dynamics including the effects of the respective intervention were modeled and then the number of prevented infections was related to the costs of the intervention [13]. Although this is an established approach in health economics of infectious diseases, it neglects that different interventions might interact in an intricate manner and, thus, should be modeled and analyzed jointly. Our work addresses this issue by integrating both TaP and PrEP into one epidemiological dynamic model and then optimizing the allocation of a limited budget to these two interventions. Hereby, the allocation is time-variant and piece-wise constant. We assume that the target population consists in the MSM population of a circumscribed geographical unit (i.e., a larger city, region, or country), exhibiting two different levels in sexual risk behavior. Furthermore, we employ two different disease stages, acute and chronic, because the acute stage is associated with a largely higher infectivity.

The paper is structured as follows. Section 2 gives some basic facts about population balance models and presents the HIV model used in the paper. Also, the problem of observing the system states is discussed. The optimal control problem along with variations thereof and the related numerical results are described in Sect. 3. Section 4 discusses some issues that occur when applying the optimal controls to real problems. Finally, we present conclusions and outlooks for further research.

2   Mathematical Model

2.1   Population Balance Model

The common modeling framework for describing infectious disease propagation is the population balance. This means that the whole population is divided into a number of homogeneous groups (compartments). That is to say, all individuals within a compartment are assumed to be identical in their evolution [6, 11]. The state variables of the model correspond to the number of people within each group and the dynamics of the ith state is described by the following differential equation:

$$\dot{X}_i = \sum_{i \neq j} \left(a_{ij}(X) - a_{ji}(X) \right) - a_{ii}(X) + w_i, \tag{1}$$

where $a_{ji}(X)$ is the flow rate from compartment i to compartment j, $a_{ii}(X)$ is the outflow out of the ith compartment, and w_i is the inflow into the ith compartment which is typically assumed to be constant. We also assume that the following regularity conditions are fulfilled:

A1. $a_{ij}(X) \geq 0$ and $w_i(X) \geq 0$ for all $X \in \mathbb{R}_{\geq 0}^n$.
A2. $X_i = 0$ implies $a_{ji}(X) = 0$ and $a_{ii}(X) = 0$, i.e. there is no flow out of an empty compartment.

It is known that the system (1) satisfying the above assumptions is *non-negative* [10]. This means that for any initial condition $x_0 \in \mathbb{R}_{\geq 0}^n$, the solution $x(t)$ of (1) belongs to $\mathbb{R}_{\geq 0}^n$ for all $t \in [0, \infty)$.

Controlled Population Balance Model While the population based models have proved to be useful in modeling various infectious diseases, less attention has been paid to controlling the propagation of the diseases within the population. In contrast to classical control systems where the control is applied at the input or the output of the system thus representing an external action, the notion of control for population based models is associated with redistribution. Typically, the set of compartments is extended by a number of new compartments and the controls are used to "modulate" the flows from the old compartments into the new ones.

In particular, a new compartment can correspond to the people on treatment, vaccinated or isolated infecteds and the respective controls define the inflows into these compartments. The controls enter the model multiplicatively, i.e., the system (1) turns into

$$\dot{X}_i = \sum_{i \neq j} \left(a_{ij}(X) - a_{ji}(X) \right) - a_{ii}(X) + w_i + \sum_{k} \sum_{i \neq j} \left(c_{ij}^k(X) - c_{ji}^k(X) \right) u^k, \tag{2}$$

where $c_{ij}^k(X)$ is the total number of people within the jth compartment that can be (potentially) assigned to a particular kind of treatment associated with the state X_i

within the unit time interval, and u^k is the fraction of this number that are enrolled into this particular treatment. Due to obvious restrictions, $u^k \in [0, 1]$. The upper bound on u^k means that the enrollment cannot exceed 100%. The functions $c_{ij}^k(X)$ are assumed to satisfy the assumptions A1.–A2. and the system (2) can be shown to be non-negative [9].

2.2   HIV Propagation Model

We consider an HIV propagation model of a general MSM population from the United States. The detailed model derivation is presented in [9].

The total population is divided into nine compartments with corresponding states. Transitions between states happen due to individuals becoming infected, progressing from acute to chronic stage, and dying of AIDS ($S. \rightarrow I_A. \rightarrow I_C. \rightarrow D$). Moreover, individuals change their risk behavior $((\cdot)_L \leftrightarrow (\cdot)_H, P \rightarrow S_L)$, thus going from high-risk to the low-risk category and back. There are two controls that correspond to putting individuals on PrEP or TaP treatment ($S_H \leftrightarrow P$, $I_C. \leftrightarrow T.$). Note that the administered treatment may fail thus leading to the (uncontrolled) flows from P and T back to the original compartments. Note that high-risk individuals on PrEP who adapt a low-risk behavior cancel their PrEP treatment and thus become low-risk susceptibles, i.e., move from P to S_L. Finally there is flow into the system due to individuals reaching an age of sexual activity and outflow from the system because of non-HIV related death or individuals becoming sexually inactive or settling in a monogamous, lifelong relationship.

The dynamics under study are described by the following system of ODEs:

$$\dot{S}_H = \alpha_H - (\phi_H(X) + \rho_H + \mu) S_H + \rho_L S_L + xP - U_P \zeta_P(X)N$$
$$\dot{S}_L = \alpha_L - (\phi_L(X) + \rho_L + \mu) S_L + \rho_H(S_H + P)$$
$$\dot{I}_{CH} = \delta_A I_{AH} - (\rho_H + \mu + \delta_C + \bar{u}_T)I_{CH} + \rho_L I_{CL} + yT_H - U_T \zeta_{T,H}(X)N$$
$$\dot{I}_{CL} = \delta_A I_{AL} - (\rho_L + \mu + \delta_C + \bar{u}_T)I_{CL} + \rho_H I_{CH} + yT_L - U_T \zeta_{T,L}(X)N$$
$$\dot{I}_{AH} = \phi_H(X)S_H - (\rho_H + \mu + \delta_A)I_{AH} + \rho_L I_{AL}$$
$$\dot{I}_{AL} = \phi_L(X)S_L - (\rho_L + \mu + \delta_A)I_{AL} + \rho_H I_{AH}$$
$$\dot{T}_H = -(y + \rho_H + \mu)T_H + \bar{u}_T I_{CH} + \rho_L T_L + U_T \zeta_{T,H}(X)N$$
$$\dot{T}_L = -(y + \rho_L + \mu)T_L + \bar{u}_T I_{CL} + \rho_H T_H + U_T \zeta_{T,L}(X)N$$
$$\dot{P} = -(x + \rho_H + \mu)P + U_P \zeta_P(X)N,$$

$$(3)$$

where $X = \begin{bmatrix} T_H & T_L & I_{CH} & I_{CL} & I_{AH} & I_{AL} & S_H & S_L & P \end{bmatrix}'$ is the vector of state variables; N is the sum of all states, $N = \langle \mathbf{1}, X \rangle$, $\mathbf{1}$ is the column of ones; $\phi_H(X)$, $\phi_L(X)$, $\zeta_{T,H}(X)$, $\zeta_{T,L}(X)$, and $\zeta_P(X)$ are the non-linear (rational) functions of X which

Table 1 Model parameters

Infection parameters	
ϕ_H, ϕ_L	Transmission rate for high-risk resp. low-risk susceptibles
δ_A	Rate that acutely infected individuals become chronically infected
δ_C	Rate that chronically infected individuals die due to AIDS
Treatment parameters	
x	Rate at which PrEP fails or is canceled
$\bar{u}_T$	Baseline enrollment rate into TaP
y	Rate at which TaP fails or is canceled
Other parameters	
α_H, α_L	Recruitment rate of new high-risk resp. low-risk susceptibles
μ	Rate that adults die of non-HIV related causes, reach an age of sexual inactivity, or settle in a lifelong relationship
ρ_H	Rate that high-risk persons become low-risk
ρ_L	Rate that low-risk persons become high-risk

take on non-negative values for any $X \in \mathbb{R}^n_{\geq 0}$. U_P and U_T are the control inputs which correspond to the fraction of total population being involved either in PrEP (U_P) or in TaP (U_T). All the remaining terms are non-negative constants[1] (see Table 1 for an explanation).

2.3 Enrollment and State Measurement

For the enrollment on PrEP and TaP, we assume that both are done by randomly sampling individuals at locations where high-risk individuals resp. chronically infecteds are prevalent. In case a sampled individual turns out to be a high-risk susceptible resp. chronically infected, he is urged to enroll in PrEP resp. TaP.

That is, we assume that there is a high-risk environment (HRE) where high-risk individuals are overrepresented, like bars or sex clubs. Then, denoting the probability that a random high resp. low risk individual is in a HRE at a random moment during the recruitment period by

$$p_H = P(\text{HRE}|R = H), \ p_L = P(\text{HRE}|R = L).$$

The probability of a random individual R encountered in a HRE at a random moment to be in one of the states corresponding to the high-risk category, i.e. to be in $X \in Z_H = \{S_H, I_{AH}, I_{CH}, T_H, P\}$, is computed using the Bayes rule:

---

[1] Here and throughout the paper we use the convention that all variables are denoted by capital letters and the constants by lowercase letters.

$$P(X|\text{HRE}) = \frac{P(\text{HRE}|R = H)P(X)}{P(\text{HRE}|R = H)P(R = H) + P(\text{HRE}|R = L)P(R = L)}$$

$$= \frac{p_H \frac{X}{N}}{p_H \frac{N_H}{N} + p_L \frac{N_L}{N}} = \frac{r_b X}{r_b N_H + N_L},$$

where $N_H = S_H + I_{AH} + I_{CH} + P + T_H$, $N_L = S_L + I_{AL} + I_{CL} + T_L$, $N = N_H + N_L$, and $r_b = p_H/p_L$. That is, r_b are the odds of a high-risk person to go to a HRE compared to a low-risk person. The probability of a random individual encountered in a HRE to be in one of the low-risk states, $X \in Z_L = \{S_L, I_{AL}, I_{CL}, T_L\}$ is computed analogously to yield

$$P(X|\text{HRE}) = \frac{X}{r_b N_H + N_L}.$$

One can easily check that the consistency condition holds, i.e.,

$$\sum_{X \in Z_H} P(X|\text{HRE}) + \sum_{X \in Z_L} P(X|\text{HRE}) = 1.$$

If, for instance, U_P is the relative rate at which individuals are sampled at HREs and then put on PrEP in case they are high risk susceptibles, the absolute rate for transition from S_H to P is

$$U_P N P(S_H|\text{HRE}) = U_P \frac{r_b S_H N}{r_b N_H + N_L},$$

where we assumed that the total number of individuals visiting a HRE during unit time interval is equal to N or to put differently, that is that the expected number of visits an individual makes at a HRE is equal to 1. Note that the term $\frac{r_b S_H N}{r_b N_H + N_L}$ in the preceding equation is just the flow rate $c_{P,S_H}^{U_P}(X)$ corresponding to the transition from the S_H compartment into the P compartment with the control U_P, cf. (2).

State Estimation An important point is that the enrollment procedure can be used to estimate the number of individuals in different states.

Let N_s be the number of individuals that were sampled while visiting a HRE during the unit time. These individuals are interviewed and tested if necessary and are divided in the respective categories according to the results. We denote the number of individuals within the respective categories by $\hat{S}_{(\cdot)}$, $\hat{I}_{(\cdot)}$, $\hat{T}_{(\cdot)}$, and $\hat{P}$; furthermore, we denote the fractions of the respective groups within the sample $\hat{s}_{(\cdot)}$, $\hat{i}_{(\cdot)}$, $\hat{t}_{(\cdot)}$, and $\hat{p}$, where, e.g., $\hat{s}_H = \frac{S_H}{N_s}$. With this, the fractions of the respective groups within the total population, denoted by $s_{(\cdot)}$, $i_{(\cdot)}$, $t_{(\cdot)}$, and p, can be computed from the respective Bayesian equations. To ensure that the problem is well defined, an additional normalization condition stating that the sum of all fractions is equal to 1 is imposed. With this, the respective fractions are found as the solution to the following set of linear equations:

$$\left[\begin{pmatrix} r_b \mathbf{I} & 0 \\ 0 & \mathbf{I} \end{pmatrix} + (1 - r_b) \begin{pmatrix} \text{diag}(\hat{x}_H) & 0 \\ 0 & \text{diag}(\hat{x}_L) \end{pmatrix} \begin{pmatrix} \mathbf{1} & 0 \\ \mathbf{1} & 0 \end{pmatrix} \right] \begin{pmatrix} x_H \\ x_L \end{pmatrix} = \begin{pmatrix} \hat{x}_H \\ \hat{x}_L \end{pmatrix} \tag{4}$$

where $\mathbf{I}$, $\mathbf{1}$, and 0 are the unitary matrix and the matrices of ones and zeros, all of appropriate dimensions; $x_H = (s_H \ i_{AH} \ i_{CH} \ t_H \ p)^T$, $x_L = (s_L \ i_{AL} \ i_{CL})^T$, $\hat{x}_H$ and $\hat{x}_L$ are defined in the same way, and $\text{diag}(\hat{x}_H)$, resp. $\text{diag}(\hat{x}_L)$ are the diagonal matrices whose entries correspond to the elements of the respective vectors. Note that the vector x_L does not contain the component t_L which is computed from the normalization constraint, i.e., $t_L = 1 - s_H - i_{AH} - i_{CH} - t_H - p - s_L - i_{AL} - i_{CL}$. Obviously, the system (4) can be reformulated to exclude any other component instead of t_L.

Proposition 2.1 *The matrix*

$$M_B = \left[\begin{pmatrix} r_b \mathbf{I} & 0 \\ 0 & \mathbf{I} \end{pmatrix} + (1 - r_b) \begin{pmatrix} \text{diag}(\hat{x}_H) & 0 \\ 0 & \text{diag}(\hat{x}_L) \end{pmatrix} \begin{pmatrix} \mathbf{1} & 0 \\ \mathbf{1} & 0 \end{pmatrix} \right],$$

defined in (4) is invertible.

Proof The proof of invertibility boils down to checking that the submatrix $r_b \mathbf{I} + (1 - r_b) \text{diag}(\hat{x}_H)$ is invertible. This matrix can be rewritten as $r_b (\mathbf{I} - \text{diag}(\hat{x}_H)) + \text{diag}(\hat{x}_H)$, which is a diagonal matrix with strictly positive elements as follows from the fact that the elements of the vector $\hat{x}_H$ (fractions) are positive and strictly less than 1. □

The above proposition implies that the system (4) is well defined and can be solved to yield the fractions x_H and x_L. If an estimate of the size of the total population, $\hat{N}$, is available, one can obtain the estimates of the state variables, e.g., $\hat{S}_H = s_H \cdot \hat{N}$ for the number of high-risk susceptibles and similarly for other states.

Remark 2.1 In practice, the members of some groups, i.e., of I_{AH} and I_{AL} cannot be recognized during the interview and merge with other groups, in this case with S_H and S_L. This results in a certain bias in the estimations that has to be resolved using additional analysis. Moreover, there might be situations in which it is too difficult to estimate r_b, rendering the approach described here infeasible.

3 Optimal Control

3.1 Problem Statement

The optimal control problem is formulated as an optimization problem, whose goal is to minimize the cost function while respecting some structural and budgetary restrictions. The cost to be minimized is defined as

$$J^C(X, U) = \int_0^{t_f} e^{-\rho t} C(X(t), U(t)) dt, \tag{5}$$

where $C(t, X(t), U(t))$ is the instantaneous cost function; $\rho > 0$ is the discount rate, and t_f is the time horizon which corresponds to the duration of the intervention. The discounting factor $e^{-\rho t}$ is typically used to describe our priorities: one may attach greater importance to decreasing the cost in the near future while paying less attention to what will happen in the farther future.

Remark 3.1 Note that the optimization problem with the cost functional (5) is well-posed if the instantaneous cost function $C(X(t), U(t))$ is nonnegative for nonnegative arguments. This follows from the fact that we are interested only in the nonnegative values of $X(t)$ and $U(t)$.

The restrictions imposed on the system are twofold:

Restriction on the Admissible Control Policies The first class of restrictions is due to the structural limitations of the decision unit. Since the intervention is performed by a state organization the control profile must be sufficiently regular. The regularity requirement boils down to assuming that the set of admissible controls consists of piecewise constant functions with a fixed interval between two consecutive switches.

Let $\mathcal{T} = \{t_i\}_{i=0}^{n_{int}}$, $0 = t_0 < t_1 < \cdots < t_{n_{int}} = t_f$ be the time instants at which control switches occur along with the initial and final time. We assume that for any $1 \leq i \leq n_{int}$, the duration of the respective interval is constant: $t_i - t_{i-1} = \delta t$. The admissible control is thus defined as a piecewise constant vector valued function that take on the discrete values $\{U^i\}_{i=1,\ldots,n_{int}}$, $U^i \in \mathbb{R}^m_{\geq 0}$, during each interval. The goal of the optimization is to determine these values in order to minimize the cost function (5) while respecting the constraints.

Note that the control values are bounded by zero from below, but there are no upper bounds.[2] The upper bounds are imposed implicitly as will be described below.

Dynamic Budget Allocation The second class of constraints is due to the budgetary limitations. In practice, an intervention incurs large expenses which are only partially compensated by the government and the insurance companies. We assume that at the beginning of each control interval $[t_{i-1}, t_i)$ a baseline budget is set by estimating the expected expenses. The money to be spent for the intervention is thus allocated starting from this baseline. This works as follows:

The total expenditures related to treating people with TaP or PrEP and the enrollment costs for TaP or PrEP are captured by the following cost function:

$$J_i^B(X, U^i) = \int_{t_{i-1}}^{t_i} B(X(s), U(s)) ds,$$

---

[2]Strictly speaking, there is an upper bound $U^i \leq 1$, but this turns out to be very loose as the real values of the control are always less than 1.

where $B(X(s), U(s))$ is a positive defined function that can be written as a sum of two components, $B(X(s), U(s)) = B^1(X(s)) + B^2(X(s))U(s)$. Here, $B^1(X(s))$ describes the expenses related to treating chronically infected patients and the costs related to carrying out the PrEP program while $B^2(X(s))$ captures the costs associated with interviewing and testing random individuals as well as the enrollment costs.

We proceed as follows: for each interval $[t_{i-1}, t_i)$, $i = 1, \ldots, n_{int}$, the value of J_i^B is computed for the uncontrolled case to determine the baseline expenses, i.e., the expenses that the state would defray if there is no enrollment. The assigned budget is allocated atop the baseline budget. Let $\tilde{X}_i(t)$, $t \in [t_{i-1}, t_i)$ be the uncontrolled ($U^i = \mathbf{0}$) solution of (3) for $t \in [t_{i-1}, t_i)$ with initial condition $\tilde{X}_i(t_{i-1}) = X(t_{i-1})$. The dynamic budget constraint (DBC) is thus formulated as follows:

$$J_i^B(X, U^i) - J_i^B(\tilde{X}_i, \mathbf{0}) \le B, \quad i = 1, \ldots, n_{int}, \tag{6}$$

where $J_i^B(\tilde{X}_i, \mathbf{0}) = \int_{t_{i-1}}^{t_i} B^1(\tilde{X}(s))ds$. The constraints (9) set an implicit limit to the control values U^i.

The resulting optimization problem can thus be formulated as follows. Determine the values of control $U^i \in \mathbb{R}^2_{\ge 0}$, $i = 1, \ldots, n_{int}$ s.t.

$$\begin{cases} J^C(X) = \int_0^{t_f} e^{-\rho t} C(X(t), U(t))dt \to \min \\ J_i^B(X, U^i) - J_i^B(\tilde{X}_i, \mathbf{0}) \le B, i = 1, \ldots, n_{int} \\ X(t), \ t \in [0, t_f], \\ \quad \text{satisfies (3) with } X(0) = X_0 \text{ and } U(t) = U^i, t \in [t_{i-1}, t_i), \\ \tilde{X}_i(t), \ t \in [t_{i-1}, t_i] \, i = 1, \ldots, n_{int} \\ \quad \text{satisfies (3) with } X_i(0) = X(t_i) \text{ and } U(t) = 0, t \in [t_{i-1}, t_i). \end{cases} \tag{7}$$

In this study, the optimal control problem consists in minimizing the incidence of HIV infection, i.e.,

$$C(X, U) = S_H \phi_H(X) + S_L \phi_L(X). \tag{8}$$

The budgetary cost function is defined as

$$B(X(t), U^i) = k_T^{(t)}[T_H(t) + T_L(t)] + k_P^{(t)} P(t) + k_T^{(e)} N(t)U_T^i(t) + k_P^{(e)} N(t)U_P^i(t) \tag{9}$$

for $i = 1, \ldots, n_{int}$ and $t \in [t_{i-1}, t_i)$. Here, $k_T^{(t)}$ and $k_P^{(t)}$ are the monthly costs for treatment with TaP and PrEP, respectively, per patient. The coefficients $k_T^{(e)}$ and $k_P^{(e)}$ represent the costs for approaching and, if necessary, enrolling one patient into TaP and PrEP, respectively.

3.2 Numerical Optimal Control

In terms of optimal control theory the budgetary constraints can be classified as
mixed integral inequality path constraints. There is in general no way to handle
such constraints analytically, but they can be treated numerically using a modified
version of the orthogonal collocation method as described in detail in [9], see also
[14] for an overview of numerical optimal control methods.

To illustrate the use of numerical optimal control we consider an outbreak
scenario in a large US city with 100,000 at-risk MSM individuals. That is, the
introduction of HIV into a subpopulation with a low prevalence (e.g., people
aged 15–25) is considered. We assume that acutely infecteds are 15 times more
contagious than chronically infecteds [4]. Furthermore, we consider a situation in
which risk is static, that is $\rho_H = \rho_L = 0$ and assume that the proportion of the MSM
population exhibiting a low risk to those with a high-risk behavior is 9:1. Hence, we
set $\alpha_L = 250$ and $\alpha_H = 28$ to ensure that a newly entering individual is high-risk
with probability 0.1.

We set $\mu = 1/360$, leading to the individuals staying 30 years in the system
on average if there was no HIV-related removal. We assume that the susceptible
get infected at the common site and that individuals from the high risk group have
ten times more sexual contacts than the ones from the low risk group based on
analysis of longitudinal sexual behavioral data [15]. The values of λ_L and $\bar{u}_T$ were
determined such that at equilibrium the prevalence was 20% and the proportion
of infected individuals on treatment was 25%, which is consistent with measured
values [16, 17]. Finally, it is assumed that treatment never fails and is never stopped,
i.e., $y = 0$.

We consider four scenarios varying in the value of x: 0, $\frac{1}{60}$, $\frac{1}{24}$, $\frac{1}{12}$. That is, in one
scenario PrEP never fails and is never canceled and in three scenarios this happens
on average after 5, 2, and 1 year, respectively. The parameters of the cost function
are given in Table 2. The cost of enrollment was estimated by considering the total
cost per enrollee of a similar intervention program implemented by the New York
City Department of Health [5] plus the cost of the labs involved in determining
infectious status. The cost of TaP and PrEP is taken from [1].

The optimal controls are shown in Fig. 1 and the ones of the number of
individuals in the various states in Figs. 2 and 3. We can observe that the lower
the value of x is, the lower is the total number of newly infecteds and the higher is
both the overall values of U_P relative to U_T and the number of individuals on PrEP.
Therefore, a lower x should render the overall intervention more effective.

Table 2 Numerical values of the coefficients in the budget functional

$k_T^{(t)}$	$k_P^{(t)}$	$k_T^{(e)}$	$k_P^{(e)}$
1299	776	266	213

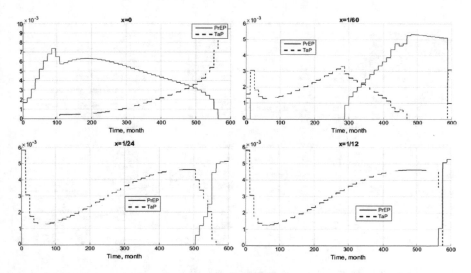

Fig. 1 The trajectories of the controls for $x = 0$, $1/60$, $1/24$, $1/12$ over 50 years

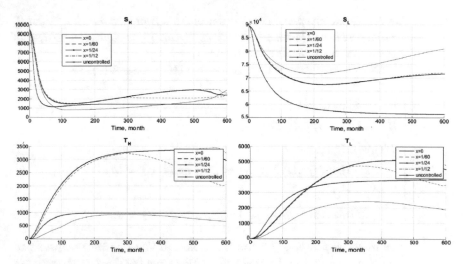

Fig. 2 The trajectories of S_H, S_L, T_H, and T_L over 50 years

3.3 Single Control Case

Albeit numerically solvable, the problem (7) remains rather computationally expensive. Typically it takes a couple of hours to get the optimal profiles. Thus it makes sense to consider a simplified problem that would be solvable with much less effort while still offering sufficient information about the problem. It turns out that the optimization problem can be tremendously simplified when being solved with respect to one control while setting another one to zero.

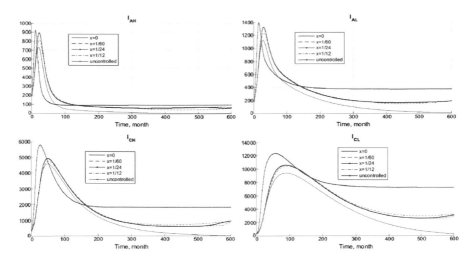

Fig. 3 The trajectories of I_{AH}, I_{AL}, I_{CH}, and I_{CL} over 50 years

Proposition 3.1 *If one component of the control U^i is kept zero for all $t \in [0, t_f]$, the optimization problem (7) with cost functions (8) and (9) is equivalent to a set of n_{int} scalar optimization problems of the form:*
    Determine the value of control $U^i \in \{0\} \times \mathbb{R}_{\geq 0}$ ($U^i \in \mathbb{R}_{\geq 0} \times \{0\}$) s.t.

$$
\begin{cases}
\|U^i\| \to \max, \\
J_i^B(X, U^i) - J_i^B(\tilde{X}_i, \mathbf{0}) \leq B, \\
X(t), \ t \in [t_{i-1}, t_i], \\
\qquad \text{satisfies (3) with } X(t_{i-1}) = X_{i-1} \text{ and } U(t) = U^i, \\
\tilde{X}_i(t), \ t \in [t_{i-1}, t_i] \\
\qquad \text{satisfies (3) with } X(t_{i-1}) = X_{i-1} \text{ and } U(t) = 0,
\end{cases}
\tag{10}
$$

which can be solved sequentially for $i = 1, \ldots, n_{int}$. At this, one has to ensure that the initial value X_i is equal to the final value of the state obtained as the result of solving the previous (partial) optimization problem.

Corollary 3.1 *The solution to the single control optimization problem is invariant with respect to the discount rate ρ.*

That is to say, when optimizing with respect to a single control, the optimization problem on each interval boils down to determining the maximal value of the respective control such that the budgetary constraint holds. In contrast to (7), the scalar optimization problems (10) can be solved within seconds. Thus, within the reasonable time one can obtain a large amount of data for different values of model parameters that can be used when making decisions about the intervention as will be discussed in Sect. 4.

4 Applications to Decision Making

As was discussed above, even the most precise optimization profile may turn out to be useless if the parameters of the model are not exactly known. Therefore, the epidemiologist who aims at designing the rules for resource allocation might wish to have a less precise result that would however have sufficiently general validity, i.e. it should be applicable to a wide class of problems. An alternative approach would be to determine a subset of model parameters such that the process under study can be controlled using the same set of rules. This issue is addressed in Sect. 4.1. On the other hand, one can attempt to estimate the system's parameters dynamically using the available data. This is discussed in Sect. 4.2.

4.1 Derivation of General Rules Using Suboptimal Solutions

Table 3 shows the values of the cost function for three cases: single TaP ($U_P = 0$), single PrEP ($U_T = 0$), and the mixed case corresponding to the results shown in Fig. 1. Note that the value of the cost function corresponding to TaP is the same for all cases as it does not depend on x.

We can see that the minimal value taken over the first two rows approximate pretty well the best possible result that is achieved when using both controls. Furthermore, the following rule can be formulated: the control policy corresponding to the smaller value of the cost function dominates in the mixed case (e.g., $x = 1/60$). If the difference between the values of the respective cost functions is small, the optimal control profile is likely to consist of two controls (e.g., $x = 0$).

To analyze the space of parameter sets, parallel coordinate plots can be employed. That is, one varies the values of different parameters, e.g. λ_H/λ_L, β_H/β_L, and ρ_L/ρ_H, and then determines for each parameter set whether exclusive allocation in TaP resp. PrEP yields a lower cost. This information can be expressed in a parallel coordinate plot as illustrated in Fig. 4. Compared with plotting histograms for each parameter separately, PCPs allow for detection of dependencies between parameters.

Table 3 Values of the cost function for different optimal control profiles and for different values of x

	$x = 0$	$x = 1/60$	$x = 1/24$	$x = 1/12$
Single TaP	6.843e+04	6.843e+04	6.843e+04	6.843e+04
Single PrEP	4.179e+04	7.753e+04	8.852e+04	9.300e+04
Mixed	4.109e+04	6.634e+04	6.829e+04	6.841e+04

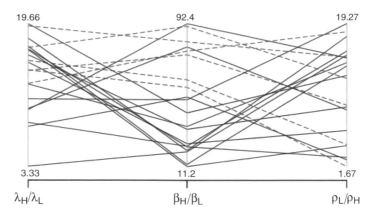

Fig. 4 For each parameter sets the cost of exclusive allocation in TaP resp. PrEP is compared. For parameter sets represented by solid lines exclusive allocation in TaP yielded lower cost, for the ones represented by dashed lines exclusive allocation in PrEP yielded lower cost

4.2 On-the-Fly Parameter Estimation

The problem of parameter estimation for epidemiological models is a crucial component of any quantitative analysis of infectious disease dynamics. This is considered to be a very challenging task as the related mathematical problems are typically non-convex and ill-conditioned [8]. There is a number of approaches among which we can mention Pseudo-Linear Regression, Prediction Error, and Bayesian Methods just to mention a few. Most of these approaches are developed for off-line identification. This means that the model parameters are identified on the basis of some historical measurements. However, fast changes in parameters make these estimations relatively useless as the obtained results may turn out to be inconsistent with the current state of things. Thus, there is a need in the methods that allow for an "on-the-fly" parameter estimation using the actually obtained data.

In this paper we present an optimization based approach to the parameter estimation that requires only the actual measurements of the system state. Below we present a schematic description of the algorithm. Let

$$\dot{x} = f(x, u, p) \tag{11}$$

be the system of differential equations describing the studied process that depends on the parameter p to be identified. We assume that the state is measured at certain discrete time instants t_i, $i = 0, \ldots, k$, which are uniformly distributed (say, we measure the state each month). We proceed as follows:

(1) Initialization:
   Set $x(t_0) = x_0$, $p = \bar{p}$, where $\bar{p}$ is an a priori estimate of the parameter. Set the index $i = 1$.
(2) Measure the state at t_i to get $\hat{x}_i$.

(3) Solve the following optimization problem:

$$\min_p \| x_p(t_i) - \hat{x}_i \|$$

(12)

$$\dot{x}_p = f(x, u, p), \quad x_p(t_{i-1}) = x_{i-1}.$$

using $\bar{p}$ as the initial guess. The optimization involves solving the DE (11) which can be performed either numerically or using the orthogonal collocations.

(4) Denote by p^o the optimal value of p found on the previous step. Set x_i to be the solution of (11) at time t_i computed for $p = p^o$ with initial condition $x(t_{i-1}) - x_{i-1}$.

(5) Set $i = i + 1$. Go to (2).

We illustrate the described algorithm for the case described in Sect. 3.2. Namely, we wish to estimate the value of the contact rate of high-risk individuals λ_H that enters the expression for $\phi_H(x)$ (see [9] for details). We assume that the system's states are measured in some way, for instance using the approach described in Sect. 2.3 and the measurements are performed each month. To make the setup more realistic, we assume that the measured values of the state have the time lag of 2 weeks. That is to say, what we expect to be the state on December 31 does actually correspond to the value of the state as on December 16.

Figure 5 shows the estimated value of the parameter λ_H at time points t_i. We see that the estimated value of λ_H enters a small neighborhood of the actual value $\lambda_H = 40.9400$ and stays there as time goes on. There is also a small bias in the estimated parameter which is due to the lag in measurements.

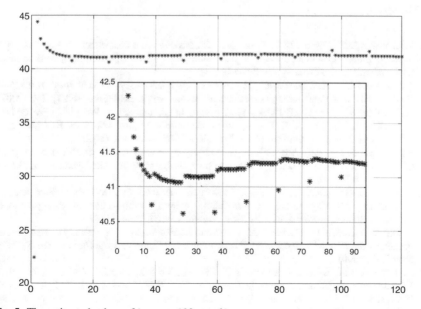

Fig. 5 The estimated values of λ_H over 120 months

5 Conclusion

In this contribution, we briefly described an application of optimal control theory to the problem of resource allocation for HIV control and treatment as well as further ramifications thereof. In particular, we addressed the issue of practical implementation of developed optimal control policies, the obstacles that one encounters when applying these policies and possible ways to overcome these difficulties.

This contribution does not address the mentioned issues in full detail as a detailed analysis of any of the outlined directions would unfold into a full-scale research project. We hope that this brief overview will serve as a catalyst for the further research along the lines presented in the paper.

Currently, we are in the process of setting up a computational framework for calculating optimal controls for general epidemiological settings described by basically any mass action model. The framework is composed of (1) clients, which define settings for which they would like to have an optimal control calculated, and (2) servers, which carry out the requested calculations. The communication between clients and servers is realized via a MySQL database.

Acknowledgements This project has been funded in whole or in part with Federal funds from the Centers for Disease Control and Prevention/OID/NCHHSTP/DSTDP, Department of Health and Human Services, under Interagency Agreement No. 17FED1710397.

Dmitry Gromov thanks to the International Union of Biological Sciences (IUBS) for partial support of living expenses in Moscow, during the 17th BIOMAT International Symposium, October 29–November 04, 2017.

References

1. S.S. Alistar, D.K. Owens, M.L. Brandeau, Effectiveness and cost effectiveness of oral pre-exposure prophylaxis in a portfolio of prevention programs for injection drug users in mixed HIV epidemics. PLoS One **9**(1), e86584 (2014)
2. A.H. Auchincloss, A.V. Diez Roux, A new tool for epidemiology: the usefulness of dynamic-agent models in understanding place effects on health. Am. J. Epidemiol. **168**(1), 1–8 (2008)
3. J.M. Baeten, D. Donnell, P. Ndase, N.R. Mugo, J.D. Campbell, J. Wangisi, J.W. Tappero, E. A. Bukusi, C.R. Cohen, E. Katabira et al. Antiretroviral prophylaxis for HIV prevention in heterosexual men and women. N. Engl. J. Med. **367**(5), 399–410 (2012)
4. S.E. Bellan, J. Dushoff, A.P. Galvani, L.A. Meyers, Reassessment of HIV-1 acute phase infectivity: accounting for heterogeneity and study design with simulated cohorts. PLoS Med. **12**(3), e1001801 (2015)
5. S. Blank, K. Gallagher, K. Washburn, M. Rogers, Reaching out to boys at bars: utilizing community partnerships to employ a wellness strategy for syphilis control among men who have sex with men in New York City. Sex. Transm. Dis. **32**, S65–S72 (2005)
6. O. Diekmann, J.A.P. Heesterbeek, *Mathematical Epidemiology of Infectious Diseases: Model Building, Analysis and Interpretation.* Mathematical and Computational Biology, vol. 5 (Wiley, New York, 2000)
7. D. Donnell, J.M. Baeten, J. Kiarie, K.K. Thomas, W. Stevens, C.R. Cohen, J. McIntyre, J.R. Lingappa, C. Celum, Partners in Prevention HSV/HIV Transmission Study Team et al., Heterosexual HIV-1 transmission after initiation of antiretroviral therapy: a prospective cohort analysis. Lancet **375**(9731), 2092–2098 (2010)

8. A. Gábor, J.R. Banga, Robust and efficient parameter estimation in dynamic models of biological systems. BMC Syst. Biol. **9**(1), 74 (2015)
9. D. Gromov, I. Bulla, O.S. Serea, E.O. Romero-Severson, Numerical optimal control for HIV prevention with dynamic budget allocation. Math. Med. Biol. J. IMA dqx015 (2017). https://doi.org/10.1093/imammb/dqx015
10. W.M. Haddad, V. Chellaboina, Stability and dissipativity theory for nonnegative dynamical systems: a unified analysis framework for biological and physiological systems. Nonlinear Anal. Real World Appl. **6**(1), 35–65 (2005)
11. J.A. Jacquez, C.P. Simon, Qualitative theory of compartmental systems. SIAM Rev. **35**(1), 43–79 (1993)
12. R.D. Moore, R.E. Chaisson, Natural history of HIV infection in the era of combination antiretroviral therapy. AIDS **13**(14), 1933–1942 (1999)
13. B.E. Nichols, C.A.B. Boucher, M. van der Valk, B.J.A. Rijnders, D.A.M.C. van de Vijver, Cost-effectiveness analysis of pre-exposure prophylaxis for HIV-1 prevention in the Netherlands: a mathematical modelling study. Lancet Infect. Dis. **16**(12), 1423–1429 (2016)
14. A.V. Rao, A survey of numerical methods for optimal control. Adv. Astronaut. Sci. **135**(1), 497–528 (2009)
15. E.O. Romero-Severson, E. Volz, J.S. Koopman, T. Leitner, E.L. Ionides, Dynamic variation in sexual contact rates in a cohort of HIV-negative gay men. Am. J. Epidemiol. **182**(3), 255–262 (2015)
16. E.S. Rosenberg, G.A. Millett, P.S. Sullivan, C. Del Rio, J.W. Curran, Modeling disparities in HIV infection between black and white men who have sex with men in the United States using the HIV care continuum. Lancet HIV **1**(3), e112–e118 (2014)
17. A. Smith, I. Miles, B. Le, T. Finlayson, A. Oster, E. DiNenno, Prevalence and awareness of HIV infection among men who have sex with men – 21 Cities, US 2008. Morb. Mortal. Wkly Rep. **59**(37), 1201–1227 (2010)

The Inverse Magnetoencephalography Problem and Its Flat Approximation

A. S. Demidov and M. A. Galchenkova

1 Unique Solution of the Inverse MEEG-Problem

The inverse MEEG-problem is the problem of finding the distribution of dipoles $\mathbf{q} : Y \to \mathbb{R}^3$ (current dipole moment) in the neurons of the brain, which occupies a domain $Y \subset \mathbb{R}^3$, according to the electric $\mathcal{D} = \varepsilon\mathcal{E}$, as well as the magnetic induction $\mathcal{B} = \mu\mathcal{H}$, measured on the surface X, which is the internal part of the helmet, with the SQUID sensors (Superconducting quantum Interference device) [1, 2]. The fields $\mathcal{E}$ and $\mathcal{H}$ are called the electric and magnetic field strengths. The parameters μ and $\varepsilon = \varepsilon(\mathbf{x}) > 0$ are magnetic and dielectric permeabilities. For the bio-medium $\mu \approx \mu_0$ magnetic permeability of the vacuum. The dielectric permittivity $\varepsilon = \varepsilon(\mathbf{x})$ is generally speaking different in $Y = Y_-,\ Y_0,\ Y_+$, where Y_0 and Y_+ as the regions corresponding to the skull and the air surrounding the head.

We shall start from the Maxwell equations

$$\mu\partial_t\mathcal{H}(\mathbf{x}, t) + \mathrm{rot}\,\mathcal{E}(\mathbf{x}, t) = 0, \quad \mathrm{div}\mathcal{B}(\mathbf{x}, t) = 0,$$

$$-\varepsilon(\mathbf{x})\partial_t\mathcal{E}(\mathbf{x}, t) + \mathrm{rot}\,\mathcal{H}(\mathbf{x}, t) = \mathbf{J}^v(\mathbf{x}) + \mathbf{J}^p(\mathbf{x}), \quad \mathrm{div}\mathcal{D}(\mathbf{x}, t) = \rho. \tag{1}$$

Here $\mathbf{J}^v = \sigma\mathcal{E}$ is the so-called volumetric or, as they say, ohmic current (more precisely, its density), because it satisfies Ohm's law associated with the coefficient of electrical conductivity $\sigma = \sigma(\mathbf{x}) \geq 0$, which is assumed to be independent of t. We note that such conditions are physically justified:

A. S. Demidov
Lomonosov Moscow State University, Moscow, Russian Federation
Moscow Institute of Physics and Technology (State University), Moscow, Russian Federation

M. A. Galchenkova (✉)
Moscow Institute of Physics and Technology (State University), Moscow, Russian Federation

© Springer International Publishing AG, part of Springer Nature 2018
R. P. Mondaini (ed.), *Trends in Biomathematics: Modeling, Optimization and Computational Problems*, https://doi.org/10.1007/978-3-319-91092-5_10

$$\sigma_+ = 0 \quad \text{on} \quad Y_+, \qquad \sigma_0 > 0 \quad \text{on} \quad Y_0, \qquad \sigma_- > \sigma_0 \quad \text{on} \quad Y_-. \tag{2}$$

The volume current is the result of the action of a macroscopic electric field on the charge carriers in the conducting medium of the brain. Neuronal same activity causes the so-called primary (principal) current $\mathbf{J}^p$. It arises as a result of dielectric polarization and it represents a movement of charges inside or near the cell. The volume density of these charges is denoted by ρ. Particles possessing these charges are part of the molecules. They are displaced from their equilibrium positions under the action of an external electric field, without leaving the molecule into which they enter.

Essential is the circumstance, especially noted in the fundamental work [2]. It is related to the frequency ratio ω of the oscillations of the electromagnetic field $\mathcal{H}(\mathbf{x}, t) = \mathbf{H}(\mathbf{x})e^{i\omega t}$, $\mathcal{E}(\mathbf{x}, t) = \mathbf{E}(\mathbf{x})e^{i\omega t}$ and the frequency of electrical oscillations in brain cells. The analysis in [2] (on page 426) shows that the quasistatic approximation for the (1) system is valid. There, on the same page, is additionally noted: "A current dipole $\mathbf{q}$, approximating a localized primary current, is a widely used concept in neuromagnetism... In EEG and MEG applications, a current dipole is used as an equivalent source for the unidirectional primary current that may extend over several square centimeters of cortex." As a result, we arrive at the following equations

$$\text{rot}\mathbf{E} = 0, \qquad \text{rot}\mathbf{B} = \mu(\sigma\mathbf{E} + \mathbf{q}), \qquad \text{div}\mathbf{B} = 0, \qquad \text{div}\mathbf{D} = \rho. \tag{3}$$

By the Stokes–Poincare theorem,[1,2]

---

[1]The vector $\mathbf{E}$, having the Cartesian coordinates (E_1, E_2, E_3) corresponds to the differential form

$$\omega_{\mathbf{E}}^1 = E_1 dx_1 + E_2 dx_2 + E_3 dx_3\,,$$

and to the vector $\text{rot}\mathbf{E} = \left(\frac{\partial E_3}{\partial x_2} - \frac{\partial E_2}{\partial x_3}, \frac{\partial E_1}{\partial x_3} - \frac{\partial E_3}{\partial x_1}, \frac{\partial E_2}{\partial x_1} - \frac{\partial E_1}{\partial x_2}\right)$—is the differential form

$$\omega_{\text{rot}\mathbf{E}}^2 = \left(\frac{\partial E_3}{\partial x_2} - \frac{\partial E_2}{\partial x_3}\right) dx_2 \wedge dx_3 + \left(\frac{\partial E_1}{\partial x_3} - \frac{\partial E_3}{\partial x_1}\right) dx_3 \wedge dx_1 + \left(\frac{\partial E_2}{\partial x_1} - \frac{\partial E_1}{\partial x_2}\right) dx_1 \wedge dx_2.$$

We have: $d\omega_{\mathbf{E}}^1 = \omega_{\text{rot}\mathbf{E}}^2$. Therefore, the condition $\text{rot}\mathbf{E} = 0$ implies $d\omega_{\mathbf{E}}^1 = 0$. Consequently, $\int_{\partial\Omega} \omega_{\mathbf{E}}^1 = \int_{\Omega} d\omega_{\mathbf{E}}^1 = 0$, where Ω is a surface in $\mathbb{R}^3$, limited by the boundary $\partial\Omega$. If $\partial\Omega$ is a curve (in other words, Ω is a simply connected surface), then the equality $\int_{\partial\Omega} \omega_{\mathbf{E}}^1 = 0$ means that the integral from some point $P_0 \in \partial\Omega$ to some other point $P \in \partial\Omega$ does not depend on which part of the curve $\partial\Omega$ it will be taken. In other words, $\omega_{\mathbf{E}}^1 = E_1 dx_1 + E_2 dx_2 + E_3 dx_3$ is the total differential: $\omega_{\mathbf{E}}^1 = -d\Phi$, i.e. $\mathbf{E} = -\nabla\Phi$. According to physical representations, at infinity the potential Φ of the field $\mathbf{E} = -\nabla\Phi$ is a constant that can be considered equal to zero.

[2]If $\omega_{\mathbf{B}}^2 = B_1 dx_2 \wedge dx_3 + B_2 dx_3 \wedge dx_1 + B_3 dx_1 \wedge dx_2$, then $d\omega_{\mathbf{B}}^2 = \omega_{\text{div}\mathbf{B}}^3$. Therefore, the condition $\text{div}\mathbf{B} = 0$ implies the equality $d\omega_{\mathbf{B}}^2 = 0$, i.e. the closed form $\omega_{\mathbf{B}}^2$. By the Poincare lemma, it is exact in a simply-connected domain, that is $\omega_{\mathbf{B}}^2 = d\omega_{\mathbf{A}}^1$, in other words $\mathbf{B} = \text{rot}\mathbf{A}$. We can assume that $\mathbf{A}\big|_\infty = 0$ as $\mathbf{B}\big|_\infty = 0$.

$$\text{rot}\mathbf{E} = 0 \quad \Leftrightarrow \quad \mathbf{E} = -\nabla\Phi, \qquad\qquad \text{div}\mathbf{B} = 0 \quad \Leftrightarrow \quad \mathbf{B} = \text{rot}\mathbf{A}. \qquad (4)$$

Since $\text{div}(\varepsilon\mathbf{E}) = \rho$, then

$$-\varepsilon\Delta\Phi - \nabla\varepsilon\nabla\Phi = \rho. \qquad (5)$$

According to physical representations, the field potential $\Phi \stackrel{(5)}{=} \Phi_\rho$ at infinity is a constant, which can be considered equal to zero. For similar reasons, the vector potential $\mathbf{A}$ of field $\mathbf{B} = \text{rot}\mathbf{A}$ is also chosen to be zero at infinity.

Since $\text{rot}(\text{rot}\mathbf{A}) = \nabla\,\text{div}\mathbf{A} - \Delta\mathbf{A}$, then $\Delta\mathbf{A} = -\text{rot}\mathbf{B} + \nabla\,\text{div}\mathbf{A}$. But $\text{rot}\mathbf{B} = \sigma\mathbf{E} + \mathbf{q}$, and $\mathbf{E} = -\nabla\Phi$. Thus

$$\Delta\mathbf{A}(\mathbf{x}) = -\mathbf{q}(\mathbf{x}) + \nabla\big[\sigma(\mathbf{x})\Phi(\mathbf{x}) + \text{div}\mathbf{A}(\mathbf{x})\big] - \Phi(\mathbf{x})\nabla\sigma(\mathbf{x}). \qquad (6)$$

The vector potential $\mathbf{A}$ is determined up to a potential field. Indeed, we have: $\text{rot}(\mathbf{A} - \mathbf{A}^*) = 0 \stackrel{(4)}{\Leftrightarrow} \mathbf{A} - \mathbf{A}^* = \nabla\varphi$, i.e. $\mathbf{A} = \mathbf{A}^* + \nabla\varphi$, where φ is a function.

Taking as φ solution of the equation[3] $\Delta\varphi = -\text{div}\mathbf{A}^* - \sigma\Phi$, subjected to condition $\varphi\big|_\infty = 0$ (because $\mathbf{A}^*\big|_\infty = 0$, $\Phi\big|_\infty = 0$), we obtain

$$\Delta\mathbf{A}(\mathbf{x}) = -\mathbf{F}(\mathbf{x}), \qquad \text{where} \qquad \mathbf{F}(\mathbf{x}) = \mathbf{q}(\mathbf{x}) + \Phi_\rho(\mathbf{x})\nabla\sigma(\mathbf{x}). \qquad (7)$$

Note that $\mathbf{A} = \mathbf{A}_\rho$, like Φ, depends on ρ.

Assuming $\mathbf{a} = (a_1, a_2, a_3)$, where $\Delta a_j(\mathbf{x}) = \delta(\mathbf{x})$, $a_j(\infty) = 0$, ie $a_j(\mathbf{x}) = -\frac{1}{4\pi}\frac{1}{|\mathbf{x}|}$, we obtain

$$\Delta\mathbf{A}(\mathbf{x}) \stackrel{(7)}{=} -\int \mathbf{F}(\mathbf{y})\Delta\mathbf{a}(\mathbf{x} - \mathbf{y})\,d\mathbf{y} = \Delta\left[-\int \mathbf{F}(\mathbf{y})\mathbf{a}(\mathbf{x} - \mathbf{y})\,d\mathbf{y}\right].$$

From here

$$\mathbf{A}(\mathbf{x}) = \frac{1}{4\pi}\int \mathbf{F}(\mathbf{y})\frac{1}{|\mathbf{x} - \mathbf{y}|}\,d\mathbf{y} \stackrel{(7)}{=} \frac{1}{4\pi}\int \Big(\mathbf{q}(\mathbf{y}) + \Phi(\mathbf{y})\nabla\sigma(\mathbf{y})\Big)\frac{1}{|\mathbf{x} - \mathbf{y}|}\,d\mathbf{y},$$

since the Laplace equation has a unique solution that vanishes at infinity (as already noted, $\mathbf{A}\big|_\infty = 0$). As a result, we obtain an integral equation of the I-kind

$$\int_Y \frac{\mathbf{q}(\mathbf{y})d\mathbf{y}}{|\mathbf{x} - \mathbf{y}|} = \mathbf{f}(\mathbf{x}), \qquad \mathbf{x} \in Y, \qquad (8)$$

---

[3]It depends on Φ_ρ and therefore on ρ.

whose right-hand side, given by the formula

$$\mathbf{f}(\mathbf{x}) = 4\pi \mathbf{A}(\mathbf{x}) - \int_Y \frac{\Phi(\mathbf{y})\nabla\sigma(\mathbf{y})}{|\mathbf{x}-\mathbf{y}|} \, d\mathbf{y}, \tag{9}$$

is completely determined by the fields $\mathbf{E} = -\nabla\Phi_\rho$ and $\mathbf{B} = \mathrm{rot}\mathbf{A}_\rho$, depending on the functional parameter ρ.

If the function σ, subject to the condition (2), is piecewise constant, then

$$\mathbf{f}(\mathbf{x}) \stackrel{(9)}{=} 4\pi \mathbf{A}(\mathbf{x}) - (\sigma_0 - \sigma_+)\mathbf{n}_X \int_X \frac{\Phi_0(\mathbf{y}_X)\, d\mathbf{y}_X}{|\mathbf{x}-\mathbf{y}_X|} - (\sigma_- - \sigma_0)\mathbf{n}_S \int_S \frac{\Phi_0(\mathbf{y}_S)\, d\mathbf{y}_S}{|\mathbf{x}-\mathbf{y}_S|},$$

where $\mathbf{n}_X$ and $\mathbf{n}_S$ are the external normals to $X = \partial Y_0 \cap \partial Y_+ = \partial Y_+$ and $S = \partial Y_0 \cap \partial Y_- = \partial Y$.

We can assume [1, 2] that the components of the vector $\mathbf{f}$ are sufficiently smooth, in any case, belongs to the Sobolev space $H^s(Y)$, where $s > 3/2$.

Theorem 1.1 (See [3]) *Equation (8) is uniquely solvable and the solution has the form*

$$\mathbf{q}(\mathbf{x}) = \mathbf{q}_0(\mathbf{x}) + \mathbf{p}_0(\mathbf{y}')\delta\Big|_{\partial Y} \in H^{s-2}(Y) + H^{s+1}(\partial Y) \otimes \delta\Big|_{\partial Y},$$

where $\delta\Big|_{\partial Y}$ —δ-function on ∂Y.

The fact that the fields here $\mathbf{E} = -\nabla\Phi_\rho$ and $\mathbf{B} = \mathrm{rot}\mathbf{A}_\rho$ depend on the functional parameter ρ allows us to apply the methods of optimal reconstruction (interpolation, see [4–6]) of these fields by their values at a finite number of points of the set X.

We also note that in [7] a connection is established between the solution $\mathbf{q}$ of the integral equation (8) and the solution $\mathbf{u}$ of an integral equation of the second kind

$$4\pi \eta^2 \mathbf{u}(\mathbf{x}) + \int_Y \frac{\mathbf{u}(\mathbf{y})}{|\mathbf{x}-\mathbf{y}|} d\mathbf{y} = \mathbf{f}(\mathbf{x}), \qquad \mathbf{x} \in Y, \quad \eta > 0. \tag{10}$$

Theorem 1.2 (See [7]) *The solution of equation (10) is representable in the form*

$$\mathbf{u}(\mathbf{x}) = \mathbf{q}_0(\mathbf{x}) + \frac{1}{\eta}\mathbf{p}_0(\mathbf{y}')\varphi\, e^{-\mathbf{y}_n/\eta} + r_0(\mathbf{x}, \eta), \tag{11}$$

where $\|r_0\|_{L^2} \le C\sqrt{\eta}$, $\mathbf{y}_n$ *is the distance along the normal from* $\mathbf{x}$ *to* $\mathbf{y}' \in \Gamma$, *and* $\varphi \in C^\infty(Y)$, $\varphi \equiv 1$ *in a small neighborhood* ∂Y *and* $\varphi \equiv 0$ *outside a slightly larger neighborhood.*

The formula (11) can serve as a basis for numerical solution of the problem (8)–(9).

2 Flat Model of the Inverse MEG-Problem

This is not a MEEG problem, since there is no data on the electric field. However, this problem is of particular interest, since it has a direct relationship to *scanning magnetic microscopes*. These tools [8] make it possible to record magnetic fields, for example, in integrated circuits, in magnetotactic bacteria. They are used in materials science, mineralogy, paleomagnetic analysis [9, 10].

In those cases it is impossible to completely find $\mathbf{q}$. However, partial information about the distribution $\mathbf{q} : Y \ni \mathbf{y} \mapsto \mathbf{q}(\mathbf{y})$ we can still get [11]. In the following model case X is the plane $\mathbb{R}^2 \ni \mathbf{x} = (x_1, x_2)$, and Y is a parallel to it a flat layer, which is kept from X at some distance $\alpha > 0$, i.e.

$$Y = \{\mathbf{y} = (\mathbf{z}, -h) \; : \; \alpha \le h \le \beta \,, \text{ where } \mathbf{z} = (z_1, z_2) \in Z = \mathbb{R}^2\}.$$

And X is a surface at the points $\mathbf{x}$ of which the magnetic field $\mathbf{B}(\mathbf{x}) = (B_1(\mathbf{x}), B_2(\mathbf{x}), B_3(\mathbf{x}))$ is measured, and Y is the set in which we have the distribution of the electric dipoles $\mathbf{q} : Y \ni \mathbf{y} = (\mathbf{z}, -h) \mapsto \mathbf{q}(\mathbf{y}) = (q_1(\mathbf{y}), q_2(\mathbf{y}), q_3(\mathbf{y}))$. In what follows we assume that the measuring system is such that $\frac{\mu}{4\pi} = 1$.

Lemma 2.1 *If $\beta - \alpha = 0$, then[4]*

$$\int_Y \mathbf{K}(\mathbf{x} - \mathbf{y}) \, \mathbf{q}(\mathbf{y}) \, d\mathbf{y} = \mathbf{B}(\mathbf{x}) \,. \tag{12}$$

Here $\mathbf{K}(\mathbf{x} - \mathbf{y}) \, \mathbf{q}(\mathbf{y}) = \frac{\mathbf{q}(\mathbf{y}) \times (\mathbf{x} - \mathbf{y})}{|\mathbf{x} - \mathbf{y}|^3}$, $\mathbf{a} \times \mathbf{b}$ is cross product $\mathbf{a}$ and $\mathbf{b}$. There by

$$\mathbf{K}(\mathbf{t}) = \begin{bmatrix} 0 & K_{12}(\mathbf{t}) & -K_{31}(\mathbf{t}) \\ -K_{12}(\mathbf{t}) & 0 & K_{23}(\mathbf{t}) \\ K_{31}(\mathbf{t}) & -K_{23}(\mathbf{t}) & 0 \end{bmatrix}, \quad \mathbf{t} = (t_1, t_2, t_3) \in \mathbb{R}^3,$$

where

$$K_{12}(\mathbf{t}) = \frac{t_3}{|\mathbf{t}|^3}, \; K_{31}(\mathbf{t}) = \frac{t_2}{|\mathbf{t}|^3}, \; K_{23}(\mathbf{t}) = \frac{t_1}{|\mathbf{t}|^3} \,, \; |\mathbf{t}| = \sqrt{t_1^2 + t_2^2 + t_3^2}. \tag{13}$$

If $\beta - \alpha > 0$, instead of (12) we will consider the following equation

---

[4]The formula (12) is an integral version of the Biot–Sawar law: $\mathbf{B}(\mathbf{x}) = \frac{\mathbf{q} \times (\mathbf{x} - \mathbf{y})}{|\mathbf{x} - \mathbf{y}|^3}$ for the field $\mathbf{B}$, that induced by a current dipole $\mathbf{q}$. This formula, which was experimentally established in 1820 by the French physicists Jean-Baptiste Biot (1774–1862) and Felix Sawar (1791–1841), in the process of observing the effect on the magnetic needle of a conductor with the current flowing along it, is a consequence of the equations Maxwell.

$$\int_Z \left[\frac{1}{\beta - \alpha} \int_\alpha^\beta \mathbf{K}(\mathbf{x} - \mathbf{y}) \Big|_{\mathbf{y} = (z_1, z_2, -h)} \, dh \right] \mathbf{Q}(\mathbf{z}) \, d\mathbf{z} = \mathbf{B}(\mathbf{x}) \,. \tag{14}$$

for the function $\mathbf{Q} : \mathbb{R}^2 \ni \mathbf{z} \mapsto \mathbf{Q}(\mathbf{z}) = \big(Q_1(\mathbf{z}), Q_2(\mathbf{z}), Q_3(\mathbf{z})\big)$, which is some "averaging" for h of the function $\mathbf{q}(\cdot, h) : \mathbf{z} \mapsto \mathbf{q}(\mathbf{z}, h)$. For $\beta - \alpha = 0$ the dipole $\mathbf{q}$ will also be denoted by $\mathbf{Q}$.

2.1 The Fourier Image of the Kernel K

Rewrite Eq. (14) in the term of pseudo-differential equation

$$Op\big(\widetilde{\mathbf{K}}(\boldsymbol{\xi})\big)\mathbf{Q} = \mathbf{B}\,, \qquad Op\big(\widetilde{\mathbf{K}}(\boldsymbol{\xi})\big) \overset{def}{=} \mathbf{F}_{\boldsymbol{\xi} \to \mathbf{x}}^{-1}\big(\widetilde{\mathbf{K}}(\boldsymbol{\xi})\big)\mathbf{F}_{\mathbf{z} \to \boldsymbol{\xi}}$$

with the following matrix symbol

$$\widetilde{\mathbf{K}} : \mathbb{R}^2 \ni \boldsymbol{\xi} = (\xi_1, \xi_2) \mapsto \widetilde{\mathbf{K}}(\boldsymbol{\xi}) = \begin{bmatrix} 0 & \widetilde{K}_{12}(\boldsymbol{\xi}) & -\widetilde{K}_{31}(\boldsymbol{\xi}) \\ -\widetilde{K}_{12}(\boldsymbol{\xi}) & 0 & \widetilde{K}_{23}(\boldsymbol{\xi}) \\ \widetilde{K}_{31}(\boldsymbol{\xi}) & -\widetilde{K}_{23}(\boldsymbol{\xi}) & 0 \end{bmatrix} \,.$$

Here $\overset{\smile}{\widetilde{K}}_{lm}(\boldsymbol{\xi}) = \frac{1}{\beta - \alpha} \int_\alpha^\beta \widetilde{K}_{lm}(\boldsymbol{\xi}, h) dh$, and

$$\widetilde{K}_{lm}(\boldsymbol{\xi}, h) = F_{\mathbf{s} \to \boldsymbol{\xi}} \, K_{lm}(\mathbf{s}, h) = \int_{\mathbb{R}^2} e^{-\overset{\circ}{i} \mathbf{s} \boldsymbol{\xi}} \, K_{lm}(\mathbf{s}, h) d\mathbf{s}\,, \quad \text{where} \quad \overset{\circ}{i} \overset{def}{=} 2\pi i \,. \tag{15}$$

Equation (15) means that

$$\widetilde{\mathbf{K}}(\boldsymbol{\xi})\widetilde{\mathbf{Q}}(\boldsymbol{\xi}) = \widetilde{\mathbf{B}}(\boldsymbol{\xi})\,, \quad \text{where } \widetilde{\mathbf{Q}} = (\widetilde{Q}_1, \widetilde{Q}_2, \widetilde{Q}_3), \ \widetilde{\mathbf{B}} = (\widetilde{B}_1, \widetilde{B}_2, \widetilde{B}_3), \tag{16}$$

and $\widetilde{Q}_j(\boldsymbol{\xi}) = \mathbf{F}_{\mathbf{z} \to \boldsymbol{\xi}} Q_j(\mathbf{z})$, $\widetilde{B}_j(\boldsymbol{\xi}) = \mathbf{F}_{\mathbf{x} \to \boldsymbol{\xi}} B_j(\mathbf{x})$.

Lemma 2.2 [5]

$$\widetilde{\mathbf{K}}(\boldsymbol{\xi}) = \begin{bmatrix} 0 & 1 & i\frac{\xi_2}{|\xi|} \\ -1 & 0 & -i\frac{\xi_1}{|\xi|} \\ -i\frac{\xi_2}{|\xi|} & i\frac{\xi_1}{|\xi|} & 0 \end{bmatrix} E(\boldsymbol{\xi})\,, \tag{17}$$

where $E(\boldsymbol{\xi}) = \left[\frac{1 - e^{-2\pi(\beta - \alpha)|\xi|}}{(\beta - \alpha)|\xi|} \right] e^{-2\pi\alpha|\xi|} \quad \Rightarrow \quad \lim_{\beta \to \alpha} E(\boldsymbol{\xi}) = 2\pi e^{-2\pi\alpha|\xi|}.$

---

[5]It has been proved by A.S. Kochurov.

Proof Let $\bar{1} = \{23\}$, $\bar{2} = \{31\}$, $\bar{3} = \{12\}$, i.e. $\bar{m}$—these are two of the three digits $\{1, 2, 3\}$ that complement the index m for a cyclic permutation: $\{1, 2, 3\} \rightarrow \{2, 3, 1\} \rightarrow \{3, 1, 2\}$. We set $s_h = (\mathbf{s}, h)$, $\mathbf{s} = (s_1, s_2)$. Note that

$$K_{\underset{m}{-}}(\mathbf{s}, h) \overset{(13)}{=} -\frac{\partial}{\partial s_m} \frac{1}{|\mathbf{s}_h|} \quad \text{when} \quad m \neq 3. \tag{18}$$

Thus for $m \neq 3$ we have

$$\tilde{K}_{\underset{m}{-}}(\xi, h) = -\lim_{N \to \infty} \int_{s_1^2 + s_2^2 \leq N^2} e^{-\overset{\circ}{i}(s_1\xi_1 + s_2\xi_2)} \frac{\partial}{\partial s_m} \frac{1}{|\mathbf{s}_h|} ds_1 ds_2 ,$$

and

$$\tilde{K}_{\underset{3}{-}}(\xi, h) \overset{\text{i.e.}}{=} \tilde{K}_{12}(\xi, h) \overset{(13)}{=} h \int_{\mathbb{R}^2} \frac{e^{-\overset{\circ}{i}(s_1\xi_1 + s_2\xi_2)} ds_1 ds_2}{[s_1^2 + s_2^2 + h^2]^{3/2}} .$$

Assuming that $re^{i\phi} = s_1 + is_2$, $\rho e^{i\psi} = \xi_1 + i\xi_2$, $|\xi| = \sqrt{\xi_1^2 + \xi_2^2}$, rewrite $\tilde{K}_{\underset{3}{-}}(\xi, h)$, using the following Hankel formula, also called the Fourier–Bessel transform:

$$h \int_{r=0}^{\infty} \int_{\phi=0}^{2\pi} \frac{e^{-\overset{\circ}{i}r|\xi|\cos(\phi - \psi)}}{[r^2 + h^2]^{3/2}} r \, dr d\phi = 2\pi h \int_0^{\infty} \frac{r J_0(2\pi|\xi|r)}{[r^2 + h^2]^{3/2}} dr ,$$

where $J_0(\zeta) = \frac{1}{2\pi} \int_0^{2\pi} e^{i\zeta \cos\theta} d\theta$ is Bessel function of zero order. Similarly, for $m \neq 3$ we have

$$-\lim_{N \to \infty} \int_{s_1^2 + s_2^2 \leq N^2} e^{-\overset{\circ}{i}(s_1\xi_1 + s_2\xi_2)} \frac{\partial}{\partial s_m} \frac{1}{\sqrt{s_1^2 + s_2^2 + h^2}} ds_1 ds_2$$

$$= -\overset{\circ}{i}\xi_m \lim_{N \to \infty} \int_{r=\sqrt{s_1^2 + s_2^2} \leq N} \frac{e^{-\overset{\circ}{i}(s_1\xi_1 + s_2\xi_2)} ds_1 ds_2}{[r^2(s_1, s_2) + h^2]^{1/2}}$$

$$= -\overset{\circ}{i}\xi_m \lim_{N \to \infty} \int_0^N \left(\int_0^{2\pi} \frac{e^{-\overset{\circ}{i}r\rho \cos\phi}}{[r^2 + h^2]^{1/2}} d\phi \right) r \, dr$$

$$= -2\pi \overset{\circ}{i}\xi_m \int_0^{\infty} \frac{r J_0(2\pi|\xi|r)}{[r^2 + h^2]^{1/2}} dr.$$

It is known (see, for example, in [12] formulas 6.554 (1 and 4)) that

$$\int_0^\infty \frac{r J_0(qr)\, dr}{(r^2 + a^2)^{3/2}} = \frac{1}{a} e^{-aq}\Big|_{q>0}, \quad \int_0^\infty \frac{r J_0(qr)\, dr}{(r^2 + a^2)^{1/2}} = \frac{1}{q} e^{-aq}\Big|_{q>0}.$$

Thus,

$$\widetilde{K}_{\underset{3}{-}}(\xi, h) = 2\pi e^{-2\pi h |\xi|}, \quad \widetilde{K}_{\underset{m}{-}}(\xi, h)\Big|_{m \neq 3} = -2\pi i \frac{\xi_m}{|\xi|} e^{-2\pi h |\xi|}.$$

□

According to Lemma 2.2, the coordinate-wise recording of Eq. (16) is as follows:

$$\left[\widetilde{Q}_2(\xi) + i \frac{\xi_2}{|\xi|} \widetilde{Q}_3(\xi) \right] E(\xi) = \widetilde{B}_1(\xi),$$

$$-\left[\widetilde{Q}_1(\xi) + i \frac{\xi_1}{|\xi|} \widetilde{Q}_3(\xi) \right] E(\xi) = \widetilde{B}_2(\xi), \tag{19}$$

$$\left[-i \frac{\xi_2}{|\xi|} \widetilde{Q}_1(\xi) + i \frac{\xi_1}{|\xi|} \widetilde{Q}_2(\xi) \right] E(\xi) = \widetilde{B}_3(\xi).$$

Lemma 2.3 *The following relations hold*

$$\widetilde{Q}_1(\xi) = -\frac{\widetilde{B}_2(\xi)}{E(\xi)} - i \frac{\xi_1}{|\xi|} \widetilde{Q}_3(\xi), \quad \widetilde{Q}_2(\xi) = \frac{\widetilde{B}_1(\xi)}{E(\xi)} - i \frac{\xi_2}{|\xi|} \widetilde{Q}_3(\xi), \tag{20}$$

$$\xi_1 \widetilde{B}_1(\xi) + \xi_2 \widetilde{B}_2(\xi) + i |\xi| \widetilde{B}_3(\xi) = 0. \tag{21}$$

The formulas (20) instantly follow from (19), and substituting $\widetilde{Q}_1(\xi)$ and $\widetilde{Q}_2(\xi)$ from (20) in $\left[-i \frac{\xi_2}{|\xi|} \widetilde{Q}_1(\xi) + i \frac{\xi_1}{|\xi|} \widetilde{Q}_2(\xi) \right] E(\xi) \overset{(19)}{=} \widetilde{B}_3(\xi)$, we get (21).

Directly from Lemma 2.3 follows[6]

Theorem 2.1 *Let $\widetilde{B}_1$ and $\widetilde{B}_2$ are such that $\widetilde{B}_k/E \in L^1$. Then the general solution of equation will look like*

$$\int_Z \left[\frac{1}{\beta - \alpha} \int_\alpha^\beta \mathbf{K}(\mathbf{x} - \mathbf{z}, h)\, dh \right] \mathbf{Q}(\mathbf{z})\, d\mathbf{z} \overset{(14)}{=} \mathbf{B}(\mathbf{x})$$

---

[6]A similar result is initiated by the problem of measuring the magnetic field by scanning magnetic microscope, was obtained in [13] using a generalization of the classic decomposition Hodge Laplace operator on a compact orientable manifold in the form of a sum $d\delta + \delta d$, where δ is the operator conjugate to the operator of exterior differentiation d.

and is representable in the form $\mathbf{Q} = \mathbf{Q}^B + \mathbf{Q}^0$. *Here* $\mathbf{Q}^0 = (Q_1^0, Q_2^0, Q_3^0)$, *where*

$$Q_1^0 = -Op\left(i\frac{\xi_1}{|\xi|}\right)Q_3, \quad Q_2^0 = -Op\left(i\frac{\xi_2}{|\xi|}\right)Q_3, \quad Q_3^0 \in L^2,$$

and $\mathbf{Q}^B = (A_1(\mathbf{y}), A_2(\mathbf{y}), 0)$, *where*

$$A_1(\mathbf{y}) \stackrel{def}{=} \mathbf{F}_{\xi \to \mathbf{y}}^{-1}\left(-\frac{\widetilde{B}_2(\xi)}{E(\xi)}\right), \quad A_2(\mathbf{y}) \stackrel{def}{=} \mathbf{F}_{\xi \to \mathbf{y}}^{-1}\left(\frac{\widetilde{B}_1(\xi)}{E(\xi)}\right). \tag{22}$$

In the next subsection we strengthen Theorem 2.1, by taking into account that the vector **B**, according to its physical meaning, is real and is given in a finite number of points $\mathbf{x}_k$.

2.2 Formulas for Numerical Calculations

The vector $\mathbf{B} = (B_1, B_2, B_3)$, according to the physical meaning, is real. It is given in a finite collection of points $\mathbf{x}_k \in X$. Therefore, the condition $\widetilde{B}_k/E \in L^1$ stipulated in Theorem 2.1 is naturally supplemented by such requirements:

(1) The functions $\mathbf{F}_{\xi \to \mathbf{x}}^{-1}\widetilde{B}_j(\xi)$ are real[7] for each $j = 1 \div 3$.
(2) Vector function $\xi \mapsto \left(\widetilde{B}_1(\xi), \widetilde{B}_2(\xi), \widetilde{B}_3(\xi)\right)\Big|_{\widetilde{B}_3(\xi) \stackrel{(21)}{=} \frac{i}{|\xi|}\left(\xi_1\widetilde{B}_1(\xi)+\xi_2\widetilde{B}_2(\xi)\right)}$ delivers

a minimum of functional

$$\Phi : \widetilde{\mathbf{B}} \mapsto \Phi(\widetilde{\mathbf{B}}) = \sum_{j=1}^{3} \sum_{k=(k_1,k_2)} \left|\mathbf{F}_{\xi \to \mathbf{x}_k}^{-1}\widetilde{B}_j(\xi) - B_j(\mathbf{x}_k)\right|^2, \tag{23}$$

where $B_j(\mathbf{x}_k)$ are the experimental values of B_j at the points $\mathbf{x}_k$.

When analyzing these requirements and numerical realization of the formulas (22) and (23), the following two sentences are useful.

Proposition 2.1 *Let* $x_1 = r\cos 2\pi\theta$, $x_2 = r\sin 2\pi\theta$ *and*

$$D(r, \theta) \stackrel{def}{=} d(x_1, x_2) = \sum_{m \in \mathbb{Z}} D_m(r)e^{\overset{\circ}{i}m\theta}, \quad D_m(r) \in \mathbb{C}.$$

---

[7]This imposes restrictions on the real and imaginary parts of the functions $\widetilde{B}_j$.

Then

$$\mathbf{F}_{\mathbf{x}\to\xi}\, d(x) = \sum_{n\in\mathbb{Z}}(-i)^n e^{\overset{\circ}{i}\omega n}\int_0^\infty r\,D_n(r)\,J_n(2\pi|\xi|r)\,dr\,, \tag{24}$$

where $\mathbf{x}=(x_1,x_2)$, $\xi=(\xi_1,\xi_2)$ *and* $\xi_1=|\xi|\cos 2\pi\omega$, $\xi_2=|\xi|\sin 2\pi\omega$.

Proof We have

$$\mathbf{F}_{\mathbf{x}\to\xi}\, d(x) = \int_0^\infty r\left(\int_0^1 D(r,\theta)e^{-i|\xi|r\cos 2\pi(\theta-\omega)}d\theta\right)dr, \text{ and}^8$$

$$e^{-\overset{\circ}{i}|\xi|r\cos 2\pi(\theta-\omega)} = \sum_{n\in\mathbb{Z}} J_n(-2\pi|\xi|r)i^n e^{\overset{\circ}{i}n(\theta-\omega)}\,. \tag{25}$$

And since $\int_0^1 e^{\overset{\circ}{i}(n-m)\theta}d\theta = \begin{cases} 0 \text{ in condition when } m\neq n \\ 1 \qquad\quad\text{and}\qquad\quad m=n, \end{cases}$ and

$$J_{-n}(-a) = J_n(a) \overset{def}{=} \frac{1}{\pi}\int_0^\pi \cos(nt - a\sin t)dt,$$

we have (24).   □

Similarly we can prove[9]

Proposition 2.2 *Let* $\xi_1 = |\xi|\cos 2\pi\omega$, $\xi_2 = |\xi|\sin 2\pi\omega$, *and* $\widetilde{C}(|\xi|,\omega) \overset{def}{=} \widetilde{c}(\xi_1,\xi_2) = \sum_{m\in\mathbb{Z}} \widetilde{C}_m(|\xi|)e^{-im\omega}$, $\widetilde{C}_m(\rho) \in \mathbb{C}$. *Then*

$$\mathbf{F}^{-1}_{\xi\to\mathbf{y}}\widetilde{c}(\xi) = \sum_{n\in\mathbb{Z}} i^n e^{-\overset{\circ}{i}\phi n}\int_0^\infty |\xi|\widetilde{C}_n(|\xi|)\,J_n(2\pi|\xi|\rho)\,d|\xi|\,, \tag{26}$$

when $\xi=(\xi_1,\xi_2)$, $\mathbf{y}=(y_1,y_2)$ *and* $y_1=\rho\cos 2\pi\phi$, $y_2=\rho\sin 2\pi\phi$.

Corollary 2.1 $\mathbf{F}^{-1}_{\xi\to\mathbf{x}}\frac{1}{|\xi|} = \int_0^\infty |\xi|\frac{1}{|\xi|}J_0(2\pi|\mathbf{x}||\xi|)\,d|\xi| = \frac{1}{2\pi|\mathbf{x}|}$.

---

[8]*Generating function* for $J_n(\mu)$, i.e. the formal power series $\sum_{n\in\mathbb{Z}} J_n(\mu)\,t^n$, is $e^{\frac{\mu}{2}\left(t-\frac{1}{t}\right)}$ (see [14]).
Assuming $t = ie^{\overset{\circ}{i}(\theta-\omega)}$, we derive (25).

[9]$\mathbf{F}^{-1}_{\xi\to\mathbf{y}}\widetilde{c}(\xi) = \int_0^\infty |\xi|\left(\int_0^1 \widetilde{C}(|\xi|,\omega)e^{\overset{\circ}{i}\rho|\xi|\cos 2\pi(\omega-\phi)}d\omega\right)d|\xi|$ and $e^{\overset{\circ}{i}\rho|\xi|\cos 2\pi(\omega-\phi)} = \sum_{n\in\mathbb{Z}} J_n(2\pi\rho|\xi|)i^n e^{\hat{i}n(\omega-\phi)}$ (cf. (25)).

Lemma 2.4 *Let* $\mathbf{x} = (x_1, x_2) = (\rho \cos 2\pi\phi, \rho \sin 2\pi\phi)$, *and* $\widetilde{B}_k(\xi)\big|_{k=1,2} =$
$\sum_{n\in\mathbb{Z}} \widetilde{C}_n^k(|\xi|) e^{-in\omega}$, *where* $\widetilde{C}_n^k(|\xi|) = p_n^k(|\xi|) + iq_n^k(|\xi|)$ *satisfy the expression:*
$\widetilde{B}_k(\xi) = 0$ *when* $|\xi| > R$ *for some* $R > 0$ *and, furthermore, suppose the following condition is satisfied*[10]:

$$\sum_{l\in\mathbb{Z}} (-1)^l \int_0^\infty |\xi| \Big\{ \Big[J_{2l}\big[q_{2l}^k \cos(4\pi l\phi) - p_{2l}^k \sin(4\pi l\phi) \big] d|\xi|$$
$$+ J_{2l+1}\big[p_{2l+1}^k \cos\big(2\pi(2l+1)\phi\big)$$
$$+ q_{2l+1}^k \sin\big(2\pi(2l+1)\phi\big) \big] \Big] \Big\} d|\xi| = 0. \tag{27}$$

Then

$$B_k(\mathbf{x})\big|_{k=1,2} = \mathbf{F}_{\xi\to\mathbf{x}}^{-1} \widetilde{B}_k(\xi) \quad \text{and} \quad B_3(\mathbf{x}) = \mathbf{F}_{\xi\to\mathbf{x}}^{-1}\left(\frac{i\big(\xi_1 \widetilde{B}_1(\xi) + \xi_2 \widetilde{B}_2(\xi)\big)}{|\xi|} \right)$$

are real, and for $k = 1, 2$ *(and* $\mathbf{x} = (\rho \cos 2\pi\phi, \rho \sin 2\pi\phi)$*)*

$$B_k(\mathbf{x}) = \sum_{l\in\mathbb{Z}} (-1)^l \int_0^\infty |\xi| \Big\{ \Big[J_{2l}\big[p_{2l}^k \cos(4\pi l\phi) + q_{2l}^k \sin(4\pi l\phi) \big]$$
$$+ J_{2l+1}\big[p_{2l+1}^k \sin\big(2\pi(2l+1)\phi\big) - q_{2l+1}^k \cos\big(2\pi(2l+1)\phi\big) \big] \Big] d|\xi|, \tag{28}$$

and

$$B_3(\mathbf{x}) = \frac{1}{4\pi^2} \int_{\mathbb{R}^2} \frac{\partial_{y_1} B_1(\mathbf{y}) + \partial_{y_2} B_2(\mathbf{y})}{|\mathbf{x} - \mathbf{y}|} d\mathbf{y}. \tag{29}$$

Proof Indeed, taking into account (26), we have

$$B_k(\mathbf{x}) = \sum_{l\in\mathbb{Z}} (-1)^l \int_0^\infty |\xi| e^{-2il\phi} \Big\{ \Big[\big(p_{2l}^k + iq_{2l}^k \big) J_{2l} \Big]$$
$$+ ie^{-i\phi} \Big[\big(p_{2l+1}^k + iq_{2l+1}^k \big) J_{2l+1} \Big] \Big\} d|\xi|$$
$$= \sum_{l\in\mathbb{Z}} (-1)^l \int_0^\infty |\xi| e^{-2il\phi} \Big\{ \Big[p_{2l}^k J_{2l} - e^{-i\phi} q_{2l+1}^k J_{2l+1} \Big]$$

---

[10] Arguments of $p_n^k(|\xi|)$, $q_n^k(|\xi|)$, $J_n(2\pi|\xi|\rho)$ for brevity are omitted.

$$+i\left[q_{2l}^k J_{2l} + e^{-\overset{\circ}{i}\phi} p_{2l+1}^k J_{2l+1}\right]\right\} d|\xi|$$

$$= \sum_{l\in\mathbb{Z}} (-1)^l \int_0^\infty \left[|\xi|\{\cos(4\pi l\phi) - i\sin(4\pi l\phi)\}\right.$$

$$\times\left\{\left[(p_{2l}^k J_{2l} - (\cos(2\pi\phi) - i\sin(2\pi\phi))q_{2l+1}^k J_{2l+1}\right]\right.$$

$$\left.\left.+i\left[(q_{2l}^k J_{2l} + (\cos(2\pi\phi) - i\sin(2\pi\phi))p_{2l+1}^k J_{2l+1}\right]\right\}\right]d|\xi|$$

$$= \sum_{l\in\mathbb{Z}} (-1)^l \int_0^\infty \left[|\xi|\{\cos(4\pi l\phi) - i\sin(4\pi l\phi)\}\right.$$

$$\times\left\{\left[p_{2l}^k J_{2l} - \cos(2\pi\phi)q_{2l+1}^k J_{2l+1} + i\sin(2\pi\phi)q_{2l+1}^k J_{2l+1}\right]\right.$$

$$\left.\left.+i\left[q_{2l}^k J_{2l} + \cos(2\pi\phi)p_{2l+1}^k J_{2l+1} - i\sin(2\pi\phi)p_{2l+1}^k J_{2l+1}\right]\right\}\right]d|\xi|$$

$$= \sum_{l\in\mathbb{Z}} (-1)^l \int_0^\infty \left[|\xi|\{\cos(4\pi l\phi) - i\sin(4\pi l\phi)\}\right.$$

$$\times\left\{\left[p_{2l}^k J_{2l} - \cos(2\pi\phi)q_{2l+1}^k J_{2l+1} + \sin(2\pi\phi)p_{2l+1}^k J_{2l+1}\right]\right.$$

$$\left.\left.+i\left[q_{2l}^k J_{2l} + \cos(2\pi\phi)p_{2l+1}^k J_{2l+1} + \sin(2\pi\phi)q_{2l+1}^k J_{2l+1}\right]\right\}\right]d|\xi|$$

$$= \sum_{l\in\mathbb{Z}} (-1)^l \int_0^\infty |\xi|\left\{\left[J_{2l}\left[p_{2l}^k \cos(4\pi l\phi) + q_{2l}^k \sin(4\pi l\phi)\right]\right.\right.$$

$$+J_{2l+1}\left[p_{2l+1}^k \sin\left(2\pi(2l+1)\phi\right) - q_{2l+1}^k \cos\left(2\pi(2l+1)\phi\right)\right]\right]$$

$$+i\left[J_{2l}\left[q_{2l}^k \cos(4\pi l\phi) - p_{2l}^k \sin(4\pi l\phi)\right]\right.$$

$$\left.\left.+J_{2l+1}\left[p_{2l+1}^k \cos\left(2\pi(2l+1)\phi\right) + q_{2l+1}^k \sin\left(2\pi(2l+1)\phi\right)\right]\right]\right\} d|\xi|.$$

Hence we obtain the formula (28) under the condition (27). The formula (29) follows directly from Corollary 2.1, since $\mathbf{F}_{\xi\to\mathbf{x}}^{-1} i\xi_k \widetilde{B}_k(\xi) = \frac{1}{2\pi}\partial_{x_k} B_k(\mathbf{x})$.

2.3 The Problem of Minimization

Let $\mathbf{x} = (x_1, x_2) = (\rho\cos 2\pi\phi, \rho\sin 2\pi\phi)$, and $\widetilde{B}_k(\xi)\Big|_{k=1,2} = \sum_{n\in\mathbb{Z}} \widetilde{C}_n^k(|\xi|)e^{-\overset{\circ}{i}n\omega}$, where $\widetilde{C}_n^k(|\xi|) = p_n^k(|\xi|) + iq_n^k(|\xi|)$ are that $\widetilde{B}_k(\xi) = 0$ when $|\xi| > R$ for some $R > 0$. Then

$$B_k(\mathbf{x})\Big|_{k=1,2} = \mathbf{F}^{-1}_{\xi\to\mathbf{x}}\widetilde{B}_k(\xi) \quad \text{and} \quad B_3(\mathbf{x}) = \mathbf{F}^{-1}_{\xi\to\mathbf{x}}\left(\frac{i\big(\xi_1\widetilde{B}_1(\xi) + \xi_2\widetilde{B}_2(\xi)\big)}{|\xi|}\right)$$

are real, and for $k = 1, 2$ (and $\mathbf{x} = (\rho\cos 2\pi\phi, \rho\sin 2\pi\phi)$)

$$B_k(\mathbf{x}) = \sum_{l\in\mathbb{Z}}(-1)^l \int_0^\infty |\xi|\Big\{\big[J_{2l}\big[p^k_{2l}\cos(4\pi l\phi) + q^k_{2l}\sin(4\pi l\phi)\big]$$

$$+J_{2l+1}\big[p^k_{2l+1}\sin\big(2\pi(2l+1)\phi\big) - q^k_{2l+1}\cos\big(2\pi(2l+1)\phi\big)\big]\Big]d|\xi|\,,$$

and

$$B_3(\mathbf{x}) = \frac{1}{4\pi^2}\int_{\mathbb{R}^2}\frac{\partial_{y_1}B_1(\mathbf{y}) + \partial_{y_2}B_2(\mathbf{y})}{|\mathbf{x} - \mathbf{y}|}d\mathbf{y}\,. \tag{30}$$

The required vector of the magnetic field must deliver the minimum to the following functional (cf. [15]):

$$\Phi(\widetilde{\mathbf{B}}(\mathbf{x})) = \sum_{j=1}^{3}\sum_{k=(k_1,k_2)}\left|\mathbf{F}^{-1}_{\xi\to x_k}\widetilde{\mathbf{B}}(x_k) - \mathbf{B}(x_k)\right|\,.$$

Acknowledgements This work is partially supported by grants of Russian Foundation for Basic Research (15-01-03576, 16-01-00781 and 17-01-00809).

References

1. T.A. Stroganova et al., EEG alpha activity in the human brain during perception of an illusory kanizsa square. Neurosci. Behav. Physiol. **41**(2), 130–139 (2011)
2. M. Hämäläinen et al., Magnetoencephalography – theory, instrumentation, and applications to noninvasive studies of the working human brain. Rev. Mod. Phys. **65**(2), 413–497 (1993)
3. A.S. Demigod Elliptic pseudodifferential boundary value problems with a small parameter in the coefficient of the leading operator. Math. USSR-Sb. **20**(3), 439–463 (1973)
4. K.Y. Osipenko, Optimal recovery of linear functionals and operators. Commun. Appl. Math. Comput. **30**(4), 459–482 (2016)
5. C.A. Micchelli et al., The optimal recovery of smooth functions. Numer. Math. **26**, 191–200 (1976)
6. P.W. Gaffney et al., Optimal interpolation Lect. Notes Math. **506**, 90–99 (1976)
7. A.S. Demigod, Asymptotics of the solution of the boundary value problem for elliptic pseudo-differential equations with a small parameter with the highest operator. Trudy Moskov. Math. obshchestva, 32, Moscow University Press, M. 119–146 (1975) [in Russian]
8. Y. Martin et al., Magnetic imaging by "force microscopy" with 1000 a resolution. Appl. Phys. Lett. **50**, 1455–1547 (1987)
9. M.H. Acuna, G. Kletetschka, J.E.P. Connerney, Mars crustal magnetization: a window into the past?, in *The Martian Surface: Composition, Mineralogy and Physical Properties*, ed. by J.F. Bell (Cambridge University Press, Cambridge, 2008), pp. 242–262

10. B.P. Weiss, E.A. Lima, L.E. Fong, F.J. Baudenbacher, Paleomagnetic analysis using SQUID microscopy. J. Geophys. Res. **112**, B09105 (2007)
11. A.S. Demidov, M.A. Galchenkova, A.S. Kochurov, On inverse problem magneto-ence-phalography, in *Quasilinear Equations, Inverse Problems and Their Applications*, Moscow, 30 Nov 2015–02 Dec 2015. Conference Handbook and Proceedings (2015), p. 22
12. I. Gradshteyn, I. Ryzhik, *Table of Integrals, Series, and Products*, 7th edn. ed. by A. Jeffrey, D. Zwillinger (2007). http://www.mathtable.com/gr/
13. L. Baratchart et al., Characterizing kernels of operators related to thin-plate magnetizations via generalizations of Hodge decompositions. Inverse Prob. **29**, 1–29 (2013)
14. M.A. Lavrentiev, B.V. Shabat, *Methods of the Theory of Functions of a Complex Variable* (Nauka, Moscow, 1965)
15. A. Demidov, I. Fedotov, M. Shatalov, Estimation des paramètres pour des équations de la cinétique chimique fourni des informations partielles sur leurs solutions, in *Proceedings of EDP-Normandie Conference 2015* (2015), pp. 199–205. [See also: Theor. Found. Chem. Technol. **50**(2), 1–11 (2016)]

Reaction–Diffusion Equations with Density Dependent Diffusion

V. N. Razzhevaikin

1 Basic Equation

Consider a reaction–diffusion equation of the form

$$u_t = D(u)\nabla_x \left(N(u)\nabla_x u\right) + F(u). \tag{1}$$

Here $x \in \Omega \subset \mathbf{R}^n$ stands for space variable, $t \in \mathbf{R}_+$ for time, $u = u(x,t) \in \mathbf{R}$ for phase variable. The density dependent diffusion is described by two parts: the internal $N(u) > N_0 > 0$ and external diffusion functions $D(u) > D_0 > 0$. The monotone change of variable u of the form $v(u) = \int_0^u N(y)\,dy$ or $w(u) = \int_0^u \frac{dy}{D(y)}$ can kill, respectively, the first or the second of them by redefining the remaining diffusion function and the reaction (source) function $F(u)$. So, for general results (see, e.g., the next section) we can leave only one of them without loss of generality. Nevertheless, we prefer to ignore this possibility for the sake of convenience in using the received results and for getting several estimates, which can't be achieved for Eq. (1) in the form with a single nonlinear diffusion function (see below Sect. 3.2). Note that in the case of the constant diffusion functions $D(u) \equiv N(u) \equiv 1$ we get the classical form of the reaction–diffusion equation.

V. N. Razzhevaikin (✉)
A.A. Dorodnicyn Computing Center of the Russian Academy of Sciences, Moscow, Russia
e-mail: evt@ccac.ru

© Springer International Publishing AG, part of Springer Nature 2018
R. P. Mondaini (ed.), *Trends in Biomathematics: Modeling, Optimization and Computational Problems*, https://doi.org/10.1007/978-3-319-91092-5_11

2 Instability and Stabilization

2.1 Instability in Bounded Convex Domain

For a convex bounded region $\Omega = \operatorname{co}\Omega \in \mathbf{R}^n$, with a smooth impenetrable boundary $\partial\Omega$ (i.e., $(\nabla_x u, v)\,|_{\partial\Omega} = 0$, $v \perp \partial\Omega$) the following result takes place.

Theorem 2.1 (Razzhevaikin [1]) *Any stationary non-constant solution of equation* (1) *on a convex bounded region with impenetrable boundary is not stable (e.g., in norm $C(\Omega)$).*

2.2 Instability on Real Line

For the whole real line $\Omega = \mathbf{R}$ the instability result is also true if there is no monotonic staying wave (see Sect. 3.1). A more strict statement is the following. Define a potential function as

$$J(U) = \int_0^U \frac{F(u)N(u)}{D(u)} du. \tag{2}$$

Theorem 2.2 (Razzhevaikin [2]) *Any non-constant stationary bounded solution of equation* (1) *on the whole real line is not stable if the source function $F(u)$ does not have two different zeros with equal potentials (i.e., $(u_1 \neq u_2)\,\&\,(F(u_1) = F(u_2) = 0) \Rightarrow (J(u_1) \neq J(u_2))$).*

2.3 Stabilization to Dominating Equilibrium

Let $\Omega = \mathbf{R}^n$, $I = [0, 1]$, $F(0) \geq 0$, $F(1) \leq 0$. The stationary solution (i.e., equilibrium) of equation (1) $\tilde{u}(x) \equiv U \in I$ (i.e., such that $F(U) = 0$) is said to be *dominating* in I, if $J(U) > J(u)$ for all $u \in I\backslash\{U\}$. Here $J(u)$ is from (2). It is evidently *stable*, i.e., $(u - U)F(u) < 0$ for all $u \in O(U)\backslash\{U\} \subset I$ ($O(U)$ is some vicinity of U).

We say that $u(x, t)$ *stabilizes* to $\tilde{u}(x)$ if for any bounded $K \in \mathbf{R}^n$ $\sup\limits_{x \in K} |u(x, t) - \tilde{u}(x)| \to 0$ as $t \to +\infty$.

The following theorem characterizes the dominating equilibrium as an "almost" global attractor for solutions with initial distributions inside I.

Theorem 2.3 (Razzhevaikin [3]) *Let $\tilde{u}(x) \equiv U$ be the dominating equilibrium of Eq.* (1) *on I, and some neighborhood $O(U) \subset I$ does not include other equilibria. Then for any segment $[\Lambda, B] \subset O(U)$, $A < U < B$, there exists $X > 0$ such* *that the solution $u(x, t)$ of equation (1) with an initial distribution $u(x, 0) \in I$ for $x \in \mathbf{R}^n$ and $u(x, 0) \in [A, B]$ for $\|x\| < X$ stabilizes to U.*

3   Single Travelling Wave

3.1   Travelling Waves

Let $\Omega = \mathbf{R}$, $U_1 < U_2$, $F(U_{1,2}) = 0$. A *travelling wave* (TW) (or a *wave solution*) between U_1 and U_2 is a solution of the form $u(x, t) = U(\xi)$ with $U(\xi) \in [U_1, U_2]$, $\xi = x + ct$ for some $c \in \mathbf{R}$, for which the boundary conditions

$$U(-\infty) = U_1, \quad U(+\infty) = U_2 \tag{3}$$

are satisfied. TW corresponds to motion of the wave having the profile $U(\xi)$ and the velocity c (from the right to the left for $c > 0$).

As far as TW satisfies the equation $cU_\xi = D(U)(N(U)U_\xi)_\xi + F(U)$, the problem of finding all of them is reduced to the one parameter boundary value problem for it and (3).

Let $P(\xi) = N(U(\xi))U_\xi(\xi)$. Instead of the last second order ordinary differential equation one can use the following system

$$
\begin{cases}
U_\xi = \dfrac{P}{N(U)} \\[2mm]
P_\xi = \dfrac{cP}{N(U)D(U)} - \dfrac{F(U)}{D(U)}.
\end{cases}
\tag{4}
$$

Theorem 3.1 (Razzhevaikin [4]) *Solutions of boundary value problem* (4), (3) *with $U(\xi) \in [U_1, U_2]$ for all $\xi \in \mathbf{R}$ satisfy the inequality $P(\xi) > 0$.*

Thus, TW is monotonic. By division the equations in (4) under condition $U_\xi > 0$ one can also get the equation for $\hat{P}(U)$ such that $\hat{P}(U(\xi)) = P(\xi)$

$$\frac{d\hat{P}}{dU} = \frac{c}{D(U)} - \frac{F(U)N(U)}{D(U)\hat{P}}. \tag{5}$$

From (3) we also get $\hat{P}(U_1) = \hat{P}(U_2) = 0$. Multiplying (5) by $\hat{P}(U)$ and integrating over $U \in [U_1, U_2]$ we get for $J(U)$ from (2) the relationship $J(U_2) - J(U_1) = c \int_{U_1}^{U_2} \frac{\hat{P}(U)}{D(U)} dU$, so $\operatorname{sign}(J(U_2) - J(U_1)) = \operatorname{sign} c$. Thus, we have a simple way to find the direction of TW motion. The case of $J(U_2) = J(U_1)$ corresponds to existence of a monotonic staying wave (see Sect. 2.2) with $c = 0$. When considered over the plane (U, E), where $E = E(U, P) = \frac{P^2}{2} + J(U)$ is an "energy" function, TW corresponds to a monotonic (as far as $\frac{dE(U,P)}{dU} = \frac{Pc}{D(U)}$) curve connecting two

extremal points of $J(U)$ (e.g., U_1 and U_2, where $J_U(U_{1,2}) = 0$ with $\hat{P}(U_{1,2}) = 0$).
So, any equilibrium $(\bar{U}, 0)$ on the tale of TW should be stable for homogeneous over
x system corresponding to (1) in rather small half-neighborhood of $\bar{U}$ onto the range
of $U(\xi)$ direction. Indeed, for $c \geq 0$ the "energy" shouldn't decrease with increasing
U and hence $J(U)$ should increase while $P(U)$ vanishes with $U \to \bar{U}-0$. Note that
without loss of generality we can suppose here and further that $c \geq 0$. Otherwise
we can use the following changes: $\hat{c} = -c$, $\hat{x} = -x$, $\hat{u} = -u$, $\hat{F}(\hat{u}) = -F(-\hat{u})$,
$\hat{D}(\hat{u}) = D(-\hat{u})$, $\hat{N}(\hat{u}) = N(-\hat{u})$.

As to the equilibrium on the front of TW, it can be either stable or unstable. The
first case is called *trigger*, whereas the second one is *Kolmogorov's* case (the first
work where it was studied is Kolmogorov[5]).

Investigation of boundary value problem (4), (3) is based upon studying of
asymptotic solutions, which meet system (4) and one of boundary conditions (3).
The comparison technique applied to them results in existence and uniqueness
theorems. The most important of them are presented below.

3.2   Kolmogorov's Case

The following result generalizes the well-known theorem from Kolmogorov [5].

Let $U_1 < U_2$, $F(U_1) = F(U_2) = 0$ and $F(u) > 0$ for $u \in (U_1, U_2)$, and let
$$c_m = 2\sqrt{N(U_1)\, D(U_1)\, F_u(U_1)}, \quad c_M = 2\sqrt{\sup_{u \in (U_1, U_2)} N(u)F(u)/\int_{U_1}^{u} \frac{dv}{D(v)}}.$$

Theorem 3.2 (Razzhevaikin [4]) *There exists some minimal* $0 < c^* \in [c_m, c_M]$
such that for all $c \geq c^*$ *TW between* U_1 *and* U_2 *exists.*

3.3   Trigger Case

When both zeros $U_1 < U_2$ of $F(u)$ are stable there exists at most one TW between
them. Moreover, if $F(U_1 + \varepsilon) < 0$, $F(U_2 - \varepsilon) > 0$ for all rather small $\varepsilon > 0$ it
really exists when the following conditions are fulfilled.

(i) $\forall u \in (U_1, U_2)$, $J(u) < J(U_2)$,
(ii) $\bar{u} \in (U_1, U_2) \& F(\bar{u}) = 0 \Rightarrow J(\bar{u}) \leq J(U_1)$.

Theorem 3.3 (Razzhevaikin [4]) *Conditions (i) and (ii) together are sufficient for
existence of TW between* U_1 *and* U_2. *Moreover, under them such TW is unique and
its velocity* $c \geq 0$.

Remarks

1. In (ii) it's enough to check zeros $\bar{u}$ with $F(\bar{u} - \varepsilon) > 0$, $\forall \varepsilon \in (0, \varepsilon_0)$ for some
   $\varepsilon_0 > 0$.

2. From Theorem 3.3 the existence of TW with some $c \in \mathbf{R}$ between two
   neighboring stable zeros of $F(u)$ follows. Indeed, the change of variables
   indicated above can bring the alternative situation to the conditions of this
   theorem.

4 Wave Chains

Let $0 = u_0 < u_1 < \cdots < u_k = 1$ be isolated zeros of $F(u)$ such that u_i for $i \geq 1$
are stable (u_0 may be unstable). Assume also that between neighboring zeros u_i and
u_{i+1} there exists at most one zero, which evidently should be unstable. We shall call
such a collection of zeros a *full set of zeros*.

According to Theorems 3.3 and 3.2 between neighboring zeros u_i and u_{i+1} in
the full set of zeros there exists a TW solution (single for $i \geq 1$) with velocity
$c_{i,i+1} \in \mathbf{R}$. One shouldn't exclude the case of existence of TW between u_i and
u_j with velocity $c_{ij} \in \mathbf{R}$ when $j \neq i + 1$. We shall denote this TW as $C(i, j)$.
When such a wave does not exist, one can observe a chain of travelling waves with
monotonic total profile and several intermediate zeros from the full set of zeros
staying for boundary conditions for each wave from the chain. It is rather interesting
that asymptotic of such chains is defined rigorously almost independently on initial
monotonic distributions (see Sect. 5 below).

4.1 Trigger Wave Chains

It proves to be that when a back wave overtakes a leading one a united wave appears
with an intermediate velocity. Successive junctions of waves result in a chain of
waves with not increasing velocities. For the trigger case it turns out to be that such
a final chain is single.

Theorem 4.1 (Razzhevaikin [4]) *For any full set of stable zeros $\{u_i,\ i = 1, \ldots, k\}$
there exists a single subset $\{i_j,\ j = 1, \ldots, \kappa\} \subset \{1, \ldots, k\}$ such that*

   i) *$i_1 = 1,\ i_\kappa = k$;*
  ii) *there exists the travelling wave $C(i_j, i_{j+1})$ with velocity $c_{i_j, i_{j+1}} \in \mathbf{R}$ for each*
      $j \in \{1, \ldots, \kappa - 1\}$;
 iii) *$c_{i_j, i_{j+1}} \geq c_{i_{j+1} i_{j+2}}$ for each $j \in \{1, \ldots, \kappa - 2\}$.*

Such a chain would be called the *true trigger chain* over the full set of stable zeros
$\{u_i,\ i = 1, \ldots, k\}$.

4.2   Kolmogorov's Wave Chains

When $u_0 = 0$ in the full set of zeros is unstable then $F(u) > 0$ for all $u \in (u_0, u_1)$.
Using Theorem 3.2 we can get a set of TW between u_0 and u_1 with velocities
$c \in [c^*, +\infty)$. TW between u_0 and u_i for $i > 1$ can also exist.

Theorem 4.2 (Razzhevaikin [4]) *In Kolmogorov's case the velocity values range*
$I_i = [c^i, \hat{c}^i)$ *for $i \geq 2$ such that TW $C(0, i)$ with velocity c exists if $c \in I_i$ can be*
constructed by the following inductive procedure.

Let $c^1 = c^*$ from Theorem 3.2 with $U_1 = 0$ and $U_2 = u_1$ and $\hat{c}^1 = +\infty$. For
$i \geq 2$ $\hat{c}^i$ is the minimal wave velocity in the true trigger chain formed over the full
set of zeros $\{u_1, \ldots, u_i\}$. If I_{i-1} is already defined, then $c^i \in I_{i-1}$. Moreover, if
$c^{i-1} > c_m$ from Theorem 3.2, then $c^i > c^{i-1}$.

Thus, I_i shrinks with increasing i up to empty set when decreasing $\hat{c}^i$ becomes
not greater than c^i. Note also that from Theorems 3.2 and 4.2 it follows that $c > 0$
even in the case of $F_u(0) = 0$.

From Theorem 4.2 it follows that behind TW one can observe a true trigger chain
formed over the remainder of the full set of zeros after ignoring those its elements
that were found inside segment $[u_0, u_i]$. Moreover, velocity of leading TW in this
remaining chain is larger than velocity of TW between u_0 and u_i. Its union with
this true trigger chain is called a *true Kolmogorov's chain* over the full set of zeros
$\{u_i, i - 0, \ldots, k\}$. It is already not unique but it is possible to select among such
chains one with minimal velocity of the leading wave. This selected chain is called
the *minimal velocity true Kolmogorov's chain*.

5   Stabilization to Wave Chains

5.1   Phase Plane Equation

Let again in Eq. (1) $x \in \mathbf{R}$, and $F(0) = F(1) = 0$. If $u(x, 0)$ is not constant and
not decreasing, then for solution $u(x, t)$ of the Cauchy problem for all $t > 0$ and
$x \in \mathbf{R}$ its space derivative exists and $u_x(x, t) > 0$. So, we can construct the function
$Q(u, t)$ such that $Q(u(x, t), t) = u_x(x, t)$. It ought to satisfy the following equation

$$Q_t = Q^2 \left(D(QN)_u + \frac{F}{Q} \right)_u \left(= Q^2(D(QN)_u)_u + F_u Q - F Q_u \right). \quad (6)$$

This equation is called the *phase plane equation* whereas its solution $Q(u, t)$ is
called the *phase plane solution* corresponding to solution $u(x, t)$ of equation (1).

If $u(\pm\infty, 0) = \alpha_\pm(0)$ and functions $\alpha_\pm(t)$ are the solutions of the Cauchy
problems $\pm\frac{\alpha_\pm(t)}{dt} = \pm F(\alpha_\pm) \geq 0$ with initials $\alpha_\pm(0)$, then the range of $Q(u, t)$ is
the increasing in time interval $(\alpha_-(t), \alpha_+(t))$. Moreover, the following result takes
place.

Theorem 5.1 (Razzhevaikin [4]) *Let $Q(u, 0)$ be an everywhere positive initial distribution defined over $(\alpha_-(0), \alpha_+(0))$ and $\alpha_\pm(t)$ as above. Then there exists a unique solution $Q(u, t) > 0$ of equation (6) with the continuous almost everywhere positive initial distribution $Q(u, 0)$, which satisfies the boundary conditions $Q(\pm\alpha_\pm(t), t) = 0$ and for which the integrals $\int_{\alpha_\pm(t)}^{\alpha_\pm(t)\mp\varepsilon} \frac{du}{Q(u,t)}$ diverge for $t > 0$.*

5.2   True Solutions and Phase Plain Convergence

Solutions of equation (6) with constant $\alpha_\pm(t) = \hat{\alpha}_\pm$ such that $F(\hat{\alpha}_\pm) = 0$ are called *true solutions*. Any of them, which is positive for $t > 0$, correspond to a *simple wave*, i.e. to a monotonic over x solution of equation (1) up to shifting along x. Those ones, which have zeros at zeros of function $F(u)$, corresponds to a "chain" of simple waves ordered in accordance with zeroes order and infinitely shifted one from other.

Among true solutions one can find *phase plain wave chains* that correspond to true trigger and true Kolmogorov's wave chains. Those ones are stationary, i.e., do not change with time. The convergence of true solutions to phase plain wave chains is called the *phase plain convergence*. Such a convergence results in convergence of $u(x, t)$ to some representative of a true wave chain.

The main technique in proofs of the phase plain convergence is based on properties of sub- and super-solutions of equation (6) that satisfy inequalities obtained from (6) by changing the first sign of equality by sign $\leq$ and $\geq$, respectively. Among these, stationary ones have a special interest. Their main property is the following. True solutions of (6), which have sub-solutions (respectively, super-solutions) as the initial distributions, monotonically increase (respectively, decrease) in time. The comparison technique allows to establish that any super-solution remains greater than any sub-solution if it takes place at the beginning. Thus, for establishing the phase plain convergence it is enough for the phase plane solution $Q(u, \hat{t})$ corresponding to solution $u(x, \hat{t})$ of equation (1) at some moment $\hat{t} \geq 0$ to find, on the one hand, such a stationary sub-solution, which is a lower bound of $Q(u, \hat{t})$, and, on the other hand, an exceeding $Q(u, \hat{t})$ stationary super-solution, between which one can find only one stationary solution of (6), which is also a phase plain wave chain. For details see Razzhevaikin [4]. Here we only formulate several theorems.

Theorem 5.2 (Razzhevaikin [4]) *If in the trigger case a continuous distribution $Q(u, 0)$ is positive over interval (α, β) with $F(\alpha) \leq 0$ and $F(\beta) \geq 0$ and takes zero values outside it, then the true solution of equation (6) converges on the phase plain to a true trigger chain. This trigger chain is the minimal chain over interval $(u_-, u_+) \supset (\alpha, \beta)$, where $u_\pm$ are stable zeros of $F(u)$, for which other its zeros between them and respectively α and β do not exist.*

5.3 Convergence to Minimal Velocity True Kolmogorov's Chains

Let $0 = u_0 < u_1 < \cdots < u_k = u_+$ be the full set of zeros of function $F(u)$ with unstable u_0, i.e., $F(u) > 0$ for $u \in (u_0, u_1)$. Let also the initial distribution $Q(u, 0)$ satisfy conditions of Theorem 5.2 with $\alpha \equiv 0$ and $\beta \in (u_{k-1}, u_k)$. The characteristic equation for Jacobian of system (4) at point $(0, 0)$ may be written as

$$c = q\lambda D(0)N(0) + \frac{F_u(0)}{\lambda} \quad (= \tilde{c}(\lambda)), \tag{7}$$

where λ is slope of $Q_\lambda(u)$ (the solution of equation (5) with the parameter c from (7)) at 0. Let $\tilde{\lambda}(c) = \min\{\lambda > 0 : \tilde{c}(\lambda) = c\}$ for $F_u(0) > 0$ and $\tilde{\lambda}(c) \equiv 0$ for $F_u(0) = 0$. The function $c = \tilde{c}(\lambda) > 0$ has minimum c_m at $\hat{\lambda} = \sqrt{\frac{F_u(0)}{(D(0)N(0))}}$. Let the minimal velocity of TW between u_0 and u_1 is equal to $c^* \geq c_m$ (see Theorem 3.2).

Theorem 5.3 (Razzhevaikin [4]) *Let a nonnegative continuous function $Q(u, 0)$ be positive over $(0, \beta)$ with $F(\beta) > 0$ or $\beta \in \{u_i\}$, $i = 1, \ldots, k$. Then under the inequality $\lim\inf_{u \to +0} \frac{Q(u,0)}{u} \geq \tilde{\lambda}(c^*)$ for $F_u(0) > 0$ (or $> \frac{1}{c^*}$ for $F_u(0) = 0$) the true solution $Q(u, t)$ with the initial distribution $Q(u, 0)$ converges on the phase plain to the minimal velocity true Kolmogorov's chain on interval $(0, u_+)$, where $u_+ = \min u_i \geq \beta$, $i - 1, \ldots, k$.*

Corollary (Razzhevaikin [4]) *In the case of a strictly increasing initial distribution $u(x, 0)$, vanishing at some finite $\bar{x}$ with $u_x(\bar{x}, 0) > 0$ or with $u_x(x, 0) > 0$ and $u_{xx}(x, 0) > -\delta$ for $x \in (\bar{x}, \bar{x} + \varepsilon)$ and some $\varepsilon, \delta > 0$ the first wave velocity would be minimal.*

5.4 The Case of Non-minimal Velocity

The conditions of Theorem 5.3 imply rather hard restrictions for $Q(u, 0)$ asymptotic under $u \to +0$. Their violation can result in the other character of asymptotic behavior of solutions of equation (6). For example, in the case of single Kolmogorov's wave ($k = 1$) under all its conditions with the exclusion of the asymptotic inequality for $Q(u, 0)$, stationary solutions of equation (6), which correspond to non-minimal velocity, also satisfy it. It appears that nevertheless, such solutions are also stable in somewhat weaker sense. Hereafter in this section we restrict our consideration with the case of a single wave, i.e., with $F(0) = F(1) = 0$ and $F(u) > 0$ for $u \in (0, 1)$.

A stationary solution of equation (6) is uniquely defined by its velocity $c > c^*$ value. We shall denote it as $C_c(u)$.

Theorem 5.4 (Razzhevaikin [4]) *Let $Q(u, 0) = u(q + g(u))$ with $q \in (0, \tilde{\lambda}(c^*))$ and a continuous over $[0, 1]$ function $g(u) = o(1)$, for which integral $\int_{+0}^{1} \frac{g(u)}{u} du$ converges. Then the true solution $Q(u, t)$ converges to the wave $C_{\tilde{c}(q)}(u)$ with $t \to +\infty$ in the following sense.*

There exist functions $Q_\pm(u, t)$ *such that*

i) $0 \le Q_-(u, t) \le Q(u, t) \le Q_+(u, t)$,

ii) $Q_-(u, t) \le C_{\tilde{c}(q)}(u) \le Q_+(u, t)$,

iii) $Q_-(u, t)$ *converges to* $C_{\tilde{c}(q)}(u)$ *uniformly on compacts* $I \Subset (0, 1)$,

iv) $\rho^t\left(Q_+(u, t), C_{\tilde{c}(q)}(u)\right) \to 0$ *with* $t \to +\infty$ *in metric* $\rho^t(Q_1, Q_2) = \int_t^{+\infty} \int_0^1 \frac{F(u)}{D(u)} \left| \frac{1}{Q_1(u,\tau)} - \frac{1}{Q_2(u,\tau)} \right| du d\tau$.

Remarks

1. Statements of Theorem 5.4 may be strengthened. The two-sided boundedness $u_-(x, t) \le u(x, t) \le u_+(x, t) \le$ for solutions of equation (1) also takes place. Here $u_\pm(x, t)$ are the properly shifted solutions of equation (1) corresponding to $Q_\pm(u, t)$, respectively.
2. Under conditions of Theorem 5.4 there exists a derivative $Q_u(u, 0)|_{u=+0} = q$. It implies particularly, an asymptotics $u(x, 0) \sim e^{qx}$ for $x \to -\infty$. Convergence of the integral takes place for any initial $Q_u(u, 0)$ with Hölder continuous at $u = 0$ derivative.

6 Leader Selection in Competing Species Diffusion Model

Consider the reaction–diffusion competition system with Lotka–Volterra reaction term. This system is assumed to describe the dynamics of a biological community distributed over space variables. Suppose for simplicity that such a distribution can be reduced to a single variable $x \in \mathbf{R}$. Let also $i = 1, \ldots, N$ stand for species numbers, $u^i(x, t) \ge 0$ is density of i-th species at (x, t). The diffusion coefficient of i-th species is $D^i > 0$ whereas $F^i(u) = u^i\left(M^i - \sum_{j=1}^{N} \gamma_{ij} u^j\right)$ is its Malthusian function, which depends on the whole vector $u = (u^1, \ldots, u^N)$ of population densities $u^j = u^j(x, t)$ at (x, t). Also, $M = (M^1, \ldots, M^N) > 0$ is the vector of Malthusian coefficients in the absence of competition. Local interactions between species are assumed to be described by coefficients γ_{ij} with $\gamma_{ii} > 0$.

The reaction–diffusion system has the form:

$$u_t^i = D^i u_{xx}^i + u^i F^i(u). \tag{8}$$

We discuss several asymptotical properties of solutions of Cauchy problems for (8) with initial distributions $u^i(x, 0) \ge 0$ that have bounded supports $S^i = \operatorname{supp} u^i(x, 0) = \operatorname{cl}\{x : u^i(x, 0) > 0\} \ne \emptyset$. The total convex envelope $S = \operatorname{co}\left(\bigcup_{i=1}^{N} S^i\right)$ is called the *seat* of the community.

The following robust assumption is called the "common ecological niche hypothesis."

(H1) $\gamma_{ij} = \alpha_i \beta_j$, and $m^i = \frac{M^i}{\alpha_i}$ are different.

For $N = 1$ we can find Kolmogorov's velocity $c^1 = 2\sqrt{D^1 M^1}$ and the asymptotical value $\hat{u}^1 = \frac{M^1}{\gamma_{11}}$.

A species with number $i_1 \in \{1, \dots, N\}$ is called a *leader* in system (8), if for any fixed collection of $\delta^l > 0$ for all $X > 0$ there exist $\hat{x} > X$ and $\hat{t} > 0$ for which the following conditions are fulfilled:

(1) $u^{i_1}(\hat{x}, \hat{t}) > \delta^i$;
(2) $u^k(x, \hat{t}) < \delta^k \; \forall k \neq i_1$ and $x \geq \hat{x}$.

We assume also the following robust assumption

(H2)   For $i_1 \in \{1, \dots, N\}$ the problem

$$\sqrt{D^{i_1} M^{i_1}} = \max_i \left\{ \sqrt{D^i M^i} \right\}$$

has a unique solution.

Theorem 6.1 (Razzhevaikin[6]) *Under hypotheses (H1) and (H2) the leader in system (8) exists and its number does not depend on initial finite distributions.*

7   Diffusion Model of Genetic Waves

As an example of application of the theory of reaction–diffusion equations with density dependent diffusion consider a model of genetic waves propagation over space [7]. It describes a population structured over space and several discrete genetic parameters. The dynamical properties of individuals depend on phenotypic characteristics whereas those ones change in accordance with genetic lows.

7.1   Basic Hypotheses

The following hypotheses are assumed to be fulfilled.

H1.   The population is distributed in space with one variable $x \in \mathbf{R}$.
H2.   Phenotypic particularities of genotypes differ in one two-allelic gene with alleles A and a.
H3.   Let $p(x, t)$ be the part of the population fertile individuals density at (x, t) recounted as fraction of the allele's A number to the number of both alleles A and a. So, if $U_A(x, t)$, $U_{Aa}(x, t)$, $U_a(x, t)$ are the densities of AA, Aa and aa pairs carriers, then $p(x, t) = \frac{2U_A(x,t)+U_{Aa}(x,t)}{2(U_A(x,t)+U_{Aa}(x,t)+U_a(x,t))}$. Let $q(x, t)$ be the corresponding value for a. Thus the total density for both alleles A and a is constant: $p(x, t) + q(x, t) = 1$.

H4.   Changes of alleles A and a numbers occur through their carriers with genotypes AA, Aa, and aa, each of which has the fitness (the probability to alive from conception to the fertile stage) equal, respectively, to α, β, γ (all are positive constants). The departure rate from the fertile stage (death-rate plus aging) is assumed to be constant.

H5.   The full panmixia locally over each x is assumed to be fulfilled.

H6.   Crossbreeding is realized via the gametes spatial carrying (e.g., via moveable sperm), which depend only on $p(x, t)$ and does not depend on other genotype properties including their densities.

7.2 Integro-Differential Model

The resulting base mathematical model has an integro-differential form. To describe it we shall take the maturation time h as the time unit.

Define the long-range genetic action operator as convolution of the form $K(p)(x, t) = \int_{-\infty}^{+\infty} k(x \cdot \xi) p(\xi, t) d\xi$ with the normal distribution with the kernel $k(x \cdot \xi) = \dfrac{e^{-\frac{(x-\xi)^2}{2\sigma^2}}}{\sqrt{2\pi\sigma^2}}$ for some $\sigma > 0$, so $K(1) = 1$. Denote $\varphi(x, t) = K(p)(x, t)$, $\psi(x, t) = K(q)(x, t) = 1 - \varphi(x, t)$ (gametes distributions, see H6); $u_A(x, t) = p(x, t)\varphi(x, t)$, $u_{Aa}(x, t) = p(x, t)\psi(x, t) + q(x, t)\varphi(x, t)$, $u_a(x, t) = q(x, t)\psi(x, t)$ (conception of AA, Aa, and aa genotypes carriers rates, see H5). Over long time intervals one can use a differential approximation of changes for $p(x, t)$ and $q(x, t)$, so in accordance with H4 it is possible to construct the following integro-differential system

$$
\begin{cases}
p_t(x, t) = 2\alpha u_A(x, t) + \beta u_{Aa}(x, t) - p(x, t)r(x, t), \\[2mm]
q_t(x, t) = \beta u_{Aa}(x, t) + 2\gamma u_a(x, t) - q(x, t)r(x, t),
\end{cases}
\tag{9}
$$

where $r(x, t)$ is defined from condition H3, so, $r(x, t) = r(x, t)(p(x, t) + q(x, t)) = 2(\alpha u_A(x, t) + \beta u_{Aa}(x, t) + \gamma u_a(x, t))$. For $q(x, t) = 1 - p(x, t)$ instead (9) we have

$$
p_t(x, t) = \varphi(x, t)R(p(x, t)) + [\beta p(x, t)(1 - 2p(x, t)) - 2p(x, t)(1 - p(x, t))\gamma],
\tag{10}
$$

with

$$
R(p) = 2(\alpha + \gamma - 2\beta)p(1 - p) + \beta > 0.
\tag{11}
$$

7.3   PDE Approximation

Via expending $p(\xi, t) = p(x, t) + \frac{\partial p(x,t)}{\partial x}(\xi - x) + \frac{1}{2}\frac{\partial^2 p(x,t)}{\partial x^2}(\xi - x)^2 + \cdots$ we have for small $\sigma > 0$ the following approximation $K(p)(x, t) = p(x, t) + \frac{\sigma^2}{2}\frac{\partial^2 p(x,t)}{\partial x^2} + \cdots$ that from (10) results in the equation

$$p_t = D(p)p_{xx} + F(p) \qquad (12)$$

with $D(p) = \frac{\sigma^2}{2}R(p)$, $R(p)$ from (11) and

$$
\begin{aligned}
F(p) &= pR(p) + \beta p(1 - 2p) - 2p(1 - p)\gamma \\
     &= 2p[(2\beta - \alpha - \gamma)p^2 + (\alpha + 2\gamma - 3\beta)p + (\beta - \gamma)].
\end{aligned} \qquad (13)
$$

7.4   Main Results

The function $F(u)$ from (13) has three zeros:

$$p_0 = 0, \quad p_1 = 1, \quad p^* = \frac{\gamma - \beta}{\alpha + \gamma - 2\beta}$$

that serves as equilibria in Eq. (12).

The direction of TW for (12) is defined by sign of integral $J = \int_0^1 \frac{F(u)}{D(u)}du$. It is easy to see that sign $J = $ sign $\left(\frac{1}{2} - p^*\right)$.

Let $\delta = \frac{\alpha + \gamma}{2}$. Without loss of generality assume that $\alpha > \gamma$. The following properties depend on β localization with respect to α and γ. Note also, that the function $R(p)$ is concave for $\delta > \beta$ and is convex for $\delta < \beta$.

7.4.1   Case $\beta < \gamma$

In this case $F'(0) = 2(\beta - \gamma) < 0$, $F'(1) = 2(\beta - \alpha) < 0$, and $p^* \in \left(0, \frac{1}{2}\right)$. Thus, it corresponds to the trigger case. As far as $J(1) > 0$ then TW goes from p_0 to p_1. In genetic waves interpretation it may be treated as spreading of the stronger allele A with the weaker allele a vanishing. Note that the initial distribution of $p(x, 0)$ should be ruther large for this wave can occur. At least it is necessary that measure of the region where $p(x, 0) > p^*$ is positive.

7.4.2 Case $\gamma < \beta < \delta$

Here $F'(0) > 0$, $F'(1) < 0$, and $p^* < 0$. Thus $F(p) > 0$ for $p \in (0, 1)$. It corresponds to Kolmogorov's case. TW solutions with velocity $c \geq c^* \geq c_K$, where c^* is some minimal value and $c_K = 2\sqrt{D(0)F'(0)} = 2\sigma\sqrt{\beta(\beta - \gamma)}$ is Kolmogorov's velocity. The genetic waves interpretation is the same as above, but the velocity of the wave depends already on the initial distribution (asymptotically minimal for initial bounded distribution of allele A). Also, there is no need to demand of massive initial distribution. For the wave to appear it's enough to have any distribution of $p(x, 0)$ with a positive integral.

7.4.3 Case $\delta < \beta < \alpha$

Here $F'(0) > 0$, $F'(1) < 0$, and $p^* > 1$. So, again $F(p) > 0$ for $p \in (0, 1)$. Since in this case the function $D(p)$ is convex (hence its maximums are localized at the ends of the interval $(0, 1)$), and for $p \in (0, 1)$ from $F'(p) > 0$ follows $F''(p) < 0$ (the inflection point of a cubic curve is between its extremums), then $c^* = c_K$. The genetic waves interpretation is just the same as above, but the minimal velocity of the wave is already known.

7.4.4 Case $\beta > \alpha$

Here again $F'(0) > 0$, $F'(1) > 0$, whereas $p^* \in \left(\frac{1}{2}, 1\right)$. For monotonic initial distribution with $p(x, 0) \rightarrow \frac{1}{2} \pm \frac{1}{2}$ for $x \rightarrow \pm\infty$ one can observe a chain of two waves scattering from p^* into opposite directions. Their velocities value spread depending on initial distributions, respectively, from $c_K^0 = 2\sqrt{D(0)F'(0)} = 2\sigma\sqrt{\beta(\beta - \gamma)}$ and $c_K^1 = 2\sqrt{D(1)F'(1)} = 2\sigma\sqrt{\beta(\beta - \alpha)}$ to $+\infty$. Note that the velocities' minimal values are exact and are reached for bounded supporters (in this case it is where $p(x, 0) \neq \frac{1}{2} \pm \frac{1}{2}$). Here it follows from implication $F''(\bar{u}) = 0 \Rightarrow F'(\bar{u}) < 0$. The interpretation in terms of genetic waves here corresponds to the heterozygotic genotype propagation over the region.

Acknowledgements This work is supported by RFFI, project No 15-07-06947.

References

1. V.N. Razzhevaikin, Instability of stationary non-monotone solutions of the reaction equation with diffusion depending on density. Differ. Equ. **42**(4), 567–575 (2006)
2. V.N. Razzhevaikin, Instability of non-constant stationary solutions to reaction-diffusion equations with density-dependent diffusion in a convex domain. Differ. Equ. **32**(2), 287–290 (1996)

3. V.N. Razzhevaikin, Stabilization of solutions of the Cauchy problem for the reaction-nonlinear diffusion equation to a dominant equilibrium. Differ. Equ. **49**(3), 320–325 (2013)
4. V.N. Razzhevaikin, Travelling-wave solutions of the nonlinear reaction-diffusion equation. Tr. Mosk. Fiz.-Tekh. Inst. **1**(4), 99–119 (2009) [in Russian]
5. A.N. Kolmogorov, I.G. Petrovskii, N.S. Piskunov, Study of diffusion equation with an increase in the quantity of matter and its application to a biological problem. Bull. Mosk. Gos. Univ. Mat. Mekh. **1**(6), 1–26 (1937)
6. V.N. Razzhevaikin, Leader in a diffusion competition model. Comput. Math. Math. Phys. **55**(3), 432–436 (2015)
7. V.N. Razzhevaikin, Diffusion model of genetic waves, in *Operations Research (Models, Systems, Solutions)* (Computing Center RAS, Moscow, 2007), pp. 23–28 [in Russian]

A Plankton-Nutrient Model with Holling Type III Response Function

Anal Chatterjee, Samares Pal, and Ezio Venturino

1 Introduction

Many models have now appeared in the literature for plantkon ecosystems, since the seminal paper [16]. The recent literature in plankton research has evolved from these early works toward more complex configurations: for instance, flip and Hopf bifurcations in a discrete predator–prey model with non-monotonic functional response are studied in [21] using the center manifold theorem.

There are several important reasons for investigating plankton ecosystems. At least three are the main grounds for basic research in this field: their role as basic food source, oxygen production, and more and more frequent insurgence of red tides. As far as the first one is concerned, plankton lies indeed at the bottom level of the trophic chain in the sea, and, in view of the oceans' extension, ultimately on Earth, so that it therefore is the food source of all the higher- level aquatic life forms. It is the basic resource on which fisheries then rely. In this context, following the early works in mathematical bioeconomics, [12], and their follow-ups for general predator–prey systems, such as [4, 13, 30], also recent efforts in plankton modeling include harvesting, especially in connection to fisheries, [14]. In [8], both in the presence and in the absence of constant harvesting rates, a detailed bifurcation study is performed for a two-species ratio-dependent model, assessing sustainability properties of the stock and the resource rent earned. Note that the optimal harvesting policy is investigated in [7] for a structured predator–prey system with stage-dependent predators. In [27] harvesting and time delay are combined, considering

A. Chatterjee · S. Pal
Department of Mathematics, University of Kalyani, 741235, India

E. Venturino (✉)
Dipartimento di Matematica "Giuseppe Peano", Università di Torino, Torino, Italy
e-mail: ezio.venturino@unito.it

© Springer International Publishing AG, part of Springer Nature 2018
R. P. Mondaini (ed.), *Trends in Biomathematics: Modeling, Optimization and Computational Problems*, https://doi.org/10.1007/978-3-319-91092-5_12

two basic models, namely the generalized Gause-type predator–prey model and the Wangersky–Cunningham model. A somewhat simpler model of Gause type is considered in [18]. It is a two population model with a sigmoidal response function, containing further an additional feature, the Allee effect. It exhibits features such as tristability and the existence of two limit cycles, one of which is unstable and the other one stable. Similarly, the Leslie–Gower situation is investigated in [19].

Further, studying plankton is important because most of the oxygen produced daily on this planet comes from marine resources, i.e., essentially from phytoplankton; for possible future dreadful scenarios due to climatic changes in this context, see the very recent investigations [35].

But nowadays one of the most relevant issues that we are facing are the so-called "red" (or "brown") tides that are responsible for the eutrophication phenomena that cause many problems to fisheries and tourism. It is being argued that these harmful algal blooms should be ascribed to the presence of toxin-releasing phytoplanktons. In such context, harvesting of a predator–prey fishery has been considered in [14]. In addition, as these are recurrent phenomena, a mathematical model for the investigation of interactions between phytoplankton and zooplankton in a periodic environment is presented in [26] that incorporates possible parameter variations mainly due to seasonal changes.

Deterministic models in ecology are not able to account for environmental fluctuations, because they neglect the effects of random parameter variations, [2]. The deterministic approach thus has some limitations in biology, being unable to deal with environmental noise in predicting the future states of the system. A stochastic model may instead describe more realistically a natural system and therefore should also be considered, in order to provide a clearer understanding of the situation. Recently, rather general deterministic and stochastic nutrient-phytoplankton-zooplankton models with toxin-producing phytoplankton have been discussed in [22], showing that increasing toxins production rates can induce chaotic system behavior, but low nutrient input may act as a control in this situation. In this context, the issue of global stability of the coexistence equilibrium becomes of paramount importance: for instance, considering epidemic models, such situations are addressed in [23] and [24]. In the former, stability of the endemic equilibrium point of the SEIR epidemic model is considered, while the latter focuses on a similar issue in the presence of delays.

In this paper, we introduce a new model, both in its deterministic and stochastic counterparts. It incorporates a feature that appears for the first time in this situation, the emergence of a Holling type III response function that has only very recently been suggested, [28]. The way of parametrizing the functional response affects food chain models. In particular the Holling type III function is theoretically known to help system stabilization, but apparently the results of laboratory experiments challenge its use. In [28] however the use of sigmoid response functions are advocated for the plankton models that do not explicitly take into account vertical heterogeneity of the population's distribution and zooplankton's active food search behavior. This is because the latter can follow phytoplankton and feed on the vertical ocean layers that are more populated by herbivores.

The use of a Holling type III functional response is however in contrast to other current models, such as [29], where the system exhibits Holling type I and type II response functions. A Holling type II with two-zooplankton one-phytoplankton system in the presence of toxicity has been discussed in [6]. The model presented here considers nutrients, as in [34], in addition to the plankton populations, but does not explicitly incorporate space. There are two trophic levels, phytoplankton, also known as autotroph, since they are able to produce their food requirement by photosynthesis from inorganic materials, and zooplankton, or herbivores, grazing on the former. The latter is assumed to be subject to harvesting which is not modeled in [5]. In contrast to many other current models, we account also for an additional food source for the autotroph biomass, represented by vitamin B_{12} as suggested in [5, 15], which helps in destabilizing the zooplankton-free equilibrium, i.e., it maintains coexistence even in the presence of a high harvesting rate. Positive invariance of the ecosystem solutions and charaterization of its equilibria are obtained, together with an analysis of the Hopf bifurcation at coexistence and of the stability of the bifurcating periodic solution.

We then move on to the stochastic counterpart of the basic model by introducing a random driving force into each system equation. This again constitutes a generalization of [29], because in the latter noise appears only in the harvest rate parameter. In our analysis we further briefly deal also with the global stability of the coexistence equilibrium, using a Lyapunov function. The stochastic stability is characterized, and conditions for the insurgence of persistent oscillations of all the system populations are investigated also via numerical simulations. Our findings show that the control proposed in the absence of the environmental disturbances retains its validity also in the presence of environmental stochasticity, provided that the intensity of the environmental fluctuation is below a critical threshold.

The paper is organized as follows. In the next section, the deterministic model is introduced, and analyzed in Sect. 3. The stochastic counterpart is presented next, results on the numerical simulations are reported in Sect. 5 and discussed in the final section.

2 The Mathematical Model

Let $N(t)$ be the concentration of the nutrient at time t. Let $P(t)$ the autotroph biomass and $Z(t)$ the number of herbivores present at time t. Let N^0 be the constant input of nutrient concentration, D is the dilution rate of nutrient. Its inverse D^{-1} represents the average time that nutrient and waste products spend in the system. Let α_1 and α_2 be the nutrient uptake rate for the autotroph biomass and conversion rate of nutrient for the growth of the autotroph biomass, respectively ($\alpha_2 \leq \alpha_1$). Further, μ and μ_2 denote respectively the mortality rates of the autotroph biomass and of the herbivore population. Let μ_3 ($\mu_3 \leq \mu$) and μ_4 ($\mu_4 \leq \mu_2$) be the nutrient recycle rates respectively coming from the dead autotroph biomass and the dead herbivore population. The maximal zooplankton's herbivore hunting rate is represented by γ_1

while γ_2 $(\gamma_2 \leq \gamma_1)$ is its maximal herbivore conversion rate. We choose Holling type II and type III functional forms to describe the grazing phenomena with K_1 and K_2 as half saturation constants. We also include harvesting of the top population in this food chain, at rate h. The harvesting is modeled via a Holling type II function with half-saturation constant E, to mimic the diminishing returns obtained via constant harvesting efforts, as it is common in fisheries models, [12].

In this subject, most current models in the literature use the Holling type I or II response function to simulate the grazing of phytoplankton by zooplankton, see, e.g., [5, 29]. In this paper, we use it to describe the uptake of nutrients by phytoplankton. Recent results in plankton modeling indicate however that a better way of modeling the zooplankton grazing is provided by a Holling type III response function [28]. Following these ideas, we propose here a system with this assumption that, to our knowledge, sets therefore this contribution apart from other previous and current works in the field.

In addition, it has been observed that other food sources are occasionally available to phytoplankton, other than the basic nutrients N. The additional food source is vitamin B$_{12}$, [5, 15]. We include the latter in our description. Thus, let rP be the phytoplankton's growth rate due to this additional vitamin supply. Then the autotroph biomass mortality rate is $\mu_1 = \mu - r \in \mathbf{R}$. With these assumptions, our deterministic system is $\dot{X} = F(X)$, with $F : \mathbf{R}_+^3 \to \mathbf{R}^3$, $X(t) = (N(t), P(t), Z(t))^T \in \mathbf{R}^3$, $F(X) = [F_1(X), F_2(X), F_3(X)]^T$. Explicitly, with its Jacobian V, it reads

$$\frac{dN}{dt} = D(N^0 - N) - \frac{\alpha_1 P N}{K_1 + N} + \mu_3 P + \mu_4 Z \equiv F_1(N, P, Z)$$

$$\frac{dP}{dt} = \frac{\alpha_2 P N}{K_1 + N} - \frac{\gamma_1 P^2 Z}{K_2^2 + P^2} - \mu_1 P \equiv F_2(N, P, Z)$$

$$\frac{dZ}{dt} = \frac{\gamma_2 P^2 Z}{K_2^2 + P^2} - \mu_2 Z - \frac{hZ}{E + Z} \equiv F_3(N, P, Z). \tag{1}$$

$$V = \begin{bmatrix} -D - \frac{\alpha_1 K_1 P}{(K_1+N)^2} & -\frac{\alpha_1 N}{K_1+N} + \mu_3 & \mu_4 \\ \frac{K_1 \alpha_2 P}{(K_1+N)^2} & \frac{\alpha_2 N}{K_1+N} - \frac{2K_2^2 \gamma_1 Z P}{(K_2^2+P^2)^2} - \mu_1 & -\frac{\gamma_1 P^2}{K_2^2+P^2} \\ 0 & \frac{2K_2^2 \gamma_2 P Z}{(K_2^2+P^2)^2} & \frac{\gamma_2 P^2}{K_2^2+P^2} - \mu_2 - \frac{Eh}{(E+Z)^2} \end{bmatrix}. \tag{2}$$

Note that some entries have a definite sign:

$$V_{11} < 0, \quad V_{13} > 0, \quad V_{21} > 0, \quad V_{23} < 0, \quad V_{32} > 0, \quad V_{33} > 0. \tag{3}$$

3 Analysis of the Deterministic System

3.1 Positive Invariance

Due to the constant supply of nutrients N^0, the system (1) is not homogeneous. Hence the origin cannot be a solution of the equilibrium equations. It is easy to check that whenever choosing $X(0) \in \mathbf{R}_+^3$ with $N = 0$, $P \neq 0$, $Z \neq 0$, then $F_1(X) > 0$. Hence on the plane $N = 0$ trajectories are entering the positive octant. The remaining coordinate planes are solutions of the respective equilibrium equations, so by the existence and uniqueness theorem, trajectories cannot cross these planes from the positive octant into the unfeasible domain.

This ensures that the solution remains within the positive octant, ensuring the biological well-posedness of the system.

3.2 Equilibria

The system (1) possesses the following three equilibria: the nutrient-only equilibrium $E_0 = (N^0, 0, 0)$, the zooplankton-free equilibrium $E_1 = (N_1, P_1, 0)$, and the coexistence of the three populations, $E^* = (N^*, P^*, Z^*)$.

The Nutrient-Only Equilibrium E_0 is always feasible; it is unstable for $\mu_1 < 0$: the eigenvalues of the Jacobian (2) evaluated at this point being $-D < 0$, $-(hE^{-1} + \mu_2) < 0$ and $\mu_1(R_0 - 1)$, and are locally asymptotically stable if and only if

$$R_0 := \frac{\alpha_2 N^0}{\mu_1(K_1 + N^0)} < 1. \tag{4}$$

The Zooplankton-Free Equilibrium At E_1 the population levels are

$$N_1 = \frac{\mu_1 K_1}{\alpha_2 - \mu_1}, \quad P_1 = \frac{D\alpha_2[N^0 \alpha_2 - (N^0 + K_1)\mu_1]}{(\alpha_2 - \mu_1)[\alpha_1 \mu_1 - \mu_3 \alpha_2]}.$$

Therefore, for $\mu_1 > 0$ this equilibrium is feasible if $\alpha_2 > \mu_1 > 0$ and either one of the two alternative conditions hold:

$$\mu_1 \frac{N^0 + K_1}{N^0} \leq \alpha_2 < \alpha_1 \frac{\mu_1}{\mu_3}; \quad \alpha_1 \frac{\mu_1}{\mu_3} < \alpha_2 \leq \mu_1 \frac{N^0 + K_1}{N^0}. \tag{5}$$

At E_1 the Jacobian (2) factorizes, to give one explicit eigenvalue $\gamma_2 P_1^2 (K_2^2 + P_1^2)^{-1} - \mu_2 - hE^{-1}$ and the quadratic equation

$$\lambda^2 + \lambda \left(D + \frac{\alpha_1 K_1 P_1}{(K_1 + N_1)^2} \right) + \frac{K_1 \alpha_2 P_1}{(K_1 + N_1)^2} \left(\frac{\alpha_1 N_1}{K_1 + N_1} - \mu_3 \right) = 0.$$

The Routh–Hurwitz conditions for the latter are easily seen to hold, but only when the first condition (5) is satisfied. Stability of E_1 is ensured by

$$R_1 = \frac{\gamma_2 E D^2 \alpha_2^2 M^2}{J^2 + D^2 \alpha_2^2 M^2 (\mu_2 E + h)} < 1, \tag{6}$$

where $J = K_2^2 (\alpha_2 - \mu_1)^2 (\alpha_1 \mu_1 - \alpha_2 \mu_3)$ and $M = N^0 \alpha_2 - \mu_1 (N^0 + K_1)$.

Remark 1 Whenever stable, E_1 is also globally asymptotically stable in the $N - P$ phase subspace, as shown by the Lyapunov function

$$W(N, P) = \int_{N_1}^{N} \frac{x - N_1}{x} dx + \frac{\alpha_1 N_1 - \mu_3 (K_1 + N_1)}{\alpha_2 N_1} \int_{P_1}^{P} \frac{x - P_1}{x} dx.$$

Estimating from above its time derivative along the trajectories of the subsystem (1) with $Z = 0$, we find using the very first inequality in (5)

$$
\begin{aligned}
\frac{dW}{dt} &= (N - N_1) \left[\frac{D(N^0 - N)}{N} - \frac{D(N^0 - N_1)}{N_1} - P \left(\frac{\alpha_1}{K_1 + N} - \frac{\mu_3}{N} \right) \right. \\
&\quad \left. + P_1 \left(\frac{\alpha_1}{K_1 + N_1} - \frac{\mu_3}{N_1} \right) + \left(\frac{\alpha_1}{K_1 + N_1} - \frac{\mu_3}{N_1} \right) \left(\frac{K_1}{K_1 + N} \right) (P - P_1) \right] \\
&\leq -(N - N_1)^2 \frac{D}{N} - \frac{(N - N_1)^2}{N N_1} [P \mu_3 + D(N^0 - N_1)] < 0.
\end{aligned}
$$

The Coexistence Equilibrium At $E^* = (N^*, P^*, Z^*)$, the levels are

$$N^* = \frac{(\gamma_1 P^* Z^* + \mu_1 (K_2^2 + P^{*2})) K_1}{(\alpha_2 - \mu_1)(K_2^2 + P^{*2}) - \gamma_1 P^* Z^*}, \qquad Z^* = \frac{(K_2^2 + P^{*2}) h}{(\gamma_2 - \mu_2) P^{*2} - K_2^2 \mu_2} - E.$$

while P^* which solves the equation with no closed form solution:

$$D(N^0 - N^*) - \frac{\alpha_1 N^* P^*}{K_1 + N^*} + \mu_3 P^* + \mu_4 Z^* = 0.$$

Thus E^* is investigated numerically. For feasibility, we certainly need

$$h > \frac{E[(\gamma_2 - \mu_2) P^{*2} - K_2^2 \mu_2]}{K_2^2 + P^{*2}}, \qquad \alpha_2 > \mu_1 + \frac{\gamma_1 P^* Z^*}{K_2^2 + P^{*2}}. \tag{7}$$

For stability, all entries of the Jacobian V but one have a definite sign. In addition to those of (3), we have $V_{12} = -(\alpha_1 N^*)(K_1 + N^*)^{-1} + \mu_3 < 0$, $V_{22} = \gamma_1 P^* Z^* [P^{*2} - K_2^2](K_2^2 + P^{*2})^{-2} \in \mathbf{R}$. The characteristic equation is

$$y^3 + \Lambda_1 y^2 + \Lambda_2 y + \Lambda_3 = 0 \tag{8}$$

where $A_1 = -tr(V)$, $A_2 = V_{11}V_{22} + V_{22}V_{33} + V_{11}V_{33} - V_{23}V_{32} - V_{12}V_{21}; A_3 = \det(V)$. Sufficient conditions for ensuring the positivity of these coefficients can be stated, which help the simulations to give bounds for P^*. In addition to the Routh–Hurwitz condition $A_1A_2 > A_3$, two cases arise.

Assume at first that $V_{22} > 0$. Now $A_1 > 0$ is implied by

$$D + \frac{\alpha_1 K_1 P^*}{(K_1 + N^*)^2} + \frac{\gamma_1 P^* Z^*(K_2^2 - P^{*2})}{(K_2^2 + P^{*2})^2} > \frac{hZ^*}{(E + Z^*)^2}; \tag{9}$$

$A_2 > 0$ is instead ensured by $V_{22}V_{33} - V_{23}V_{32} - V_{12}V_{21} > -(V_{11}V_{22} + V_{11}V_{33})$; $A_3 > 0$ holds whenever $V_{11}V_{23}V_{32} - V_{33}V_{11}V_{22} > V_{13}V_{21}V_{32} - V_{33}V_{12}V_{21}$.

For $V_{22} < 0$, we find instead that $A_1 > 0$ is satisfied if the opposite inequality (9) is verified while $A_2 > 0$ holds if $V_{11}V_{22} - V_{23}V_{32} - V_{12}V_{21} > -(V_{11}V_{33} + V_{22}V_{33})$; and finally $A_3 > 0$ is implied by $V_{11}V_{23}V_{32} - V_{33}V_{11}V_{22} > V_{13}V_{21}V_{32} - V_{33}V_{12}V_{21}$.

Remark 2 The system could have a Hopf bifurcation at the coexistence equilibrium if the following two conditions are satisfied:

$$A_1(h^*)A_2(h^*) - A_3(h^*) = 0, \quad A_1'(h^*)A_2(h^*) + A_1(h^*)A_2'(h^*) - A_3'(h^*) \neq 0. \tag{10}$$

Hopf Bifurcation at Coexistence The Hopf bifurcation occurs when the characteristic equation (8) has two purely imaginary roots $\pm \eta_2 i$, where $i = \sqrt{-1}$ denotes the imaginary unit, in addition to the one real root η_1. It follows that it can be rewritten as $(y^2 + \eta_2^2)(y - \eta_1) = 0$. Expanding, we find $y^3 - \eta_1 y^2 + \eta_2^2 y - \eta_1 \eta_2 = 0$. Comparing the coefficients with those of (8), the first condition (10) follows. Let the eigenvalues of the characteristic equation be written as $\lambda_i = u_i + i\eta_i$. The second condition (10) is obtained by observing that these eigenvalues must satisfy the transversality condition

$$\left.\frac{du_i}{dh}\right|_{h=h^*} \neq 0.$$

Let us also write $S = A_1A_2 - A_3$. Substituting λ_i into the characteristic equation, separating the real and imaginary parts, and eliminating η, we get $8u^3 + 8A_1u^2 + 2u(A_1^2 + A_2) + S = 0$. Differentiating it with respect to h:

$$24u^2\frac{du}{dh} + 16A_1u\frac{du}{dh} + 2(A_1^2 + A_2)\frac{du}{dh} + 2u\left[2A_1\frac{dA_1}{dh} + \frac{dA_2}{dh}\right] + \frac{dS}{dh} = 0.$$

At $h = h^*$, we have $u(h^*) = 0$, so that the above equation becomes

$$\left[\frac{du}{dh}\right]_{h=h^*} = -\frac{1}{2(A_1^2 + A_2)}\left[\frac{dS}{dh}\right]_{h=h^*} \neq 0$$

providing the second condition (10). An analytic verification of (10) appears in these conditions a very difficult task, as the coefficients depend on the population levels at equilibrium E^*. But the numerical simulations reveal the existence of sustained population oscillations, see, e.g., Fig. 1.

Theorem 1 *The parameter μ_{22} determines the direction of the Hopf bifurcation. If $\mu_{22} > 0 \ (< 0)$ then the Hopf bifurcation is supercritical (subcritical) and the bifurcating periodic solutions exist for $h > h^*$.*

The stability and the period of the bifurcating periodic solutions are respectively determined by the parameters β_2 and τ_2 defined in the proof. The solutions are orbitally stable (unstable) if $\beta_2 < 0 \ (> 0)$ and the period increases (decreases) if $\tau_2 > 0 \ (< 0)$.

Proof We just outline the proof, the method being based on the normal form theory [20]. Assuming (10), let us denote by a bar the system parameters; the eigenvector corresponding to the eigenvalue $\sigma = \mathbf{i}\upsilon$ is

$$\omega \left[\frac{(\mathbf{i}\upsilon)^2 - (V_{22} + V_{33})\mathbf{i}\upsilon + V_{22}V_{33} - V_{32}V_{23}}{V_{21}V_{32}}, \frac{\sigma - V_{33}}{V_{32}}, 1 \right]^T, \quad \omega \in \mathbf{R}.$$

Let us define the following quantities $b_{31} = 1, b_{32} = 0, b_{33} = 1$,

$$b_{11} = -\frac{\upsilon^2 + V_{22}V_{33} - V_{32}V_{23}}{V_{21}V_{32}}, \quad b_{12} = \frac{\upsilon(V_{22} + V_{33})}{V_{21}V_{32}}, \quad b_{21} = -\frac{V_{33}}{V_{32}},$$

$$b_{13} = \frac{\upsilon^2 - (V_{22} + V_{33})\upsilon + V_{22}V_{33} - V_{32}V_{23}}{V_{21}V_{32}}, \quad b_{22} = -\frac{\upsilon}{V_{32}}, \quad b_{23} = \frac{\upsilon - V_{33}}{V_{32}}.$$

Using the transformation $N = N^* + b_{11}x_1 + b_{12}y_1 + b_{13}p_1$, $P = P^* + b_{21}x_1 + b_{22}y_1 + b_{23}p_1$, $Z = Z^* + b_{31}x_1 + b_{32}y_1 + b_{33}p_1$, system (1) then reduces to

$$\frac{dx_1}{dt} = \frac{W_3 B_2 - W_1 b_{22} + W_2 b_{12}}{B_2 - B_1} := G^1, \tag{11}$$

$$\frac{dy_1}{dt} = \frac{(B_2 - B_1 + B_3)W_1 + B_4 W_2 + (b_{13}B_1 - b_{11}B_2)W_3}{b_{12}(B_2 - B_1)} := G^2,$$

$$\frac{dp_1}{dt} = \frac{W_1 b_{22} - W_2 b_{12} - W_3 B_1}{B_2 - B_1} := G^3,$$

where as a shorthand we have introduced the following quantities:

$$B_1 = b_{11}b_{22} - b_{21}b_{12}, \quad B_2 = b_{13}b_{22} - b_{23}b_{12}, \quad B_3 = b_{11}b_{22} - b_{13}b_{22},$$

$$B_4 = b_{13}b_{12} - b_{11}b_{12}, \quad W_1 = \bar{D}(\bar{N^0} - N^* - b_{11}x_1 - b_{12}y_1 - b_{13}p_1)$$

$$- \frac{\bar{\alpha}_1(N^* + b_{11}x_1 + b_{12}y_1 + b_{13}p_1)(P^* + b_{21}x_1 + b_{22}y_1 + b_{23}p_1)}{\bar{K}_1 + N^* + b_{11}x_1 + b_{12}y_1 + b_{13}p_1}$$

$$+ \bar{\mu}_3(P^* + b_{21}x_1 + b_{22}y_1 + b_{23}p_1) + \bar{\mu}_4(Z^* + b_{31}x_1 + b_{32}y_1 + b_{33}p_1),$$

$$W_2 = \frac{\bar{\alpha}_2(N^* + b_{11}x_1 + b_{12}y_1 + b_{13}p_1)(P^* + b_{21}x_1 + b_{22}y_1 + b_{23}p_1)}{\bar{K}_1 + N^* + b_{11}x_1 + b_{12}y_1 + b_{13}p_1}$$

$$- \frac{\bar{\gamma}_1(P^* + b_{21}x_1 + b_{22}y_1 + b_{23}p_1)^2(Z^* + b_{31}x_1 + b_{32}y_1 + b_{33}p_1)}{K_2^2 + (P^* + b_{21}x_1 + b_{22}y_1 + b_{23}p_1)^2}$$

$$- \bar{\mu}_1(P^* + b_{21}x_1 + b_{22}y_1 + b_{23}p_1),$$

$$W_3 = \frac{\bar{\gamma}_2(P^* + b_{21}x_1 + b_{22}y_1 + b_{23}p_1)^2((Z^* + b_{31}x_1 + b_{32}y_1 + b_{33}p_1))}{K_2^2 + (P^* + b_{21}x_1 + b_{22}y_1 + b_{23}p_1)^2}$$

$$- \bar{\mu}_2(Z^* + b_{31}x_1 + b_{32}y_1 + b_{33}p_1) - \frac{\bar{h}(Z^* + b_{31}x_1 + b_{32}y_1 + b_{33}p_1)}{\bar{E} + (Z^* + b_{31}x_1 + b_{32}y_1 + b_{33}p_1)}.$$

The equilibrium point of the new system (11) is now the origin. At it, the Jacobian of (11) simplifies since several entries vanish:

$$\frac{\partial G^1}{\partial x_1} = \frac{\partial G^2}{\partial y_1} = \frac{\partial G^1}{\partial p_1} = \frac{\partial G^3}{\partial x_1} = \frac{\partial G^3}{\partial y_1} = \frac{\partial G^2}{\partial p_1} = 0.$$

Further, the following auxiliary quantities can be explicitly calculated in terms of the system parameters, but we omit the explicit formulae:

$$D_1 = \frac{\partial G^3}{\partial p_1}, \quad g_{11} = \frac{1}{4}\left[\frac{\partial^2 G^1}{\partial x_1^2} + \frac{\partial^2 G^2}{\partial y_1^2} + i\left(\frac{\partial^2 G^2}{\partial x_1^2} + \frac{\partial^2 G^1}{\partial y_1^2}\right)\right],$$

$$g_{02} = \frac{1}{4}\left[\frac{\partial^2 G^1}{\partial x_1^2} - \frac{\partial^2 G^2}{\partial y_1^2} - 2\frac{\partial^2 G^2}{\partial x_1 \partial y_1} + i\left(\frac{\partial^2 G^2}{\partial x_1^2} - \frac{\partial^2 G^1}{\partial y_1^2}\right) + 2\frac{\partial^2 G^1}{\partial x_1 \partial y_1}\right],$$

$$g_{20} = \frac{1}{4}\left[\frac{\partial^2 G^1}{\partial x_1^2} - \frac{\partial^2 G^2}{\partial y_1^2} + 2\frac{\partial^2 G^2}{\partial x_1 \partial y_1} + i\left(\frac{\partial^2 G^2}{\partial x_1^2} - \frac{\partial^2 G^1}{\partial y_1^2}\right) - 2\frac{\partial^2 G^1}{\partial x_1 \partial y_1}\right],$$

$$G_{21} = \frac{1}{8}\left[\frac{\partial^3 G^1}{\partial x_1^3} + \frac{\partial^3 G^1}{\partial x_1 \partial y_1^2} + \frac{\partial^3 G^2}{\partial^2 x_1 \partial y_1} + \frac{\partial^3 G^2}{\partial y_1^3}\right]$$

$$+ \frac{i}{8}\left[\frac{\partial^3 G^2}{\partial x_1^3} + \frac{\partial^3 G^2}{\partial x_1 \partial y_1^2} - \frac{\partial^3 G^1}{\partial^2 x_1 \partial y_1} - \frac{\partial^3 G^1}{\partial y_1^3}\right],$$

$$G_{110} = \frac{1}{2}\left[\frac{\partial^2 G^1}{\partial x_1 \partial p_1} + \frac{\partial^2 G^1}{\partial y_1 \partial p_1} + i\left(\frac{\partial^2 G^2}{\partial x_1 \partial p_1} - \frac{\partial^2 G^2}{\partial y_1 \partial p_1}\right)\right],$$

$$G_{101} = \frac{1}{2} \left[\frac{\partial^2 G^1}{\partial x_1 \partial p_1} - \frac{\partial^2 G^1}{\partial y_1 \partial p_1} + i \left(\frac{\partial^2 G^2}{\partial x_1 \partial p_1} + \frac{\partial^2 G^2}{\partial y_1 \partial p_1} \right) \right],$$

$$h_{11} = \frac{1}{4} \left[\frac{\partial^2 G^3}{\partial x_1^2} + \frac{\partial^2 G^3}{\partial y_1^2} \right], \quad h_{20} = \frac{1}{4} \left[\frac{\partial^2 G^3}{\partial x_1^2} - \frac{\partial^2 G^3}{\partial y_1^2} - 2i \frac{\partial^2 G^3}{\partial x_1 \partial y_1} \right],$$

$$\omega_{11} = -\frac{h_{11}}{D_1}, \quad \omega_{20} = -\frac{h_{20}}{D_1 - 2i\omega_0}, \quad g_{21} = G_{21} + 2G_{110}\omega_{11} + G_{101}\omega_{20}.$$

The values of μ_{22} and τ_2, obtained from [20, 36], can now be calculated:

$$c_1(0) = \frac{i}{2\omega_0} \left[g_{20}g_{11} - 2|g_{11}|^2 - \frac{1}{3}|g_{02}|^2 \right] + \frac{g_{21}}{2},$$

$$\mu_{22} = -\frac{\text{Re}[c_1(0)]}{u'(0)}, \quad \tau_2 = -\frac{\text{Im}[c_1(0)] + \mu_{22}\omega'(0)}{\omega(0)}, \quad \beta_2 = 2\text{Re}[c_1(0)].$$

Recalling finally [20], if the root of the characteristic equation increases for increasing values of the bifurcation parameter h, namely $u'(0) > 0$, the cycle for $\mu_{22} > 0$ is supercritical while it is subcritical for $\mu_{22} < 0$.

The analytical results are summarized in Table 1.

4 The Stochastic Model

The above discussion rests on the assumption that the environmental parameters involved with the model are all constants irrespective of time and environmental fluctuations. In this section, we introduce the effect of environmental fluctuation on the system and consider the stochastic stability of the coexistence equilibrium.

There are two ways to develop the stochastic model from an existing deterministic system. Firstly, one can replace the environmental parameters in the deterministic model by some random parameters. For instance, the growth rate parameter 'r' could be replaced by $r_0 + \epsilon\gamma(t)$, where r_0 is the average growth rate, $\gamma(t)$ is the noise function, and ϵ is the intensity of fluctuation. Secondly, one can add a randomly

Table 1 Thresholds and stability of steady states

Thresholds (R_0, R_1)	$(N_0, 0, 0)$	$(N_1, P_1, 0)$	(N^*, P^*, Z^*)
$R_0 < 1$	Asymptotically stable	Not feasible	Not feasible
$R_0 > 1$, $R_1 < 1$	Unstable	Asymptotically stable	Not feasible
$R_1 > 1$	Unstable	Unstable	Asymptotically stable

fluctuating driving force directly into the deterministic dynamic equations without altering any particular parameter [37]. Here we follow the latter approach. We assume that the stochastic perturbations of the state variables around their steady-state values E^* are of Gaussian white noise type, which is extremely effective to model rapidly fluctuating phenomena. Thus the stochastic perturbations are proportional to the distances of each population from its equilibrium value, [3]. From the deterministic system (1), we get the stochastic model

$$dN = F_1(N, P, Z)dt + \sigma_1(N - N^*)d\xi_t^1, \tag{12}$$

$$dP = F_2(N, P, Z)dt + \sigma_2(P - P^*)d\xi_t^2,$$

$$dZ = F_3(N, P, Z)dt + \sigma_3(Z - Z^*)d\xi_t^3$$

where the intensities of environmental fluctuations $\sigma_1, \sigma_2, \sigma_3 \in \mathbf{R}$ $\xi_t^i = \xi_i(t), i = 1, 2, 3$ are standard Wiener processes independent of each other.

We consider (12) as an Itō stochastic differential system of the type

$$dX_t = F(t, X_t)dt + g(t, X_t)d\xi_t, \quad X_{t0} = X_0, \tag{13}$$

where the solution $X_t = (N, P, Z)^T$, for $t > 0$ is a Itō process, F is the slowly varying continuous component or drift coefficient, the diagonal matrix $g = \text{diag}[\sigma_1(N - N^*), \sigma_2(P - P^*), \sigma_3(Z - Z^*)]$ expresses the rapidly varying continuous random component or diffusion coefficient and $\xi_t = (\xi_t^1, \xi_t^2, \xi_t^3)^T$ is a three-dimensional stochastic process having scalar Wiener process components with increments $\Delta\xi_t^j = \xi_j(t + \Delta t) - \xi_j(t)$ that are independent Gaussian random variables $\mathbf{N}(0, \Delta t)$. Since the diffusion matrix g depends upon the solution of X_t, (12) has multiplicative noise.

4.1 Stochastic Stability of the Coexistence Equilibrium

The stochastic differential system (12) can be centered at its coexistence equilibrium E^* by introducing the perturbation vector $U(t) = (u_1(t), u_2(t), u_3(t))^T$, with $u_1 = N - N^*$, $u_2 = P - P^*$, $u_3 = Z - Z^*$. To derive the asymptotic stability in the mean square sense by the Lyapunov functions method, working on the complete nonlinear equation (12) could be attempted. But for simplicity we deal with the stochastic differential equations obtained by linearizing (12) about coexistence E^* and we assume that there is no additional food source, i.e., $r = 0$, or, equivalently, $\mu_1 = \mu$. The linearized version of (13) around E^* is given by

$$dU(t) = F_L(U(t))dt + g(U(t))d\xi(t), \tag{14}$$

where now $g(U(t)) = \text{diag}[\sigma_1 u_1, \sigma_2 u_2, \sigma_3 u_3]$ and

$$F_L(U(t)) = \begin{bmatrix} V_{11}u_1 + V_{12}u_2 + V_{13}u_3 \\ V_{21}u_1 + V_{22}u_2 + V_{23}u_3 \\ V_{31}u_1 + V_{32}u_2 + V_{33}u_3 \end{bmatrix} = VU,$$

and the coexistence equilibrium corresponds now to the origin $(u_1, u_2, u_3) = (0, 0, 0)$. Let $\Omega = \left[(t \geq t_0) \times R^3, t_0 \in R^+ \right]$ and let $\Theta(t, X) \in C^{(1,2)}(\Omega)$ be a differentiable function of time t and twice differentiable function of X. Let us use set further

$$
L\Theta(t, u) = \frac{\partial \Theta(t, u(t))}{\partial t} + f^T(u(t)) \frac{\partial \Theta(t, u)}{\partial u}
$$

$$
+ \frac{1}{2} \mathrm{tr} \left[g^T(u(t)) \frac{\partial^2 \Theta(t, u)}{\partial u^2} g(u(t)) \right],
\tag{15}
$$

where

$$
\frac{\partial \Theta}{\partial u} = \left(\frac{\partial \Theta}{\partial u_1}, \frac{\partial \Theta}{\partial u_2}, \frac{\partial \Theta}{\partial u_3} \right)^T, \quad \frac{\partial^2 \Theta(t, u)}{\partial u^2} = \left(\frac{\partial^2 \Theta}{\partial u_j \partial u_i} \right)_{i, j = 1, 2, 3}.
$$

With these positions, we now recall the following result, [1].

Theorem 2 *Assume that the functions $\Theta(U, t) \in C_3(\Omega)$ and $L\Theta$ for $\alpha > 0$ satisfy the inequalities*

$$
r_1 |U|^\alpha \leq \Theta(U, t) \leq r_2 |U|^\alpha, \quad L\Theta(U, t) \leq -r_3 |U|^\alpha, r_i > 0, \quad i = 1, 2, 3. \tag{16}
$$

Then the trivial solution of (14) is exponentially α-stable for all time $t \geq 0$.

Remark 3 For $\alpha = 2$ in (16), the trivial solution of (14) is exponentially mean square stable; furthermore, the trivial solution of (14) is globally asymptotically stable in probability, [1].

Theorem 3 *Assume $V_{ij} < 0$, $i, j = 1, 2, 3$, and that for some positive real values of ω_k, $k = 1, 2$, the following inequality holds*

$$
\left[2(1 + \omega_2) V_{22} + 2V_{32}\omega_2 + (1 + \omega_2)\sigma_2^2 \right] \left[2V_{13}\omega_1 + 2V_{23}\omega_1 + 2V_{33}(\omega_1 + \omega_2) \right.
$$

$$
\left. + (\omega_1 + \omega_2)\sigma_3^2 \right] > [V_{12}\omega_1 + V_{22}\omega_2 + V_{23}(1 + \omega_2) + V_{32}(\omega_1 + \omega_2)
$$

$$
+ V_{33}\omega_2]^2.
\tag{17}
$$

Then if $\sigma_1^2 < -2V_{11}$, it follows that

$$
\sigma_2^2 < -\frac{2V_{22}(1 + \omega_2) + 2V_{32}\omega_2}{1 + \omega_2}, \quad \sigma_3^2 < -\frac{2V_{13}\omega_1 + 2V_{23}\omega_1 + 2V_{33}(\omega_1 + \omega_2)}{\omega_1 + \omega_2},
\tag{18}
$$

$$
\omega_1^* = \frac{V_{21}}{V_{13} + V_{11} + V_{33} - V_{12} - V_{32}}, \quad \omega_2^* = \frac{V_{11} + V_{13} + V_{33}}{V_{12} - (V_{13} + V_{11} + V_{33}) + V_{32}},
\tag{19}
$$

and the zero solution of system (12) is asymptotically mean square stable.

Proof Consider $\Theta(u(t)) = \frac{1}{2}\left[\omega_1(u_1 + u_3)^2 + u_2{}^2 + \omega_2(u_2 + u_3)^2\right]$, the Lyapunov function with ω_k real positive constants to be chosen later.

It is easy to check that the left inequalities (16) hold for $\alpha = 2$. Then

$$L_\Theta(u(t)) = \left[V_{22}(1 + \omega_2) + V_{32}\omega_2\right]u_2^2 + \left[V_{13}\omega_1 + V_{23}\omega_2 + V_{33}(\omega_1 + \omega_2)\right]u_3^2$$
$$+ u_1 u_2 \left[V_{12}\omega_1 - V_{21}(1 + \omega_2) + V_{32}\omega_1\right]$$
$$+ u_2 u_3 \left[V_{12}\omega_1 + V_{22}\omega_2 + V_{23}(1 + \omega_2)\right.$$
$$+ V_{32}(\omega_1 + \omega_2) + V_{33}\omega_2] + u_3 u_1 \left[V_{13}\omega_1 + V_{11}\omega_1 - V_{21}\omega_2 + V_{33}\omega_1\right]$$
$$+ V_{11}\omega_1 u_1^2 + \frac{1}{2}\mathrm{tr}\left[g^T(u(t))\frac{\partial^2\Theta}{\partial u^2}g(u(t))\right].$$

Now we can evaluate the trace of the matrix

$$\frac{\partial^2\Theta}{\partial u^2} = \begin{vmatrix} \omega_1 & 0 & \omega_1 \\ 0 & 1 + \omega_2 & \omega_2 \\ \omega_1 & \omega_2 & \omega_1 + \omega_2 \end{vmatrix},$$

$$\mathrm{tr}\left[g^T(u(t))\frac{\partial^2\Theta}{\partial u^2}g(u(t))\right] = \omega_1\sigma_1{}^2 u_1{}^2 + (1 + \omega_2)\sigma_2{}^2 u_2{}^2 + (\omega_1 + \omega_2)\sigma_3{}^2 u_3{}^2.$$

Using then (19), the Lyapunov function becomes $L_\Theta(u(t)) = -u^T Q u$, with

$$Q = \begin{vmatrix} -V_{11}\omega_1 - \frac{1}{2}\omega_1\sigma_1^2 & 0 & 0 \\ 0 & -(1 + \omega_2)V_{22} - \omega_2 V_{32} - \frac{1}{2}(1 + \omega_2)\sigma_2^2 & Q_{23} \\ 0 & Q_{23} & Q_{33} \end{vmatrix}.$$

the real symmetric matrix where $Q_{23} = -\frac{1}{2}[V_{12}\omega_1 + V_{22}\omega_2 + V_{23}(1+\omega_2) + V_{32}(\omega_1 + \omega_2) + V_{33}\omega_2]$ and $Q_{33} = -V_{13}\omega_1 - V_{23}\omega_2 - V_{33}(\omega_1 + \omega_2) - \frac{1}{2}(\omega_1 + \omega_2)\sigma_3^2$. Easily, the inequality $L_\Theta(u(t)) \le -u^T Q u$ holds. On the other hand, (17) and (18) imply that Q is positive definite and therefore all its eigenvalues $\lambda_i(Q)$, $i = 1, 2, 3$, are positive real numbers. Let $\lambda_m = \min\{\lambda_i(Q), \ i = 1, 2, 3\} > 0$. From the previous inequality for $L_\Theta(u(t))$ we thus get $L_\Theta(u(t)) \le -\lambda_m |u(t)|^2 < 0$.

Remark 4 Theorem 3 provides the necessary conditions for the stochastic stability of the coexistence equilibrium E^* under environmental fluctuations, [2]. Thus the model parameters and environmental fluctuations intensities help in maintaining the stability of the stochastic system.

Table 2 Parameter values

Parameter	Definition	Unit	Source
$N^0 = 2.0$	Constant input of nutrient	mg ml^{-1}	[11]
$D = 0.5$	Dilution rate of nutrient	day^{-1}	[17]
$\alpha_1 = 1.6$	Nutrient uptake rate for the autotroph biomass	day^{-1}	–
$\alpha_2 = 1.2$	Nutrient conversion rate for the growth of autotroph	day^{-1}	–
$\gamma_1 = 1$	Autotroph biomass uptake rate for the herbivore	day^{-1}	[16]
$\gamma_2 = 0.9$	Autotroph biomass conversion rate for the herbivore	day^{-1}	[16]
$\mu_1 = 0.6$	Mortality rate of autotroph biomass	day^{-1}	[11]
$\mu_2 = 0.4$	Mortality rate of herbivore	day^{-1}	–
$\mu_3 = 0.1$	Nutrient recycle rate due to the death autotroph	day^{-1}	[25]
$\mu_4 = 0.1$	Nutrient recycle rate due to the death of herbivore	day^{-1}	[25]
$K_1 = 0.3$	Half-saturation constant for autotroph	mg ml^{-1}	–
$K_2 = 0.3$	Half-saturation constant for herbivore	mg ml^{-1}	–
$h = 0.4$	Harvesting rate of herbivore population	day^{-1}	[31]
$E = 1.0$	Effort required to harvest the herbivores	mg ml^{-1}	[31]

5   Numerical Simulations

In this section, we focus our attention on the occurrence and termination of the populations fluctuations. We take as the reference parameter set the one given in Table 2, see also [29] and [11], for which the existence condition of the coexistence equilibrium point $E^* = (0.7297, 0.6331, 0.1941)$ is feasible and locally asymptotically stable, namely a stable focus in view of the eigenvalues $-0.5081, -0.0327 \pm i0.3106$, see Fig. 1.

In each of the following subsections, we vary just one or two of the parameters at the time and keep the remaining ones at the reference level. Not all figures will be shown. In these simulations, we do not consider the additional food source, thus setting $r = 0$ and $\mu_1 > 0$. In general, we observe that when we use the same set of parameter values of Table 2, i.e., taking $r = 0.14$, the supplementary food destabilizes the whole system. Moreover, changing $\mu = 0.6$ into $\mu = 0.06$, $\mu_3 = 0.1$ into $\mu_3 = 0.01$, and $r = 0$ into $r = 0.14$, we observe the oscillatory behavior for $\mu_1 < 0$, and we obtain similar results for N^0, D, and h; but in the latter case for $\mu_1 > 0$, we must exclude from these considerations the value $h = 0.5$.

Effects of D   For $D = 0.55$, leaving all other parameters unaltered, the system exhibits persistent oscillations around the positive interior equilibrium E^* with eigenvalues $-0.5698, 0.0092 \pm i0.3232$, see Fig. 1.

Effects of N^0   Increasing N^0 from 2 to 2.2, the system again exhibits persistent oscillations; eigenvalues at E^* are $-0.5068, 0.0232 \pm i0.3391$.

Effects of h   Decreasing the value of h from 0.4 to 0.2, once again the system exhibits oscillatory behavior. The eigenvalues at E^* now are $-0.5028, 0.0152 \pm i0.4172$. This is in line with our analytical results: when h crosses the value h^*, the system has a Hopf bifurcation, see Fig. 2 left.

Fig. 1 The plot is obtained for the reference parameter values given in Table 2, but with $D = 0.55$

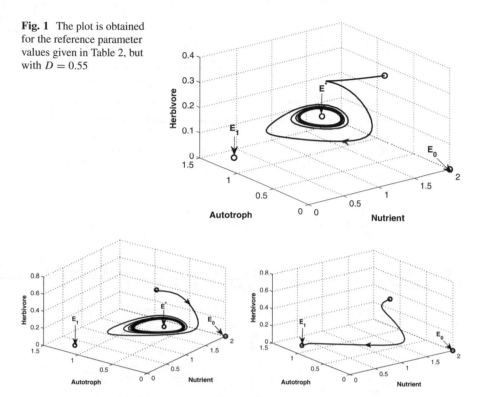

Fig. 2 The plot is obtained for the reference parameter values given in Table 2, on the left we however take $h = 0.2$, but on the right $h = 0.5$

On the other hand, the system shifts to the herbivore-free stable node $E_1 = (0.3000, 1.2143, 0)$, when h increases from 0.4 to 0.5, see Fig. 2 right.

Combined Effect of h and D Starting from the cycles for $h = 0.2$ but decreasing D from 0.5 to 0.4, coexistence returns to be stable.

Combined Effect of h and N^0 Once again starting from the cycles for $h = 0.2$, decreasing N^0 from 2 to 1.6, coexistence restabilizes.

Combined Effect of h and r As noted above, when $h = 0.5$ and in the absence of additional food, the system settles to the herbivore-free equilibrium E_1. In the presence of the additional food, taking $r = 0.14$ preserves the herbivores, since the system trajectories tend once again to the coexistence equilibrium E^*, see Fig. 3.

Hopf Bifurcations For a clear understanding of the dynamical implications of some parameter changes, we plot suitable bifurcation diagrams. We report the findings in Fig. 4 respectively in terms of the bifurcation parameters h and N^0, keeping the remaining parameters always at their reference values. Observe that $f_1(h) = A_1(h)A_2(h)$ and $f_2(h) = A_3$ intersect at $h^* = 0.135$ and $h^{**} = 0.368$

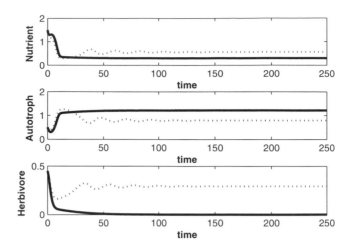

Fig. 3 The plot is obtained for the reference parameter values given in Table 2, but with $h = 0.5$; in addition, we take $r = 0$, the continuous line, and $r = 0.14$, the dotted line

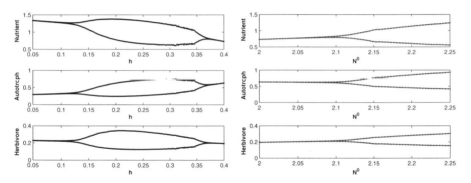

Fig. 4 Bifurcation diagrams in terms of h (left), N^0 (right), with the other parameters at the reference values given in Table 2. In each frame, top to bottom, we have the Nutrient, Autotroph, and Herbivore populations

indicating that the system (1) changes its stability when the parameter h crosses the thresholds h^* and h^{**}. Moreover, for $h > h^*$ we see that $f_1(h) < f_2(h)$ the system (1) becomes unstable at E^*. On the other hand, for $h > h^{**}$ we observe that $f_1(h) > f_2(h)$, satisfying the condition of stability at E^* (cf. Fig. 5a). More specifically, it is found that the tangent to $g_1(h) = f_1(h) - f_2(h)$ both at h^* and at h^{**} is not parallel to the h axis, satisfying the second condition of (10) (cf. Fig. 5b). Also Fig. 6a indicates the stability behavior of the system (1) for N^0. Moreover, Fig. 6b shows that the second condition of (10) for N^0 holds. For the dilution rate of the nutrient D, a diagram similar to the one of N^0 is obtained, not shown, with the threshold value for the bifurcation given by $D^* = 0.53$.

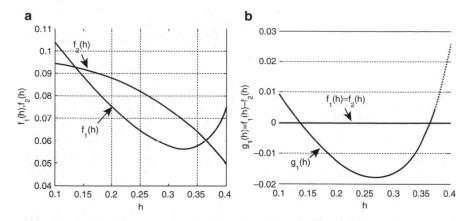

Fig. 5 (**a**) The two curves, $f_1(h)$, $f_2(h)$, intersect at the $h = h^*$ (the star). (**b**) The tangent to the curve $g_1(h) = f_1(h) - f_2(h)$ at $h = h^*$ is not parallel to the h axis

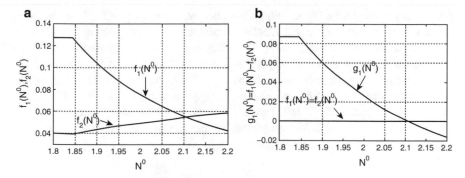

Fig. 6 (**a**) The two curves, $f_1(N^0)$, $f_2(N^0)$, intersect at the $N^0 = N^{0*}$ (the star). (**b**) The tangent to the curve $g_1(N^0) = f_1(N^0) - f_2(N^0)$ at $N^0 = N^{0*}$ is not parallel to the N^0 axis

Changes in Mortalities and Recycle Rates We now perform numerical simulations in terms of $\mu_1 = \mu - r$, taking $\mu_1 < 0$. When $r = 0.14$, keeping all the other parameters at their fixed reference value, but changing μ from 0.6 to 0.06, μ_2 from 0.4 to 0.04, μ_3 from 0.1 to 0.01, and μ_4 from 0.1 to 0.01, the system shows stable behavior. Thus when the additional food supply exceeds the mortality rate of the autotroph biomass, the system exhibits a stable behavior. Instead, taking $r = 0.14$, and keeping all the parameters at their reference value, but changing μ from 0.6 to 0.06, μ_2 from 0.4 to 0.2, μ_3 from 0.1 to 0.01, and μ_4 from 0.1 to 0.01, the system oscillates and similarly for $\mu_1 < 0$.

Environmental Fluctuations Finally, we investigate the dynamical behavior of the system in the presence of environmental disturbances. For the numerical simulation of the stochastic differential equation (12), we use the Euler–Maruyama method and MATLAB software. For the stochastic model, we have obtained the condition

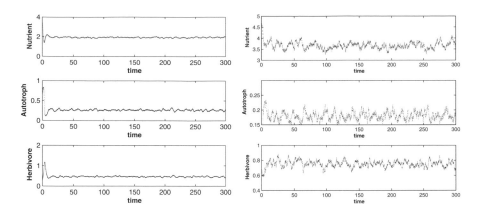

Fig. 7 Effects of environmental fluctuations. Left: $\sigma_1 = 0.09$, $\sigma_2 = 0.08$, $\sigma_3 = 0.09$; Right: $\sigma_1 = 0.25$, $\sigma_2 = 0.15$, $\sigma_3 = 0.12$

for the asymptotic stability of the coexistence equilibrium point E^* in the mean square sense by using a suitable function. These conditions depend upon σ_1, σ_2, σ_3 and the parameters of the model (12). Using the reference values for the model parameters in Table 2 and taking the following values for the intensities of the environmental perturbations, $\sigma_1 = 0.09$, $\sigma_2 = 0.08$, and $\sigma_3 = 0.09$, values that satisfy condition (18). In these conditions, the system is stochastically stable, all the three species coexist, Fig. 7 left. Instead, increasing the perturbations, $\sigma_1 = 0.25$, $\sigma_2 = 0.15$, $\sigma_3 = 0.12$, the amplitude of the fluctuations increases, implying instability of the coexistence equilibrium, Fig. 7 right.

6   Discussion

Two autotroph–herbivore interaction models are considered, a deterministic one and its corresponding stochastic version. Their main features are the use of general nutrient uptake functions and instantaneous nutrient recycling. Note that this model differs from [11] because in it the HTII response function is used, but above all because the main assumption is that the phytoplankton produces harmful toxins for the zooplankton, a fact that here is not taken into account. A similar model with HTI response was earlier introduced in [32]. Also in [9] a model similar to the present one is considered, using however an HTII response function. The other major assumption that distinguishes it from this ecosystem is that zooplankton is assumed to feed not only on phytoplankton but on nutrients as well. In view of this assumption, contrary to our findings here, zooplankton can survive also in the absence of phytoplankton. Qualitatively the other exhibited features are quite similar, differing clearly however in the quantitative expressions of the equilibria population values. The ecosystem modeled in [10] instead contains additionally detritus, which degrades into nutrients. But in this case zooplankton does not feed directly on nutrients, hence the phytoplankton-free state is not allowed. Indeed the situation with no phytoplankton, while all the other populations thrive, does not exist because zooplankton would not be able to thrive without it. Also, detritus is present only if plankton is also. This is evident, because if there is nothing to degrade into detritus, the latter will slowly decompose and turn into basic nutrients.

This paper is mainly based on the earlier one [29], but the functional response forms used here are the Holling II and III in place of their Holling types I and II counterparts. Further, we have considered stochastic perturbation terms built directly into the growth equations, in order to incorporate the effects of a randomly fluctuating environment. In [29] instead noise was introduced just in the harvest rate h and does not incorporate additional food sources for the autotroph biomass, that we instead further take into consideration. For these reasons, both our deterministic and stochastic results differ from those of [29]. In what follows in particular we further outline the investigations performed on the roles that varying the model parameters has, either just considering changes in one of them or instead by combining some of them.

Firstly, the deterministic model is studied analytically and the threshold values for the feasibility and stability of the three possible steady states are assessed, see Table 1. These are the nutrient-only, the herbivore-free, and the coexistence equilibria. They are related to each other by transcritical bifurcations, when the system parameter values satisfy suitable threshold conditions. In particular, the nutrient-only point can be obtained from the herbivore-free state if the nutrients fall below a critical value. Note that in [17] also, low nutrient concentrations imply the disappearance of plankton.

Furthermore, analytical results for a Hopf bifurcation at the coexistence equilibrium are determined. Numerical evidence supports this finding. The persistent oscillations are shown to occur for changes in various parameters. The bifurcation diagrams of Fig. 4 show that increasing values of the nutrient input as well as of its dilution rate lead to sustained population oscillations. Similar results are also found in the literature, but simulated eutrophication is obtained with further increase, for instance in the nutrient input. These phenomena are also found in other simpler models and in their delayed counterparts, [34]; they induce oscillations ultimately leading to ecosystem collapse, [17]. The same outcome occurs if instead we decrease the harvesting rate. High exploitation values of the herbivore population instead tend to deplete the system of this resource. Thriving of the latter in these conditions can be ensured if additional food for autotroph biomass becomes available. Thus increasing the intrinsic growth rate of the autotroph trophic level may prevent the herbivore population extinction.

These results in practice imply that to control the plankton population and to maintain stability of the coexistence equilibrium, we have to control the constant nutrient input, the dilution rate of the nutrients, and the harvesting rate of the herbivores. These are especially influenced by artificial eutrophication. But note that other considerable agents responsible for nutrient addition to the aquatic ecosystem are geological weathering and inputs from ocean upwelling.

We now compare the deterministic results of this research with those of the investigations involving Holling-type II response functions, e.g., [33], where toxin-producing phytoplankton and zooplankton are considered. Environmental noise for the growth rates of toxic-phytoplankton and zooplankton is assumed. Low values of noise intensities are found to lead to stochastic asymptotic stability of the system. On the other hand, slightly higher values of noise intensities give rise to oscillations with large amplitudes that become larger with increasingly higher noise intensities ultimately entailing extinction of both plankton populations. Comparison with our results seems to suggest that the different assumptions on the response functions imply drastically different system's behaviors.

We have also found that stochastically our system is stable in suitable conditions involving both the model parameters as well as the maximum size of the environmental random fluctuations. In particular, the latter must remain within certain threshold values to prevent possible unpredictable consequences in the system's behavior. Exogenous influences such as weather harsh conditions and climate changes may deeply affect these ecosystems and should therefore be monitored, to assess suitable harvesting policies that at least in the medium term contrast the ecosystem collapse.

In this respect, our conclusions agree with those in [26], where nonautonomous models are proposed in which a seasonal change in temperature is taken into account. Specifically, the phytoplankton growth rate, two zooplanktons' relevant parameters, the mortality and the ratio of grazed biomass to new herbivores are all assumed to be time-dependent. This model predicts indeed that low nutrient concentrations prevent harmful algal blooms. Also, in [26] the "predator's average growth rate" threshold is introduced, depending on nutrients availability and on the predator's viability, which helps in assessing whether the ecosystem will persist. It essentially unveils the role of zooplankton mortality on the ecosystem behavior. Our simulations reveal a similar behavior in the deterministic version of our model, because Hopf bifurcations are triggered by increasing zooplankton harvesting, although their sizes are apparently not too large and the minima keep away from zero. In [22], similar conclusions are obtained by higher toxin production rates, which induce higher zooplankton mortality. The stochastic variant of our system instead shows equilibrium destabilization and increasing oscillations amplitudes. The latter could become dangerous for the ecosystem survival if their minima achieve very small values. This is in line with the findings of a delayed stochastic model, [37], involving two plankton populations that release toxins only in the presence of the other population, in which competition is modeled via a simple Holling-type I interaction and white noise stochastic environmental parameter fluctuations are introduced: ecosystem extinction has indeed been assessed. Similar results hold considering as bifurcation parameter the nutrient input rate, and are matched by similar observations in [26]. This holds both qualitatively and quantitatively, as the bifurcation in both cases occurs for almost the same value of the nutrient input rate.

One final word comes from the various plankton models that appeared in the literature. Since many of them have a similar structure but mainly differ on the equilibria values, it would be of help if from laboratory and field experiments carried out by ecologists and biologists these equilibrium values may be assessed. From them the missing parameter values might then be evaluated. In this way, it would perhaps be possible to decide the most suitable form of the model for the ecosystem under consideration.

Acknowledgements The research of Samares Pal is supported by UGC, New Delhi, India Ref. No. MRP-MAJ-MATH-2013-609. The research of Ezio Venturino has been partially supported by the project "Metodi numerici nelle scienze applicate" of the Dipartimento di Matematica "Giuseppe Peano".

References

1. V.N. Afanas'ev, V.B. Kolmanowskii, V.R. Nosov, *Mathematical Theory of Control Systems Design* (Kluwer Academic, Dordrecht, 1996)
2. M. Bandyopadhyay, J. Chattopadhyay, Ratio-dependent predator-prey model: Effect of environmental fluctuation and stability. Nonlinearity **18**, 913–936 (2005)
3. E. Beretta, V.B. Kolmanowskii, L. Shaikhet, Stability of epidemic model with time delays influenced by stochastic perturbations. Math. Comput. Simul. **45**(3–4), 269–277 (1998)
4. F. Brauer, A.C. Soudack, Stability regions in predator-prey systems with constant rate prey harvesting. J. Math. Biol. **8**, 55–71 (1979)
5. S. Chakraborty, J. Chattopadhyay, Nutrient-phytoplankton-zooplankton dynamics in the presence of additional food source — A mathematical study. J. Biol. Syst. **16**(4), 547–564 (2008)
6. K. Chakraborty, K. Das, Modeling and analysis of a two-zooplankton one-phytoplankton system in the presence of toxicity. Appl. Math. Model. **39**(3–4), 1241–1265 (2015)
7. K. Chakraborty, S. Das, T.K. Kar, Optimal control of effort of a stage structured prey-predator fishery model with harvesting. Nonlinear Anal Real World Appl. **12**(6), 3452–3467 (2011)
8. K. Chakraborty, M. Chakraborty, T.K. Kar, Optimal control of harvest and bifurcation of a prey-predator model with stage structure. Appl. Math. Comput. **217**(21), 8778–8792 (2011)
9. A. Chatterjee, S. Pal, Effect of dilution rate on the predictability of a realistic ecosystem model with instantaneous nutrient recycling. J. Biol. Syst. **19**, 629 (2011)
10. A. Chatterjee, S. Pal, Role of constant nutrient input in a detritus based open marine plankton ecosystem model. Contemp. Math. Stat. **2**, 71–91 (2013)
11. A. Chatterjee, S. Pal, S. Chatterjee, Bottom up and top down effect on toxin producing phytoplankton and its consequence on the formation of plankton bloom. Appl. Math. Comput. **218**, 3387–3398 (2011)
12. C.W. Clark, *Mathematical Bioeconomics: The Optimal Management of Renewable Resources*, 2nd edn. (Wiley Interscience, New York, 1990)
13. G. Dai, M. Tang, Coexistence region and global dynamics of a harvested predator-prey system. SIAM J. Appl. Math. **58**, 193–210 (1998)
14. T. Das, R.N. Mukherjee, K.S. Chaudhuri, Harvesting of a prey-predator fishery in the presence of toxicity. Appl. Math. Model. **33**(5), 2282–2292 (2009)
15. M.R. Droop, Vitamin B12 in marine ecology. Nature **180**, 1041–1042 (1957)
16. A.M. Edwards, J.Brindley, Oscillatory behaviour in a three-component plankton population model. Dyn. Stab. Syst. **11**(4), 347–370 (1996)
17. A. Fan, P. Han, K. Wang, Global dynamics of a nutrient-plankton system in the water ecosystem. Appl. Math. Comput. **219**, 8269–8276 (2013)
18. E. González-Olivares, A. Rojas-Palma, Multiple limit cycles in a Gause type predator-prey model with holling Type III functional response and Allee effect on prey. Bull. Math. Biol. **73**, 1378–1397 (2011)

19. E. González-Olivares, P.C. Tintinago-Ruiz, A. Rojas-Palma, A Leslie-Gower type predator-prey model with sigmoid functional response. Int. J. Comput. Math. **92**, 1895–1909 (2015)
20. B.D. Hassard, N.D. Kazarinoff, Y.H. Wan, *Theory and Application of Hopf Bifurcation* (Cambridge University Press, Cambridge, 1981)
21. Z. Hu, Z. Teng, L. Zhang, Stability and bifurcation analysis of a discrete predator-prey model with nonmonotonic functional response. Nonlinear Anal Real World Appl. **12**(4), 2356–2377 (2011)
22. S.R.J. Jang, E.J. Allen, Deterministic and stochastic nutrient-phytoplankton-zooplankton models with periodic toxin producing phytoplankton. Appl. Math. Comput. **271**, 52–67 (2015)
23. M.Y. Li, J.S. Muldowney, Global Stability for the SEIR model in epidemiology. Math. BioSci. **125**, 155–164 (1995)
24. M.Y. Li, H. Shu, Global dynamics of an in-host viral model with intracellular delay. Bull. Math. Biol. **72**, 1492–1505 (2010)
25. Y. Li, D. Xie, J. A. Cui, The effect of continuous and pulse input nutrient on a lake model. J. Appl. Math. **2014**, Article ID 462946 (2014)
26. J. Luo, Phytoplankton-zooplankton dynamics in periodic environments taking into account eutrophication. Math. BioSci. **245**, 126–136 (2013)
27. A. Martin, S. Ruan, Predator-prey models with delay and prey harvesting. J. Math. Biol. **43**, 247–267 (2001)
28. A.Y. Morozov, Emergence of Holling type III zooplankton functional response: bringing together field evidence and mathematical modelling. J. Theor. Biol. **265**, 45–54 (2010)
29. B. Mukhopadhyay, R. Bhattacharyya, On a three-tier ecological food chain model with deterministic and random harvesting: a mathematical study. Nonlinear Anal Model. Control **16**(1), 77–88 (2011)
30. M.R. Myerscough, B.F. Gray, W.L. Hogarth, J. Norbury, An analysis of an ordinary differential equation model for a two-species predator-prey system with harvesting and stocking. J. Math. Biol. **30**, 389–411 (1992)
31. S. Pal, A. Chatterjee, Coexistence of plankton model with essential multiple nutrient in chemostat. Int. J. Biomath. **6**, 28–42 (2013)
32. S. Pal, S. Chatterjee, J. Chattopadhyay, Role of toxin and nutrient for the occurrence and termination of plankton bloom – Results drawn from field observations and a mathematical model. Biosystems **90**, 87–100 (2007)
33. F. Rao, The complex dynamics of a stochastic toxic- phytoplankton- zooplankton model. Adv. Difference Equ. **2014**, 22 (2014)
34. S. Ruan S, Oscillations in Plankton models with nutrient recycling. J. Theor. Biol. **208**, 15–26 (2001)
35. Y. Sekerci, S. Petrovskii, Mathematical modelling of spatiotemporal dynamics of oxygen in a plankton system. Math. Model. Nat. Phenom. **10**(2), 96–114 (2015)
36. A. Sen, D. Mukherjee, B.C. Giri, P. Das, Stability of limit cycle in a prey-predator system with pollutant. Appl. Math. Sci. **5**(21), 1025–1036 (2011)
37. P.K. Tapaswi, A. Mukhopadhyay, Effects of environmental fluctuation on plankton allelopathy. J. Math. Biol. **39**, 39–58 (1999)

Modeling the Endophytic Fungus *Epicoccum nigrum* Action to Fight the "Olive Knot" Disease Caused by *Pseudomonas savastanoi* pv. *savastanoi* (*Psv*) Bacteria in *Olea europaea* L. Trees

Cecilia Berardo, Iulia Martina Bulai, Ezio Venturino, Paula Baptista, and Teresa Gomes

1 Introduction

Olive knot, caused by *Pseudomonas savastanoi pv. savastanoi* (*Psv*), is one of the most important diseases of olive crop [11]. This bacterial species produces tumorous galls or knots, mostly on stems and branches of olive trees, causing their death and loss of tree vigor, and thus endangering olive harvest [11]. *Psv* does not survive for long in soil, being usually found in olive tree phyllosphere as an epiphyte [10] and/or endophyte [7]. Disease development is shown to be dependent on several factors, such as concentration of *Psv* at infection sites and their interaction with other microorganisms [11]. These interacting microorganisms are usually found as epiphytic but can also occur in the knots [6, 8, 9]. When found in olive knots together with *Psv*, some of them have been shown either to depress *Psv* growth or to produce an increase in knot size [6]. Since olive knot is difficult to control, prevention being the only reliable strategy, information of naturally occurring antagonistic microorganisms with capacity to suppress *Psv* is of great importance. In fact, the use of these antagonists as biological control agents against *Psv* could be a promising tool to reduce olive knot incidence on olive crops. The use of such biocontrol agents meets many of the European policies aiming at moving toward more sustainable crop production systems (Directive 2009/128/EC) and follows the "Guidelines for Integrated Production of Olives" published by IOBC/WPRS [5]. In line with this, we have started studying the endophytic fungal community associated to *Psv* in the phyllosphere of Portuguese olive tree cultivars, and their capacity to antagonize

C. Berardo · I. M. Bulai · E. Venturino (✉)
Dipartimento di Matematica "Giuseppe Peano", Università di Torino, Torino, Italy
e-mail: iuliam@live.it; ezio.venturino@unito.it

P. Baptista · T. Gomes
CIMO, School of Agriculture, Polytechnic Institute of Bragança, Bragança, Portugal
e-mail: pbaptista@ipb.pt

© Springer International Publishing AG, part of Springer Nature 2018
R. P. Mondaini (ed.), *Trends in Biomathematics: Modeling, Optimization and Computational Problems*, https://doi.org/10.1007/978-3-319-91092-5_13

Psv under in vitro conditions. Antagonism tested on solid media with agar overlays showed an inhibition zone for the endophyte *Epicoccum nigrum* and its supernatant was showed to reduce *Psv* growth/biomass around 96%, after 48 h of incubation (unpublished data).

Mathematical models in biology and ecology have become of widespread use nowadays. In particular, it is possible to model population interactions. Two most common features in these interactions are the classical predator–prey systems introduced by Lotka and Volterra, about a century ago, [4, 13], and competition systems, first studied experimentally in [2], see also the classical monography [14]. However, population communities also can benefit from their associations, leading to diverse phenomena such as mutualism, amensalism, and symbiosis. Some investigations of mathematical nature in the latter framework have recently been performed [1, 3, 12]. The ecological setting described above shows that in the context of the olive tree all the three types of mutual relationships described in this paragraph simultaneously occur. In order to begin to better understand the effect of the resident fungus (*E. nigrum*) in the *Psv* development, we formulate and develop a mathematical model, taking into account the interactions olive tree–*Psv*–*E. nigrum*. This model helps in elucidating the role of the endophytic fungi on the spread and severity of olive knot disease.

The paper is organized as follows. In the next section, we formulate the model, while the following section contains the equilibria analysis. Numerical simulations are then carried out and a final discussion concludes the paper. The appendices contain some more technical mathematical details.

2   The Model

We consider a single olive tree (*Olea europaea*) that is affected by the olive knot disease caused by the bacterium *Psv*. We assume the presence of the endophytic fungus *E. nigrum* on the olive tree, having a double effect: a positive one on the tree, and another one at large on the environment, removing bacteria. In doing so, the endophytic fungus also receives a benefit, by gaining indirectly more space on the plant and directly more food from the plant that they inhabit.

The four populations that are modeled in the ecosystem, all measurable by biomass (or extent of their surface), are:

- S: the healthy branches of the olive tree;
- I: the branches of the same olive tree that are infected by bacteria;
- B: the pathogenic bacterium *Psv*, attacking and infecting the olive branches;
- N: the endophytic fungus *E. nigrum*, that removes the *Psv* bacterium B, with this action benefiting by getting more space for their own growth and also more nutrients from the plant; with this behavior, they also exert a beneficial effect on the healthy parts of the plant, S.

The model, in which all the parameters represent nonnegative quantities, reads:

$$\frac{dS}{dt} = s \left(1 - \frac{S+I}{K} \right) S - \lambda SB + bNS \tag{1}$$

$$\frac{dI}{dt} = \lambda SB - qIB - s\frac{S+I}{K}I - gI$$

$$\frac{dB}{dt} = hqIB - aNB - mB - rB^2$$

$$\frac{dN}{dt} = ebNS + uaNB - nN - pN^2.$$

In the first equation, the evolution of the tree's healthy branches is modeled. They reproduce following a logistic growth, with net biomass production rate s and carrying capacity K and become infected at rate λ by the action of the bacterium. We assume that there is well mixing of the bacteria in all parts of the tree because they are transported by the wind and the rain. In this way, their interaction with the healthy parts of the tree is modeled via a mass action term, this fact in epidemiological terms corresponding to the homogeneous mixing λSB. The last term expresses the fact that they get benefit at rate b from the endophytic fungi, with which they have a beneficial relationship.

The second equation for the infected branches shows firstly that they become so when they are attacked by the bacterium, secondly they also suffer the action of the bacteria B at rate q, then they are also subject to intraspecific competition for space and nutrients from other healthy and infected branches, and finally experience an additional, disease-related, mortality at rate g. However, the pathogenic bacteria and endophytic fungus remain in dead branches and thus are still part of the tree ecosystem.

In the third equation, we model the bacterium. It gets nutrients from the infected branches, as already mentioned at rate q, with a conversion factor $h < 1$. The second term indicates that they are killed by the *E. nigrum* at rate a, and the third one that their natural mortality rate is m. Bacteria can die also by intraspecific competition at rate r.

The endophytic fungus is modeled in the fourth equation. They get benefit from their relationship with the healthy branches, at rate b, scaled by a factor $e < 1$, and also by killing the bacterium at rate a, because in this way they get more space for growth, and indirectly also more nutrients. Here, $u < 1$ represents another conversion factor. In the last two terms, we model their removal from the ecosystem: *E. nigrum* naturally die at rate n and experience also intraspecific competition at rate p.

The Jacobian of (1) is

$$J = \begin{bmatrix} J_{11} & -\frac{s}{K}S & -\lambda S & bS \\ \lambda B - \frac{s}{K}I & J_{22} & \lambda S - qI & 0 \\ 0 & hqB & J_{33} & -aB \\ ebN & 0 & uaN & J_{44} \end{bmatrix} \tag{2}$$

where

$$J_{11} = s - \frac{2s}{K}S - \frac{s}{K}I - \lambda B + bN, \quad J_{22} = -qB - g - \frac{s}{K}S - \frac{2s}{K}I,$$

$J_{33} = hqI - aN - m - 2rB$, $J_{44} = ebS + uaB - n - 2pN$.

To study the local stability of the equilibria, we evaluate (2) at each equilibrium point and compute the sign of the eigenvalues of the characteristic polynomials.

3 The Ecosystem's Steady States

By solving the system equilibrium equations, we find the following five possible feasible equilibria.

There are the origin $E_0 = (0, 0, 0, 0)$ and the healthy-tree only equilibrium $E_1 = (K, 0, 0, 0)$, which are both always feasible.

The eigenvalues for E_0 are $-g < 0$, $-m < 0$, $-n < 0$, and $s > 0$, showing its unconditional instability.

The corresponding eigenvalues of E_1 are $-s < 0$, $-m < 0$, $-g - s < 0$, and $ebK - n$, giving the following condition ensuring the stability of the equilibrium:

$$ebK < n. \tag{3}$$

Then, there is the equilibrium in which the endophytic fungi are absent:

$$E_2 = \left(K - \left(1 + \frac{\lambda hqK}{rs} \right) I_2 + \frac{\lambda mK}{rs}, I_2, \frac{hq}{r}I_2 - \frac{m}{r}, 0 \right) \tag{4}$$

where I_2 is the positive root of the quadratic equation:

$$U_2 I_2^2 + V_2 I_2 + Z_2 = 0, \quad U_2 = -\frac{\lambda^2 h^2 q^2 K}{r^2 s} - \frac{hq^2}{r},$$

$$V_2 = \frac{\lambda K hq}{r} + 2\frac{\lambda^2 hqKm}{r^2 s} + \frac{mq}{r} - g - s,$$

$$Z_2 = -\frac{\lambda K m}{r} - \frac{\lambda^2 m^2 K}{r^2 s} \tag{5}$$

with $U_2 < 0$ and $Z_2 < 0$. For a nonnegative root, by Descartes' rule of signs, we need the conditions:

$$V_2 > 0, \quad V_2^2 \geq 4U_2 Z_2.$$

Moreover, the conditions:

$$hqI_2 \geq \frac{m}{r}, \quad K + \frac{\lambda mK}{rs} > \left(1 + \frac{\lambda hqK}{rs} \right) I_2$$

ensure the nonnegativity of the populations B_2 and S_2 at this point and therefore impose an interval in which the infected branches must lie. The feasibility conditions are then given by:

$$V_2 > 0, \quad V_2^2 \geq 4U_2Z_2, \quad \frac{m}{hq} < I_2 < \frac{K(rs + \lambda m)}{rs + \lambda hqK}. \tag{6}$$

For E_2, one eigenvalue is explicit, given by $-\frac{sI_2}{K} - \lambda S_2 B_2$. It is always negative, with S_2, I_2, and B_2 given in (4), while the other three eigenvalues are the roots of the cubic equation $\mu^3 + R_1\mu^2 + R_2\mu + R_3 = 0$, with R_1, R_2, and R_3 given in Appendix 5. Here, stability is ensured by the Routh–Hurwitz conditions:

$$R_1 > 0, \quad R_3 > 0, \quad R_1R_2 > R_3. \tag{7}$$

Next, there is the point with no infection, i.e., where the tree is healthy and the phyllosphere microorganisms thrive in it,

$$E_3 = \left(\frac{K(bn - sp)}{Keb^2 - sp}, 0, 0, \frac{s(n - Keb)}{Keb^2 - sp} \right).$$

It is feasible if either one of the following sets of conditions holds:

$$\max \left\{ \frac{sp}{n}, \sqrt{\frac{sp}{Ke}} \right\} < b < \frac{n}{Ke} \tag{8}$$

or

$$\frac{n}{Ke} < b < \min \left\{ \frac{sp}{n}, \sqrt{\frac{sp}{Ke}} \right\}. \tag{9}$$

These conditions ensure the nonnegativity of the populations at this point.

Two of the eigenvalues of $J(E_3)$ are explicitly evaluated as $-g - sK^{-1}S_3 < 0$, $-m - aN_3 < 0$. The remaining two eigenvalues are the roots of the quadratic equation:

$$\mu^2 - \text{tr}(Q) + \det(Q) = 0$$

with

$$-\text{tr}(Q) = \frac{s}{K}S_3 + pN_3 > 0, \quad \det(Q) = \frac{s}{K}pS_3N_3 - b^2eS_3N_3.$$

By the Routh–Hurwitz criterion, the stability of E_3 reduces to ensuring that $\det(Q) > 0$, that is:

$$sp > Kb^2e. \tag{10}$$

Note that condition (8) is the opposite of the stability condition (10) for E_3. Thus, whenever (8) holds, the equilibrium E_3 is unstable. Conversely, in case of (9), condition (10) is verified and E_3 is stable.

Further, by comparing the stability condition (3) of E_1 with the feasibility condition (9) of E_3, a transcritical bifurcation is seen to connect these two equilibria.

Finally, the coexistence equilibrium is also a possible outcome:

$$E^* = \left(\frac{sK - sI^* - k\lambda B^* + KbN^*}{s}, I^*, B^*, \frac{hqI^* - rB^* - m}{a} \right). \qquad (11)$$

It is obtained with the following steps:

- Step 1: S^* in (11) is obtained solving the first equilibrium equation of (1). The population is nonnegative if $sK + KbN^* > sI^* + K\lambda B^*$.
- Step 2: From the third equation in (1), we obtain N^*. This population is nonnegative if $hqI^* > rB^* + m$.
- Step 3: We evaluate the fourth equation in (1) using the values of S^* and N^* obtained in Steps 1 and 2, to obtain an equation for I^* and B^*, that we consider a quadratic equation in I^*:

$$eb \cdot \frac{hqI^* - rB^* - m}{a} \cdot \frac{sK - sI^* - K\lambda B^* + KbN}{s}$$

$$+ uaB^* \cdot \frac{hqI^* - rB^* - m}{a} - \frac{nhqI^* - nrB^* - nm}{a}$$

$$- p \cdot \left(\frac{hqI^* - rB^* - m}{a} \right)^2 = 0. \qquad (12)$$

Its roots provide the following two values for I^*:

$$I_1^* = \frac{b^*r + m}{hq}, \quad I_2^* = \frac{A_B B^* + A_C}{A_I}, \qquad (13)$$

with

$$A_B = -abeK\lambda - b^2 eKr + a^2 su + pru, \qquad (14)$$

$$A_C = abeKs - b^2 eKm - ans + mps,$$

$$A_I = -b^2 ehKq + abes + hpqs.$$

We concentrate only on the case of I_1^*, for which, in view of the nonnegativity of the parameters involved in (13), clearly no additional conditions need to be imposed to guarantee its nonnegativity.

- Step 4: We evaluate the second equilibrium equation of (1) using S^* and I_1^* to get the following second degree equation in B^*:

$$U_1 B^{*2} + V_1 B^* + Z_1 = 0, \qquad (15)$$

where the coefficients U_1, V_1, and Z_1, in view of their complexity, are explicitly deferred to the Appendix 5.

The quadratic equation (15) has real roots if and only if $V_1^2 - 4U_1Z_1 \geq 0$. Further, for feasibility of just one value of B^* we need exactly one sign variation, which is ensured by one of the alternative situations: $U_1V_1 < 0$ or $V_1Z_1 < 0$ or $U_1Z_1 < 0$. If instead U_1 and Z_1 have the same sign and V_1 the opposite one, both roots of (15) are positive. To sum up, the following conditions ensure the feasibility of E^*:

$$sK + KbN^* > sI^* + K\lambda B^*, \quad hqI^* > rB^* + m, \quad V_1^2 - 4U_1Z_1 \geq 0. \quad (16)$$

Stability of the coexistence equilibrium E^* is guaranteed by the Routh–Hurwitz conditions

$$P_3 > 0, \quad P_4 > 0, \quad P_1P_2P_3 > P_3^3 + P_1^2P_4, \quad (17)$$

where the above quantities are the coefficients of the characteristic polynomial of $J(E^*)$

$$\mu^4 + P_1\mu^3 + P_2\mu^2 + P_3\mu + P_4. \quad (18)$$

Although these coefficients are known explicitly, they are not reported here because their representations are rather complicated.

4 Numerical Simulations

The ecosystem (1) embeds a symbiotic subsystem among the S and N populations. It is well known that this classical two-populations' mathematical model can lead to unbounded, ecologically unrealistic, growth, if its isoclines are suitably chosen. We now investigate the assumptions that need to be made on the ecosystem parameters in order to avoid this phenomenon, that might obscure our findings in the simulations that we will perform for the larger envisaged ecosystem (1).

From the first and the fourth equilibrium equations of the model, we obtain the symbiotic subsystem isoclines by imposing the parameters λ and h to vanish:

$$s - \frac{s}{K}S + bN = 0 \quad (19)$$
$$ebS - n - pN = 0.$$

They intersect at the coexistence point

$$\left(\frac{K(nb - sp)}{Keb^2 - sp}, \quad \frac{s}{b}\left(-1 + \frac{nb - sp}{Keb^2 - sp} \right) \right), \quad (20)$$

for which the nonnegativity of the coordinates of the intersection of the two curves gives the two sets of alternative feasibility conditions, (8) and (9), already found for the feasibility of E_3, and that are assumed from now on.

We want to control the spread of the disease in the olive tree and pay attention to the scenarios obtained varying mainly the disease transmission rate λ. Most importantly, for the parameters a and m, we use the fixed values $m = 0.183$ (the time unit is taken to be hours) and $a = 0.021$ (time unit in hours) provided by our laboratory experiments, yet unpublished. The parameter b is not experimentally known and thus it is chosen to satisfy the conditions (8) and (9). To run the simulations, the intrinsic ode45 routine of Matlab2016a and the bifurcation software XPPAUT are used.

For the remaining set of parameters, we take hypothetical reference values

$$s = 4.24476, \quad b = 0.2, \quad q = 9.24772, \quad g = 6.42079, \tag{21}$$

$$h = 0.226653, \quad m = 0.183, \quad a = 0.021, \quad e = 0.9,$$

$$n = 1.045, \quad u = 0.214479, \quad K = 50, \quad r = 3.95804, \quad p = 0.8 .$$

Using the initial conditions,

$$S(0) = 7.0300, \quad I(0) = 7.5415, \quad B(0) = 5.4729, \quad N(0) = 5.5348, \tag{22}$$

we construct the one-parameter bifurcation diagrams for the four populations S, I, B, and N, reported in Figs. 1 and 2. In view of the constraints (8) and (9), the parameter b can lie only in the rather narrow approximate range $[0.023, 0.2747]$.

In the first diagram for the population S, we can observe that for $0.5 \lesssim \lambda \lesssim 11.86$ five equilibrium points are feasible: for $0.5 \lesssim \lambda \lesssim 4.83$, one of the two coexistence equilibria is stable, given by the line (b), while the lines (a), (c), and (d) denote the unstable equilibria. The same equilibrium points can be noted in the other bifurcation diagrams for I, B, and N, denoted with the same letters. In the second diagram of Fig. 2, for $\lambda \gtrsim 4.83$, the population N collapses to 0 and E_2, the line (d), becomes stable after a transcritical bifurcation (TCB), which occurs between (b) and (d) for $\lambda \approx 4.83$. A saddle-node bifurcation (SNB) occurring at $\lambda \approx 0.5$ separates the stable coexistence equilibrium (b) from the unstable one, denoted with (a). For values of λ larger than 11.86, the three nonvanishing populations S, I, and B in the solution of the system (1) oscillate, due to the presence of a stable limit cycle, whose maximum and minimum values are both plotted (line (e)). The Hopf bifurcation from which the limit cycle originates occurs for $\lambda \approx 11.9$ (HB). At $\lambda \approx 11.86$, another Hopf bifurcation leads to the presence of an unstable limit cycle, surrounding the stable one. It is however not represented in the bifurcation diagrams in order not to clutter the figures.

Figure 3 shows the results for higher values of λ, namely for $\lambda = 55$. The solutions for S, I, and N according to continuous time are characterized by high amplitude oscillations, approaching values dangerously close to zero, while the N population rapidly decays and ultimately vanishes.

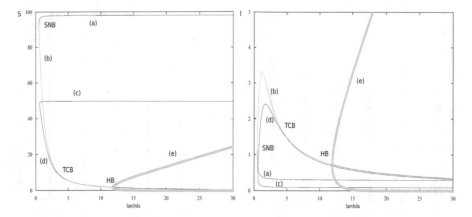

Fig. 1 Left: bifurcation diagram (λ, S). Right: bifurcation diagram (λ, I). Parameter values: $s = 4.24476$, $b = 0.2$, $q = 9.24772$, $g = 6.42079$, $h = 0.226653$, $m = 0.183$, $a = 0.021$, $e = 0.9$, $u = 0.214479$, $n = 1.045$, $K = 50$, $r = 3.95804$, $p = 0.8$. Initial conditions: $S(0) = 7.0300$, $I(0) = 7.5415$, $B(0) = 5.4729$, $N(0) = 5.5348$. HB: Hopf bifurcation point. TCB: transcritical bifurcation point. SNB: saddle-node bifurcation point. (a), (b), (c), (d): equilibrium points. (e): stable limit cycle

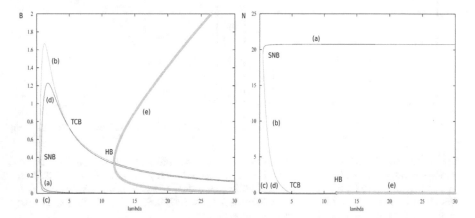

Fig. 2 Left: bifurcation diagram (λ, B). Right: bifurcation diagram (λ, N). Parameter values: $s = 4.24476$, $b = 0.2$, $q = 9.24772$, $g = 6.42079$, $h = 0.226653$, $m = 0.183$, $a = 0.021$, $e = 0.9$, $u = 0.214479$, $n = 1.045$, $K = 50$, $r = 3.95804$, $p = 0.8$. Initial conditions: $S(0) = 7.0300$, $I(0) = 7.5415$, $B(0) = 5.4729$, $N(0) = 5.5348$. HB: Hopf bifurcation point. TCB: transcritical bifurcation point. SNB: saddle-node bifurcation point. (a), (b), (c), (d): equilibrium points. (e): stable limit cycle

In Fig. 4, the trajectories for $\lambda = 0.2$ are shown: we note that for values of λ smaller than about 0.5 the system converges to the equilibrium point E_3. Consequently, the disease is eradicated due to the low value of the transmission rate. The healthy branches' population thrives to very high levels, higher than the carrying capacity, because of the beneficial effect of the endophytic fungus.

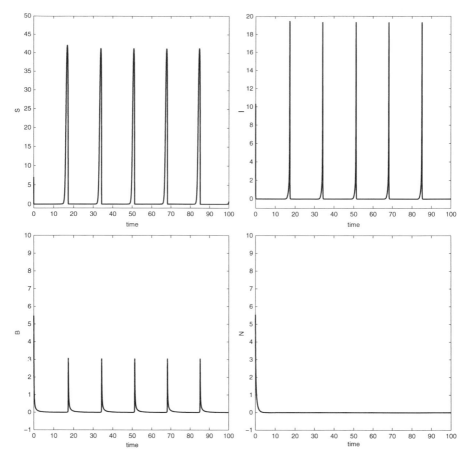

Fig. 3 The populations S, I, N, and B, shown in clockwise order as functions of time for the parameter values: $\lambda = 55$, $s = 4.24476$, $b = 0.2$, $q = 9.24772$, $g = 6.42079$, $h = 0.226653$, $m = 0.183$, $a = 0.021$, $e = 0.9$, $u = 0.214479$, $n = 1.045$, $K = 50$, $r = 3.95804$, $p = 0.8$. Initial conditions: $S(0) = 7.0300$, $I(0) = 7.5415$, $B(0) = 5.4729$, $N(0) = 5.5348$. Note that the zero value is located at higher level than the frames' bottom

Now, we perform some equilibria sensitivity analysis to variations in two parameters. This makes sense also in view of possible effects due to climate changes that may induce substantial changes in the ecosystem.

Firstly, Fig. 5 considers the pair $\lambda - g$, where the parameter g is the additional disease-related mortality rate of the infected part of the plant. For the same set of parameters and initial conditions used in Figs. 1 and 2, namely (21) and (22), but allowing the parameters to vary in the ranges $\lambda \in [0, 10]$ and $g \in [0, 10]$, we obtain the surfaces indicating the equilibrium value of each population. Observe that for low values of λ and values of g approximately higher than 6, the populations I and B vanish, while S and N thrive at rather high levels. In this region, the equilibrium

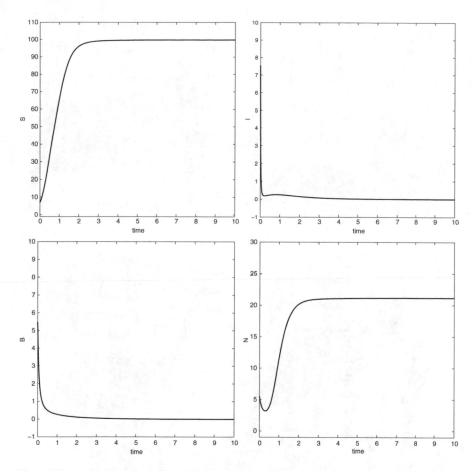

Fig. 4 The populations S, I, N, and B, shown in clockwise order as functions of time for the parameter values: $\lambda = 0.2$, $s = 4.24476$, $b = 0.2$, $q = 9.24772$, $g = 6.42079$, $h = 0.226653$, $m = 0.183$, $a = 0.021$, $e = 0.9$, $u = 0.214479$, $n = 1.045$, $K = 50$, $r = 3.95804$, $p = 0.8$. Initial conditions: $S(0) = 7.0300$, $I(0) = 7.5415$, $B(0) = 5.4729$, $N(0) = 5.5348$. Note that both infected branches and bacteria disappear, the zero value is indeed located at higher level than the frames' bottom

E_3 is stable and the disease is thus eradicated from the plant. This appears to be sensible, because the rate at which the disease spreads among the branches of the olive tree is low and the additional rate at which the infected part of the plant dies is high, in turn damaging the bacteria in the plant and causing the infected branches to disappear. For low values of both g and λ, instead the system converges to the coexistence equilibrium and when $\lambda \gtrsim 4.83$, the population of the endophytic fungi, N, disappears. Then, in such case the ecosystem converges to the endophytic fungi-free equilibrium E_2, but with alarmingly small values of the healthy parts of the plant.

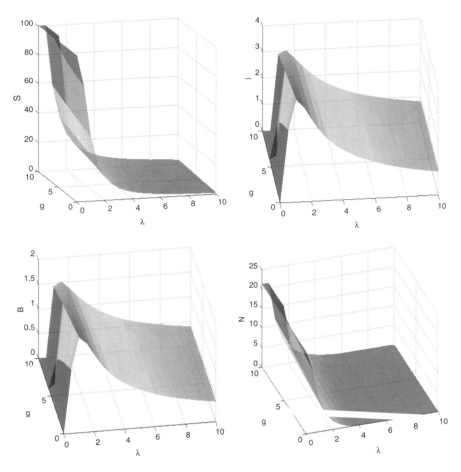

Fig. 5 The populations S, I, N, and B, shown in clockwise order as functions of the parameters λ and g. Remaining parameter values: $s = 4.24476$, $b = 0.2$, $q = 9.24772$, $h = 0.226653$, $m = 0.183$, $a = 0.021$, $e = 0.9$, $u = 0.214479$, $n = 1.045$, $K = 50$, $r = 3.95804$, $p = 0.8$. Initial conditions: $S(0) = 7.0300$, $I(0) = 7.5415$, $B(0) = 5.4729$, $N(0) = 5.5348$

In the two-parameters' bifurcation diagrams of Fig. 6, we consider the system's equilibria as function of the transmission rate and the rate at which the antagonistic fungi produce a beneficial effect on the plant, i.e., the pair of parameters $\lambda - b$, in the respective ranges $[0, 11]$ and $[0.023, 0.2747]$. For b, we impose the feasibility conditions for E_3 to be satisfied, given by (8) and (9). The system settles to the equilibrium point E_3 when the values of λ lie below about 0.5. This matches the analytical results for the equilibria stability, since b varies in the range in which the feasibility and stability conditions for E_3 are satisfied, and also the results of the previously analyzed one-parameter bifurcation diagram in Figs. 1 and 2. In this case, the bacteria and the infected parts of the plant disappear and only the healthy branches and the antagonistic fungi thrive in the ecosystem. Moreover, for low

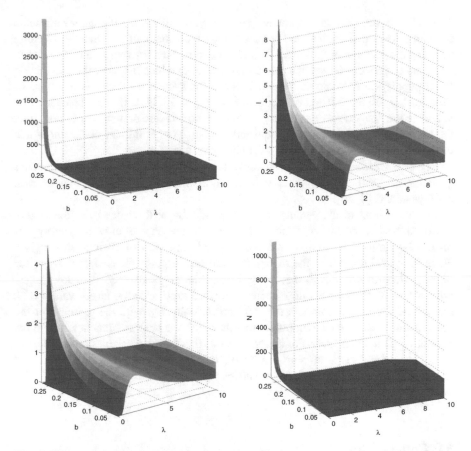

Fig. 6 The populations S, I, N, and B, shown in clockwise order as functions of the parameters λ and b. Remaining parameter values: $s = 4.24476$, $q = 9.24772$, $g = 6.42079$, $h = 0.226653$, $m = 0.183$, $a = 0.021$, $e = 0.9$, $u = 0.214479$, $n = 1.045$, $K = 50$, $r = 3.95804$, $p = 0.8$. Initial conditions: $S = 7.0300$, $I = 7.5415$, $B = 5.4729$, $N = 5.5348$

values of b the two species present in the system assume low values, while for values of b higher than approximately 0.25 they increase to biologically considerable high values. Figure 6 provides information on the range for the parameter b leading the system to settle onto the disease-free equilibrium E_3, favorable for the olive tree, where both the disease and the infected branches are eradicated. For $0.5 \lesssim \lambda \lesssim 4.83$, the system converges to a coexistence equilibrium point: N assumes low values, very close to zero. They appear a bit difficult to be clearly shown graphically, in the chosen range of the two-parameters' bifurcation diagram. For $\lambda \gtrsim 4.83$, instead the equilibrium E_2 with no endophytic fungus *E. nigrum* becomes stable. When λ is close to 4.83 and b approaches 0.2747, the infected branches and bacteria populations I and B attain their highest values, while S and N rapidly drop to values close to zero.

We finally consider the parameter pair $\lambda - q$, recalling that the latter represents the mortality rate of the infected branches due to the attack of the bacteria. Keeping on using the previously chosen set (21) of the parameters, for values of λ bigger than 11.9, the system converges to a stable limit cycle, in which however the endophytic fungi population N vanishes. Figure 7 shows the system solutions as functions of time for $\lambda = 20 > 11.9$ and $q = 5$, $q = 8$, and $q = 25$. Note that the change in the values of the disease-related mortality rate q of the infected branches does not affect the feasibility and stability of the equilibrium, but it rather influences the amplitude of the oscillations, which decreases for higher values of q.

Figure 8 shows the system behavior in terms of the variations of the parameter q in the range $[0, 20]$, together with the influence of the disease transmission coefficient $\lambda \in [0, 20]$.

For low values of the parameter q, independently of the value of λ, the system settles to the disease-free equilibrium E_3, with the presence of only the healthy part of the plant and of the antagonistic endophytic fungi. We can distinguish a region, close to the origin, where the favorable equilibrium point E_3 is stable. Then, for higher values of q and $\lambda \gtrsim 0.5$, the system achieves coexistence at steady values, with S and N showing a quick decrease toward zero. Finally, larger values of λ and q lead the system to attain the equilibrium point E_2, with the presence of the disease and of the infected branches, together with the healthy branches, where the endophytic fungi disappear. For larger values of the disease transmission coefficient, $\lambda > 11.9$, the ecosystem starts to oscillate. The surfaces with the highest and lowest peaks in the limit cycles are shown for the populations S, I, and B while N disappears.

5   Conclusions

We have proposed a model for studying how to fight by natural means the harmful bacteria that harbor in the olive trees.

From a biological point of view, the disease-free equilibrium point E_3 is the most relevant one. If the system settles at E_3, the healthy branches and the endophytic fungi survive, while the infected parts of the plant and the bacterium Psv disappear. Note that the plant benefits from the symbiotic action of the fungi, in that it thrives better, at higher levels, as shown in Fig. 4. However, for the healthy thriving of the tree, from the qualitative analysis, we obtain that the parameter b must satisfy the condition (9), to have E_3 both feasible and stable, but this implies a very narrow range of allowed values for b. In other words, the benefit that the endophytic fungi exert on the tree must be confined to a certain appropriate range.

The disease-and-endophytic fungi-free equilibrium point E_1 is also good for the plant, since it ensures anyway the survival of the healthy part of the olive tree plant. Here, without the helpful action of the endophytic fungi, the branches attain just their own carrying capacity.

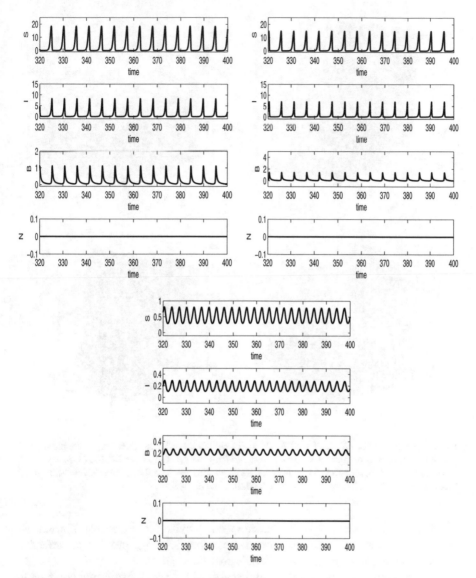

Fig. 7 Persistent limit cycles of the system. Top, left: $q = 5$. Top, right: $q = 8$. Bottom: $q = 25$. Remaining parameter values: $\lambda = 20$, $s = 4.24476$, $b = 0.2$, $g = 6.42079$, $h = 0.226653$, $m = 0.183$, $a = 0.021$, $e = 0.9$, $u = 0.214479$, $n = 1.045$, $K = 50$, $r = 3.95804$, $p = 0.8$. Initial conditions: $S = 7.0300$, $I = 7.5415$, $B = 5.4729$, $N = 5.5348$. Note that the solutions are plotted for time values in the range $[320, 400]$

On the other hand, there are equilibria that must possibly be avoided, E_2 and E^*, because they still harbor the harmful bacteria.

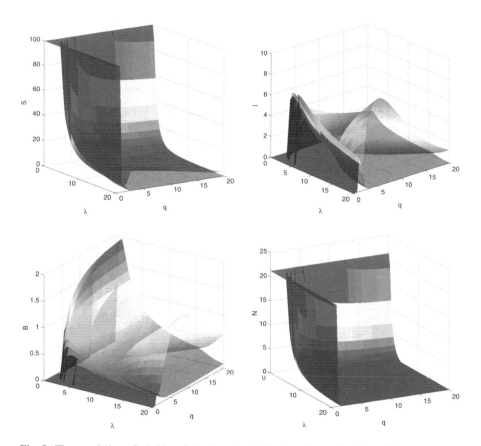

Fig. 8 The populations S, I, N, and B, shown in clockwise order as functions of the parameters λ and q. Remaining parameter values: $s = 4.24476$, $g = 6.42079$, $h = 0.226653$, $m = 0.183$, $a = 0.021$, $e = 0.9$, $u = 0.214479$, $n = 1.045$, $K = 50$, $r = 3.95804$, $p = 0.8$, $b = 0.2$. Initial conditions: $S = 7.0300$, $I = 7.5415$, $B = 5.4729$, $N = 5.5348$

First of all, observe that the endemic endophytic fungi-free equilibrium E_2 is dangerous for the plant, because the endophytic fungi provide an effective suppression mechanism for *Psv* bacteria's growth.

Further, note that a highly virulent disease, with large transmission coefficient, leads to populations' oscillations that are extremely close to vanishing levels, see Fig. 3. It is known that stochastic environmental fluctuations in such situations may lead to the ecosystem collapse and therefore ultimately to the tree's death.

An additional difficulty for obtaining a stable disease-free equilibrium is the fact that the disease transmission rate λ must be low and further it must assume values in a very small range. As shown in the two-parameters' bifurcation diagrams in

Fig. 5, the additional disease-related mortality rate must also be high, because in this way infected branches are bound to die fast and thus prevent the replication of the bacteria on the tree.

A final point is that also the disease-and-endophytic fungi-free equilibrium point E_1 is difficult to be achieved in practice, due to the fact that the stability condition hinges on high values of the parameter n, the fungi mortality, which is however known in general to be quite small.

If at all possible, field measurements of at least some of the above parameters might clarify whether they fall within the required ranges for disease eradication. Should this not be the case, our investigation shows that an alternative exists, although perhaps quite expensive, as it requires a rather continuous surveillance of the situation. The findings on the sensitivity analysis indeed indicate that a high infected branch mortality g helps in controlling the disease, providing a guideline on how to proceed for fighting the pathogens. This indeed can also be achieved by human-related external means, like pruning of the leaves and branches that appear to be disease-affected. This remark indicates that a constant close monitoring of the tree is necessary in order to keep it healthy.

Acknowledgements This research has been partially supported by The European COST Action: FA 1405—Food and Agriculture: Using three-way interactions between plants, microbes, and arthropods to enhance crop protection and production. The research of Ezio Venturino has been partially supported by the project "Metodi numerici nelle scienze applicate" of the Dipartimento di Matematica "Giuseppe Peano."

Appendix 1

$$R_1 = \frac{Khq I_2 + K\lambda S_2 B_2 - Km + sS_2 + sI_2}{K}$$

$$R_2 = -\frac{h^2\lambda q^2 S_2 I_2 - h^2 q^3 I_2^2 - h\lambda qr S_2 I_2 B_2 - h\lambda mq S_2 + hmq^2 I_2 + \lambda mr S_2 B_2}{r}$$

$$-\frac{h\lambda qs S_2 I_2 - hqrs S_2 I_2 - hqrs I_2^2 - \lambda rs S_2^2 B_2 + \lambda ms S_2 + mrs S_2 + mrs I_2}{rK}$$

$$R_3 = \frac{S_2(hq I_2 - m)(Kh^2\lambda^2 mq - h\lambda qrs S_2 + hq^2 rs I_2 + \lambda r^2 s S_2 B_2 - \lambda mrs)}{Kr^2}$$

Appendix 2

$$U_1 = \frac{\lambda\,(-aK\lambda - rbK)}{as} - \frac{r}{h} - s\left(\frac{-aK\lambda - rbK}{Khqas} + \frac{r}{h^2Kq^2}\right)r$$

$$V_1 = \frac{\lambda\left((hqx - m)bK + asK - asI_1^*\right)}{as} - \frac{m}{h} - \frac{gr}{hq}$$
$$-s\left(\frac{(hqI_1^* - m)bK + asK - asI_1^*}{hKqas} + \frac{m}{h^2Kq^2q}\right)r$$
$$-s\left(\frac{-aK\lambda - rbK}{hKqas} + \frac{r}{h^2Kq^2}\right)m$$

$$Z_1 = -\frac{gm}{hq} - s\left(\frac{(hqI_1^* - m)bK + sKa - asI_1^*}{saKhq} + \frac{m}{h^2Kq^2}\right)m$$

References

1. E. Caccherano, S. Chatterjee, L. Costa Giani, L. Il Grande, T. Romano, G. Visconti, E. Venturino, Models of symbiotic associations in food chains, in *Symbiosis: Evolution, Biology and Ecological Effects*, ed. by A.F. Camisão, C.C. Pedroso (Nova Science Publishers, Hauppauge, 2012), pp. 189–234
2. G.F. Gause, Experimental studies on the struggle for existence. J. Exp. Biol. **9**, 389–402 (1932)
3. M. Haque, E. Venturino, Mathematical models of diseases spreading in symbiotic communities, in *Wildlife: Destruction, Conservation and Biodiversity*, ed. by J.D. Harris, P.L. Brown (Nova Science Publishers, New York, 2009), pp. 135–179
4. A.J. Lotka, *Elements of Mathematical Biology* (Dover, New York, 1956)
5. C. Malavolta, D. Perdikis, IOBC technical guidelines III. Guidelines for Integrated Production of Olives. IOBC/WPRS Bulletin 77, pp. 1–19 (2012)
6. G. Marchi, A. Sisto, A. Cimmino, A. Andolfi, M. G. Cipriani, A. Evidente, G. Surico. Interaction between *Pseudomonas savastanoi pv. savastanoi* and *Pantoea agglomerans* in olive knots. Plant Pathol. **55**, 614–624 (2006)
7. G. Marchi, B. Mori, P. Pollacci, M. Mencuccini, G. Surico Systemic spread of *Pseudomonas savastanoi pv. savastanoi* in olive explants. Plant Pathol. **58**, 152–158 (2009)
8. H. Ouzari, A. Khsairi, N. Raddadi, L. Jaoua, A. Hassen, M. Zarrouk, D. Daffonchio, A. Boudabous, Diversity of auxin-producing bacteria associated to *Pseudomonas savastanoi*-induced olive knots. J. Basic Microbiol. **48**, 370–377 (2008)
9. J. M. Quesada, A. García, E. Bertolini, M.M. López, R. Penyalver, Recovery of *Pseudomonas savastanoi pv. savastanoi* from symptomless shoots of naturally infected olive trees. Int. Microbiol. **10**, 77–84 (2007)
10. J.M. Quesada, R. Penyalver, J. Pérez-Panadés, C.I. Salcedo, E.A. Carbonell, M.M. López, Dissemination of *Pseudomonas savastanoi pv. savastanoi* populations and subsequent appearance of olive knot disease. Plant Pathol. **59**, 262–269 (2010)

11. J.M. Quesada, R. Penyalver, M.M. López, Epidemiology and control of plant diseases caused by phytopathogenic bacteria: the case of olive knot disease caused by *Pseudomonas savastanoi pv. savastanoi*, in *Plant Pathology*, ed. by C.J. Cumagun (InTech, 2012). https://doi.org/10.5772/32544. ISBN: 978-953-51-0489-6
12. E. Venturino, How diseases affect symbiotic communities. Math. Biosci. **206**, 11–30 (2007)
13. V. Volterra, U. D'Ancona, La concorrenza vitale tra le specie dell'ambiente marino, VIIe Congr. Int. acquicult et de pêche, Paris 1–14 (1931)
14. P. Waltman, *Competition Models in Population Biology*. SIAM CBMS-NSF Regional Conference Series in Applied Mathematics (SIAM, Philadelphia, 1983)

Numerical Modeling of Transcranial Ultrasound

I. B. Petrov, A. V. Vasyukov, K. A. Beklemysheva, A. S. Ermakov,
A. O. Kazakov, Y. V. Vassilevski, V. Y. Salamatova, A. A. Danilov,
G. K. Grigoriev, and N. S. Kulberg

Arterial aneurysm is one of the most dangerous diseases of cerebral vessels. In total, about 70% of patients with ruptured arterial aneurysms of cerebral vessels die from primary or repeated hemorrhages. A vast majority of arterial aneurysms are located on the arteries of the base of the brain. In a pathoanatomic study of corpses of people who died from various causes, arterial aneurysms of cerebral vessels are found in 1–5% of cases.

It is considered that the first arterial aneurysm of cerebral vessels was discovered by Morgagni more than 200 years ago (1761), but only after the introduction of the clinical practice of cerebral angiography [13] this disease of the brain vessels was well studied and began to be diagnosed in living patients. Attempts were made to surgically treat arterial cerebral aneurysms [5, 8].

The ultrasound is capable of determining both the shape of vessels and the direction of the blood flow in real time. It is a very cheap, common, and efficient tool for detecting the early treatable stages of aneurysm, but its application for the cerebral vessels is very limited at the moment. The human skull is mostly impenetrable by ultrasound waves due to the cancellous bone tissue which consumes

I. B. Petrov · A. V. Vasyukov · K. A. Beklemysheva (✉) · A. S. Ermakov · A. O. Kazakov
Moscow Institute of Physics and Technology, Dolgoprudny, Russia
e-mail: petrov@mipt.ru; amisto@ya.ru

Y. V. Vassilevski · V. Y. Salamatova · A. A. Danilov
Institute of Numerical Mathematics of the RAS, Moscow, Russia

Moscow Institute of Physics and Technology, Dolgoprudny, Russia

G. K. Grigoriev
MGTS Medical and Health Center, Moscow, Russia

N. S. Kulberg
Moscow Scientific and Practical Center of Medical Radiology, Moscow, Russia

Institute of Educational Informatics, Federal Research Center, Computer Science and Control of the RAS, Moscow, Russia

© Springer International Publishing AG, part of Springer Nature 2018
R. P. Mondaini (ed.), *Trends in Biomathematics: Modeling, Optimization and Computational Problems*, https://doi.org/10.1007/978-3-319-91092-5_14

209

elastic waves. Only several gaps are available for the ultrasound, and most of them give a small aperture, which is insufficient for medical observations. The most promising way at the moment is using the temple gap. It is a place on the human skull that doesn't contain the cancellous bone tissue, and ultrasound waves can reach cerebral vessels, reflect, and return to the sensor. The only problem is the thin layer of cortical bone, which acts as an irregularly shaped lens, distorting the final image. Making the ultrasound applicable for the cerebral vessels is the final goal of our research, and the first step is to develop a reliable mathematical model of transcranial ultrasound.

1 Mathematical Model and Numerical Method

The ultrasound imaging inherently implies that the ultrasound speed is homogeneous in the human body. It means that all the tissues are assumed to have the same speed of longitudinal waves and there are no shear waves. This assumption allows to obtain B-scans in real time, allowing the ultrasound operator to determine size and position of internal organs.

Mechanical studies show that human tissues behave like nonlinear viscoelastic media with a great variability of material parameters [4, 9–12]. The assumption about the ultrasound speed homogeneity is applicable because of three reasons. Firstly, nonlinear effects in soft tissues mainly manifest themselves in static and quasi-static tests, especially strength tests. Ultrasound pressure is very low in comparison with the tissue strength, and the material behaves almost according to Hooke's law—though, certain attenuation effects can still be observed. Secondly, shear waves may not be taken into consideration because they attenuate too fast to leave a significant trace on an ultrasound sensor. Thirdly, though soft tissue parameters may vary from one person to another, they tend to have close values in a single organism. They are close enough that it is possible to determine the size of internal organs within a reasonable error.

We used the acoustic material model with the Maxwell viscosity model to reproduce both the longitudinal elastic wave pattern and its attenuation. Ultrasound waves reflect from a boundary between tissues only if their material properties are at least slightly different, so we couldn't follow the assumption about the sound speed homogeneity. The details about material properties, governing equations, and the numerical method are thoroughly described in Ref. [16].

The grid-characteristic method [3, 14, 16] was selected because it is well adapted for modeling wave processes. The method takes characteristic properties of the governing equations system into consideration, and gives us the possibility to easily implement any possible type of border or contact condition.

For the calculations, we used a 3D segmented model of a human head [6, 7]. The general view of the tetrahedral grid can be observed in Fig. 1. All the tissues of the human head were grouped into five segments with different rheological parameters: fat, muscle, bone, brain, and vessels.

Fig. 1 Tetrahedral grid for the human head model: (**a**) general view, (**b**) bone (white) and vessels (black) tissues

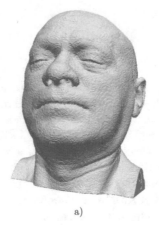
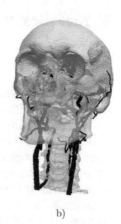

a)   b)

Although previous studies [3] showed that it is necessary to distinguish blood in vessels as a separate segment to obtain reliable results, in this study we used a homogenous vessel model with averaged properties. The problem is that the distinguishing of vessel walls as a separate segment requires a considerable refinement of the mesh, leading to a considerable increase of the calculation time. Current ultrasound technology requires several (up to 200) beams to obtain a single B-scan. In terms of numerical modeling, it means conducting several separate calculations (one for each ultrasound beam), and reducing the calculation time is one of the priority development directions for the current study. The problem can be solved by using hierarchical meshes and time steps, combined with calculations on large supercomputers.

Another development direction is implementing a hybrid method to use an elastic model for bones and an acoustic model for soft tissues. Shear waves don't attenuate in bones as fast as in soft tissues, and these shear waves in bones can generate longitudinal waves in soft tissues. It happens due to their complex shape—when an elastic wave falls on an inclined boundary, it generates waves of both types, shear and longitudinal [1]. Thus, though shear waves attenuate rapidly in soft tissues and don't affect sensor data, they still must be modeled in bones. The acoustic model is insufficient in that case, so we need an elastic model at least in bones segment. At the same time, using the elastic model for the whole craniocerebral area is very ineffective in terms of calculation time. Using the grid-characteristic method, we can easily assign different attenuation coefficients for longitudinal and shear waves, obtaining a wave pattern which will be even closer to the experimental data than an acoustic one. We plan to use this elastic model for calculations of a small area directly beneath the sensor, but applying it to the whole human head model will also considerably increase calculation time.

We used an ultrasound scanner Sonomed-500. The transducer has a phased linear array of 64 elements, its sizes are 1.5×2.0 cm, and operational frequencies range from 2 to 7.5 MHz. For the virtual transcranial ultrasound examination, we use the frequency 3.5 MHz to reduce the attenuation. The emitter phase is modeled as a boundary condition (external pressure), the receiver is modeled by recording the signal as the pressure in the same boundary nodes.

2 Numerical Results

2.1 *Phantom*

Modeling the transcranial ultrasound is a complex task, consisting of several important steps—from building a reliable human heal model to conducting numerous large calculations. To verify the numerical method and the signal processing model, we used an ultrasound phantom Gammex 1430 LE. It contains different types of inhomogeneities, which can be seen on a B-scan in Fig. 2. Phantoms are used to teach ultrasound techniques, to adjust instruments, and to determine instrument parameters: penetration depth, axial and transverse resolution, distance measurement accuracy, "dead zone" size. Parameters of their inhomogeneities are documented, which facilitate the development of their model.

Diffuse areas—areas with different concentrations of contrasting powder—exhibit behavior similar to the behavior of pipes with fluid (in the case of real biological objects—vessels). Both of them give distinct artifacts—bright spots at points located on a straight line connecting the sensor and the center of the tube. These artifacts are usually a hindrance to the ultrasound study, but they allow us to estimate the distance between objects. Diffuse areas are located at a depth of 43 mm and at a distance of 19 mm from each other. The radius of each area is 4 mm. If the sensor is aligned parallel to the pipes, a line can be seen on the screen.

Point reflectors in the phantom are thin nylon threads—about 0.1 mm in diameter. If the sensor is aligned parallel to the thread, a line can be seen on the screen, otherwise it is a bright point. The threads are located at a depth of 16 mm and at a distance of 15 mm from each other. Simulation of such small objects would require an extremely fine mesh (or at least local mesh refinement) and a very large calculation time. Thus, it was decided to simulate wider threads—about 0.5 mm thick.

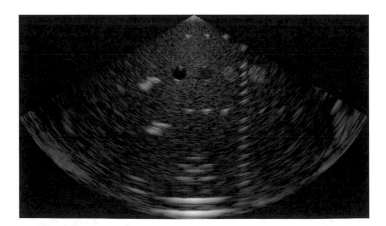

Fig. 2 B-scan of the ultrasound phantom Gammex 1430 LE

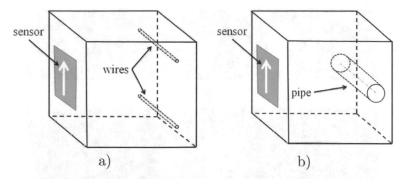

Fig. 3 A scheme of calculation area for the phantom Gammex 1430 LE: (**a**) two threads, (**b**) single pipe

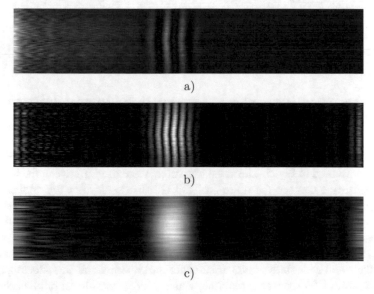

Fig. 4 Numerical raw ultrasound B-scan for a single pipe: (**a**) before processing, (**b**) after narrowband filtering, (**c**) after Hilbert transformation

A scheme of calculation area is shown in Fig. 3. All borders except for the sensor are consuming, effectively modeling a huge bulk of material—a signal from its back side doesn't reach the sensor in the considered time.

Calculation results for a single water pipe are presented in Fig. 4. Without a detailed documentation on signal processing in the ultrasound scanner, we could only assume certain processing steps. The first one is a narrowband filtering, which reduces amount of background noise and allows us to apply the Hilbert transform to obtain the signal envelope and eliminate the carrier frequency [2, 15].

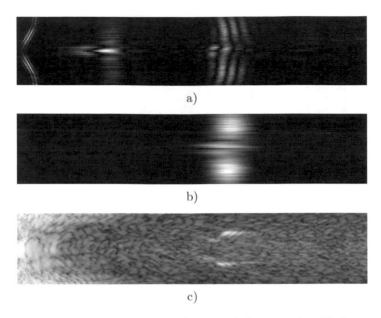

Fig. 5 Raw ultrasound B-scans for two nylon threads: (**a**) before processing, (**b**) after narrowband filtering and Hilbert transformation, (**c**) experimental data

Calculation results for two nylon threads and a comparison with the experimental data are presented in Fig. 5. The same two filters were used. Between the images of the threads, one can see the noise that did not disappear after the application of these filters. This is due to the fact that the size of the threads in the calculation was significantly larger than in the real phantom. Also, in a real sensor, other filters can be applied. They presumably compress the "significant" portions of the signal and reduce the size of the point source in the resulting image. For the correct modeling of these filters and quantitative comparison with the experimental data, documentation on the Sonomed-500 sensor is required, which is not available for public access.

2.2 Human Head Model

The pressure distribution at different times in the frontal plane is depicted in Fig. 6.

The distribution of the velocity modulus at different times on the cerebral vessels is depicted in Fig. 7. Two cases are considered—with the aneurysm and without one. The differences arise due to the changed geometry of the vascular tissue; however, their quantification requires other methods of analyzing the results.

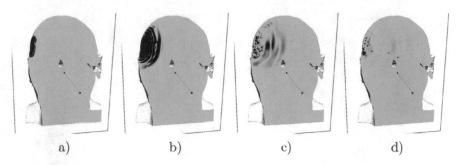

a)   b)   c)   d)

Fig. 6 The pressure distribution at different times in the frontal plane

3 Conclusions

The grid-characteristic numerical method for a viscoelastic material model on irregular tetrahedral grids was implemented as a parallel set of programs. A 3D segmented model for the human head and a model of signal processing in an ultrasound sensor were developed.

A set of calculations was performed for the medical ultrasound phantom model and the human head model. B-scans were obtained and compared to the results of the ultrasound sensor Sonomed-500. The results of the comparison show that the signal processing requires further development, but we can already judge distances to the obstacles and their size.

Acknowledgements The research was supported by the Russian Science Foundation grant 14-31-00024.

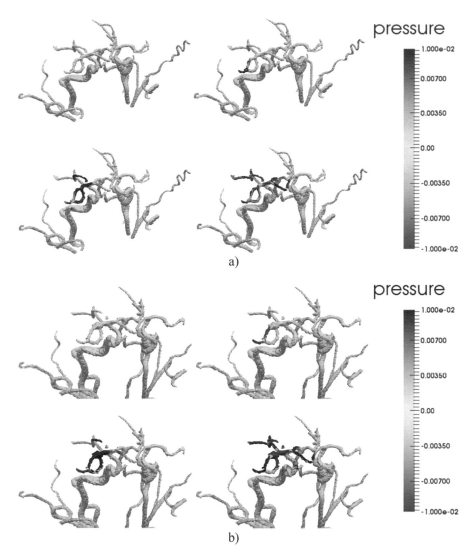

Fig. 7 The wave pattern on cerebral vessels: (**a**) without the aneurysm, (**b**) with the aneurysm

References

1. K. Aki, P.G. Richards, *Quantitative Seismology*, 2nd edn. (University Science Books, Mill Valley, 2002)
2. S.I. Baskakov, *Radio Engineering Circuits and Signals: Textbook for High Schools* (Vysshaya Shkola, Moscow, 1988) (in Russian)
3. K.A. Beklemysheva, A.A. Danilov, I.B. Petrov, V.Y. Salamatova, Y.V. Vassilevski, A.V. Vasyukov, Virtual blunt injury of human thorax: age-dependent response of vascular system. Russ. J. Numer. Anal. Math. Model. **30**(5), 259–268 (2015)

4. J. Black, G. Hastings, *Handbook of Biomaterial Properties* (Springer Science and Business Media, New York, 2013)
5. W.E. Dandy. *Intracranial Arterial Aneurysms* (Comstock Publishing, Ithaca, 1944)
6. A.A. Danilov, D.V. Nikolaev, S.G. Rudnev, V.Y. Salamatova, Y.V. Vassilevski, Modelling of bioimpedance measurements: unstructured mesh application to real human anatomy. Russ. J. Numer. Anal. Math. Model. **27**, 431–440 (2012)
7. A.A. Danilov, V.Y. Salamatova, Y.V. Vassilevski, Mesh generation and computational modeling techniques for bioimpedance measurements: an example using the vhp data. J. Phys. Conf. Series **407**(1), 02004 (2012)
8. N.M. Dott, Intracranial aneurysms: cerebral arterioradiography: surgical treatmen. Edinb. Med. J. **47**, 219–234 (1933)
9. S.A. Goss, R.L. Johnston, F. Dunn, Comprehensive compilation of empirical ultrasonic properties of mammalian tissues. J. Acoust. Soc. Am. **64**(2), 423–457 (1978)
10. P.R. Hoskins, Physical properties of tissues relevant to arterial ultrasound imaging and blood velocity measurement. Ultrasound Med. Biol. **33**(10), 1527–1539 (2007)
11. E.L. Madsen, H.J. Sathoff, J.A. Zagzebski, Ultrasonic shear wave properties of soft tissues and tissuelike materials. J. Acoust. Soc. Am. **74**(5), 1346–1355 (1983)
12. D. Mohan, J.W. Melvin, Failure properties of passive human aortic tissue. I - uniaxial tension tests. J. Biomech. **15**(11), 887–902 (1982)
13. E. Moniz, Diagnostic des tumeurs crbrales et preuve de l'encphalographie artrielle (Diagnostics of cerebral tumours and application of arterial encephalography), Paris (1931)
14. I.B. Petrov, A.V. Favorskaya, A.V. Vasyukov, A.S. Ermakov, K.A. Beklemysheva, A.O. Kazakov, A.V. Novikov, Numerical simulation of wave propagation in anisotropic media Dokl. Math. **90**(3), 778–780 (2015)
15. A.B. Sergienko, *Digital Signal Processing* (SPB, St. Petersburg, 2006) (in Russian)
16. Y.V. Vassilevsky, K.A. Beklemysheva, G.K. Grigoriev, A.O. Kazakov, N.S. Kulberg, I.B. Petrov, V.Y. Salamatova, A.V. Vasyukov, Transcranial ultrasound of cerebral vessels in silico: proof of concept. Russ. J. Numer. Anal. Math. Model. **31**(5), 317–328 (2016)

On the Dynamics of a Discrete Predator–Prey Model

Priyanka Saha, Nandadulal Bairagi, and Milan Biswas

1 Introduction

Nonlinear system of differential equations play very important role in studying different physical, chemical and biological phenomena. However, in general, nonlinear differential equations cannot be solved analytically and therefore discretization is inevitable for good approximation of the solutions [1]. Another reason of constructing discrete models, at least in case of population model, is that it permits arbitrary time-step units [2, 3]. Unfortunately, conventional discretization schemes, such as Euler method, Runge-Kutta method, show dynamic inconsistency [4]. It produces spurious solutions which are not observed in its parent model and its dynamics depend on the step-size. For example, consider the simple logistic model in continuous system:

$$\frac{dx}{dt} = rx \left(1 - \frac{x}{K}\right), \quad x(0) = x_0 > 0, \tag{1}$$

where r and K are positive constants. The system (1) has two equilibrium points with the following dynamical properties:

1. the trivial equilibrium point $x = 0$ is always unstable.
2. the nontrivial equilibrium point $x = K$ is always stable.

P. Saha · N. Bairagi (✉)
Centre for Mathematical Biology and Ecology, Department of Mathematics, Jadavpur University, Kolkata, India
e-mail: nbairagi@math.jdvu.ac.in

M. Biswas
A. J. C. Bose College, Kolkata, India

© Springer International Publishing AG, part of Springer Nature 2018
R. P. Mondaini (ed.), *Trends in Biomathematics: Modeling, Optimization and Computational Problems*, https://doi.org/10.1007/978-3-319-91092-5_15

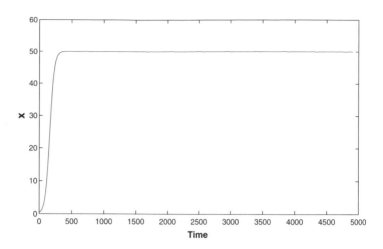

Fig. 1 Time series of the continuous system (1). It shows that the system (1) is stable around the interior equilibrium point $x = K$. Initial point and parameters are taken as $x(0) = 0.4$, $r = 3$ and $K = 50$

Figure 1 shows that even if we start very close to zero ($x_0 = 0.3$) the solution goes to $x = K = 50$, implying that the system is stable around the equilibrium point $x = K$ and unstable around $x = 0$.

The corresponding discrete model formulated by standard finite difference schemes (such as Euler forward method) is given by Anguelov and Lubuma [1]

$$\frac{x_{n+1} - x_n}{h} = rx_n\left(1 - \frac{x_n}{K}\right). \tag{2}$$

This equation can be transformed into logistic difference equation

$$x_{n+1} = x_n + hrx_n\left(1 - \frac{x_n}{K}\right), \tag{3}$$

where h is the step-size. The system (3) also has same equilibrium points with the following dynamic properties:

1. the trivial equilibrium point $x = 0$ is always unstable.
2. the nontrivial equilibrium point $x = K$ is stable if $h < \frac{2}{r}$.

The bifurcation diagram of Euler model (3) (Fig. 2) with h as the bifurcating parameter shows that the fixed point $x = K$ changes its stability as the step-size h crosses the value $\frac{2}{r} = 0.666$. The fixed point is stable for $h < 0.666$ and shows more complex behaviors (period doubling bifurcation) as the step-size is further increased. Thus, dynamics of Euler forward model (3) depends on the step-size and exhibits spurious dynamics which are not observed in the corresponding continuous system (1).

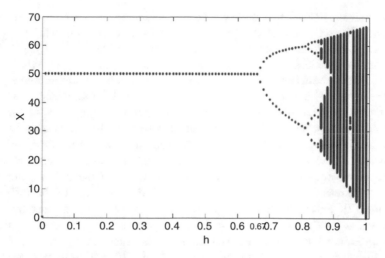

Fig. 2 Bifurcation diagram of the model (3) with h as the bifurcating parameter. It shows that the system is stable till the step-size h is less than 0.666 and unstable for higher values of h. Parameters and initial point are as in Fig. 1

Let us consider another simple example (decay equation)

$$\frac{dx}{dt} = -\lambda x, \ \lambda > 0, x(0) = x_0 > 0. \tag{4}$$

Its solution, given by

$$x(t) = x_0 e^{-\lambda t},$$

is always positive. The corresponding discrete model constructed by Euler forward method is given by

$$x_{n+1} = (1 - \lambda h)x_n. \tag{5}$$

Note that its solution will not be positive if λh is sufficiently large and therefore supposed to show numerical instability.

These examples demonstrate that the discrete systems constructed by standard finite difference scheme is unable to preserve some properties of its corresponding continuous systems. Dynamic behaviors of the discrete model depend strongly on the step-size. However, on principles, the corresponding discrete system should have same properties to that of the original continuous system. It is therefore of immense importance to construct discrete model which will preserve the properties of its constituent continuous models. In the recent past, a considerable effort has been given in the construction of discrete-time model to preserve dynamic consistency of the corresponding continuous-time model without any limitation on the step-size. Mickens first proved that corresponding to any ODE, there exists an exact difference equation which has zero local truncation error [3, 4] and proposed a non-standard finite difference scheme (NSFD) in 1989 [2]. Later in 1994, he introduced the concept of elementary stability, the property which brings correspondence between the local stability at equilibria of the differential equation and the numerical method [5]. Anguelov and Lubuma [6] formalized some of the foundations of Micken's rules, including convergence properties of non-standard finite difference schemes. They defined qualitative stability, which means that the constructed discrete system satisfies some properties like positivity of solutions, conservation laws and equilibria for any step-size. In 2005, Micken coined the term dynamic consistency, which means that a numerical method is qualitatively stable with respect to all desired properties of the solutions to the differential equation [7]. NSFD scheme has gained lot of attention in the last few years because it generally does not show spurious behavior as compared to other standard finite difference methods. NSFD scheme has been successfully used in different fields like economics [8], physiology [9], epidemic [10–12], ecology [13–15] and physics [16, 17]. Here we shall discretize a nonlinear continuous-time predator–prey system following dynamics preserving non-standard finite difference (NSFD) method introduced by Mickens [2].

The paper is arranged in the following sequence. In the next section we describe the considered continuous-time model. Section 3 contains some definitions and general technique of constructing an NSFD model. Section 4 contains the analysis of NSFD and Euler models. Extensive simulations are presented in Sect. 5. The paper ends with the summary in Sect. 6.

2   The Model

Celik [18] has investigated the following dimensionless Holling-Tanner predator–prey system with ratio-dependent functional response:

$$\frac{dN}{dt} = N(1 - N) - \frac{NP}{N + \alpha P}, \tag{6}$$

$$\frac{dP}{dt} = \beta P \left(\delta - \frac{P}{N} \right).$$

The state variables N and P represent, respectively, the density of prey and predator populations at time t, and $N(t) > 0$, $P(t) \geq 0$ for all t. Here α, β and δ are positive constants. For more description of the model, readers are referred to the work of Celik [18].

Celik [18] discussed about the existence and stability of the coexistence interior equilibrium $E^* = (N^*, P^*)$, where

$$N^* = \frac{1 + \alpha\delta - \delta}{1 + \alpha\delta}, \quad P^* = \delta N^*.$$

The following results are known.

Theorem 1.1 *The interior equilibrium point* E^* *of the system* (6) *exists and becomes stable if*

$$(i)\alpha\delta + 1 > \delta, \quad (ii)\delta(2 + \alpha\delta) < (1 + \alpha\delta)^2(1 + \beta\delta).$$

Here we seek to construct a discrete model of the corresponding continuous model (6) that preserves the qualitative properties of the continuous system and maintains dynamic consistency. We also construct the corresponding Euler discrete model and compare its results with the results of NSFD model.

3 Some Definitions

Consider the differential equation

$$\frac{dx}{dt} = f(x, t, \lambda), \tag{7}$$

where λ represents the parameter defining the system (7). Assume that a finite difference scheme corresponding to the continuous system (7) is described by

$$x_{k+1} = F(x_k, t_k, h, \lambda). \tag{8}$$

We assume that $F(., ., ., .)$ is such that the proper uniqueness-existence properties holds; the step size is $h = \nabla t$ with $t_k = hk, k = $ integer; and x_k is an approximation to $x(t_k)$.

Definition 3.1 ([7]) Let the differential equation (7) and/or its solutions have a property P. The discrete model (8) is said to be dynamically consistent with Eq. (7) if it and/or its solutions also have the property P.

Definition 3.2 ([7, 19, 20]) The NSFD procedures are based on just two fundamental rules:

(i) the discrete first–derivative has the representation
$\frac{dx}{dt} \to \frac{x_{k+1} - \psi(h)x_k}{\phi(h)}, h = \Delta t$, where $\phi(h), \psi(h)$ satisfy the conditions
$\psi(h) = 1 + O(h^2), \phi(h) = h + O(h^2)$;

(ii) both linear and nonlinear terms may require a nonlocal representation on the discrete computational lattice; for example,
$$x \to 2x_k - x_{k+1}, \quad x^3 \to (\tfrac{x_{k+1} + x_{k-1}}{2})x_k^2,$$
$$x^3 \to 2x_k^3 - x_k^2 x_{k+1}, \quad x^2 \to (\tfrac{x_{k+1} + x_k + x_{k-1}}{3})x_k.$$

While no general principles currently exist for selecting the functions $\psi(h)$ and $\phi(h)$, particular forms for a specific equation can easily be determined. Functional forms commonly used for $\psi(h)$ and $\phi(h)$ are

$$\phi(h) = \frac{1 - e^{-\lambda h}}{\lambda}, \quad \psi(h) = cos(\lambda h),$$

where λ is some parameter appearing in the differential equation.

Definition 3.3 The finite difference method (8) is called positive if for any value of the step size h, solution of the discrete system remains positive for all positive initial values.

Definition 3.4 The finite difference method (8) is called elementary stable if for any value of the step size h, the fixed points of the difference equation are those of the differential system and the linear stability properties of each fixed point being the same for both the differential system and the discrete system.

Definition 3.5 ([21]) A method that follows the Mickens rules (given in the Definition 3.2) and preserves the positivity of the solutions is called positive and elementary stable nonstandard (PESN) method.

4 Nonstandard Finite Difference (NSFD) Model

For convenience, at first we can write the continuous system (6) as

$$\frac{dN}{dt} = N - N^2 - \frac{NP}{(N + \alpha P)} + (N - N)(N + \alpha P), \tag{9}$$

$$\frac{dP}{dt} = \beta \delta P - \frac{\beta P^2}{N}.$$

Now we express the above system as follows:

$$\frac{dN}{dt} = N - N^2 - NA(N, P) + (N - N)B(N, P), \tag{10}$$

$$\frac{dP}{dt} = \beta \delta P - \beta PC(N, P),$$

where $A(N, P) = \frac{P}{N + \alpha P}$, $B(N, P) = (N + \alpha P)$ and $C(N, P) = \frac{P}{N}$.

We employ the following non-local approximations termwise for the system (10):

$$\begin{cases} \frac{dN}{dt} \to \frac{N_{n+1} - N_n}{h}, & \frac{dP}{dt} \to \frac{P_{n+1} - P_n}{h} \\ N \to N_n, & P \to P_n, \\ N^2 \to N_n N_{n+1}, & \\ PC(N, P) \to P_{n+1} C(N_n, P_n), & \\ NA(N, P) \to N_{n+1} A(N_n, P_n), & \\ (N - N)B(N, P) \to (N_n - N_{n+1})B(N_n, P_n), & \end{cases} \tag{11}$$

where $h \ (> 0)$ is the step-size.

By these transformations, the continuous-time system (9) is converted to

$$\frac{N_{n+1} - N_n}{h} = N_n - N_n N_{n+1} - \frac{N_{n+1} P_n}{N_n + \alpha P_n} + (N_n - N_{n+1})(N_n + \alpha P_n),$$

$$\frac{P_{n+1} - P_n}{h} = \beta \delta P_n - \frac{\beta P_{n+1} P_n}{N_n}. \tag{12}$$

System (12) can be simplified to

$$N_{n+1} = \frac{N_n \{1 + h + h(N_n + \alpha P_n)\}(N_n + \alpha P_n)}{(1 + 2h N_n + \alpha h P_n)(N_n + \alpha P_n) + h P_n}, \tag{13}$$

$$P_{n+1} = \frac{P_n N_n (1 + \beta \delta h)}{N_n + \beta h P_n}.$$

Note that all solutions of the discrete-time system (13) remains positive for any step-size if they start with positive initial values. Therefore, the system (13) is positive.

4.1 Existence of Fixed Points

Fixed points of the system (13) are the solutions of the coupled algebraic equations obtained by putting $N_{n+1} = N_n = N$ and $P_{n+1} = P_n = P$ in (13). However, the fixed points can be obtained more easily from (12) with the same substitutions. Thus, fixed points are the solutions of the following nonlinear algebraic equations:

$$N - N^2 - \frac{NP}{N + \alpha P} = 0, \tag{14}$$

$$\beta \delta P - \frac{\beta P^2}{N} = 0.$$

It is easy to observe that $E_1 = (1, 0)$ is the predator-free fixed point. The interior fixed point $E^* = (N^*, P^*)$ satisfies

$$1 - N^* - \frac{P^*}{N^* + \alpha P^*} = 0 \text{ and } \delta - \frac{P^*}{N^*} = 0. \tag{15}$$

From the second equation of (15), we have $P^* = \delta N^*$. Substituting P^* in the first equation of (15), we find $N^* = \frac{1 + \alpha \delta - \delta}{1 + \alpha \delta}$, which is always positive if $1 + \alpha \delta > \delta$. Thus the positive fixed point E^* exists if $1 + \alpha \delta > \delta$.

4.2 Stability Analysis of Fixed Points

The variational matrix of system (13) evaluated at an arbitrary fixed point (N, P) is given by

$$J(N, P) = \begin{pmatrix} a_{11} & a_{12} \\ a_{21} & a_{22} \end{pmatrix}, \tag{16}$$

where

$$\begin{cases} a_{11} = \frac{\{1+h+h(N_n+\alpha P_n)\}(N_n+\alpha P_n)}{(1+2hN_n+\alpha h P_n)(N_n+\alpha P_n)+h P_n} + \frac{hN_n(N_n+\alpha P_n)}{(1+2hN_n+\alpha h P_n)(N_n+\alpha P_n)+h P_n} \\ \qquad + \frac{N_n\{1+h+h(N_n+\alpha P_n)\}}{(1+2hN_n+\alpha h P_n)(N_n+\alpha P_n)+h P_n} \\ \qquad - \frac{N_n\{1+h+h(N_n+\alpha P_n)\}(N_n+\alpha P_n)\{2h(N_n+\alpha P_n)+(1+2hN_n+\alpha h P_n)\}}{\{(1+2hN_n+\alpha h P_n)(N_n+\alpha P_n)+h P_n\}^2}, \\[4pt] a_{12} = \frac{\alpha h N_n(N_n+\alpha P_n)}{(1+2hN_n+\alpha h P_n)(N_n+\alpha P_n)+h P_n} + \frac{\alpha N_n\{1+h+h(N_n+\alpha P_n)\}}{(1+2hN_n+\alpha h P_n)(N_n+\alpha P_n)+h P_n} \\ \qquad - \frac{N_n\{1+h+h(N_n+\alpha P_n)\}(N_n+\alpha P_n)\{\alpha h(N_n+\alpha P_n)+\alpha(1+2hN_n+\alpha h P_n)+h\}}{\{(1+2hN_n+\alpha h P_n)(N_n+\alpha P_n)+h P_n\}^2}, \\[4pt] a_{21} = \frac{P_n(1+\beta\delta h)}{N_n+\beta h P_n} - \frac{P_n N_n(1+\beta\delta h)}{(N_n+\beta h P_n)^2}, \\[4pt] a_{22} = \frac{(1+\beta\delta h)N_n}{N_n+\beta h P_n} - \frac{\beta h P_n N_n(1+\beta\delta h)}{(N_n+\beta h P_n)^2}. \end{cases}$$

Let λ_1 and λ_2 be the eigenvalues of the variational matrix (16) then we give the following definition in relation to the stability of the system (13).

Definition 4.1 A fixed point (x, y) of the system (13) is called stable if $|\lambda_1| < 1$, $|\lambda_2| < 1$ and a source if $|\lambda_1| > 1$, $|\lambda_2| > 1$. It is called a saddle if $|\lambda_1| < 1$, $|\lambda_2| > 1$ or $|\lambda_1| > 1$, $|\lambda_2| < 1$ and a non-hyperbolic fixed point if either $|\lambda_1| = 1$ or $|\lambda_2| = 1$.

Lemma 4.1 ([22]) Let λ_1 and λ_2 be the eigenvalues of the variational matrix (16). Then $|\lambda_1| < 1$ and $|\lambda_2| < 1$ iff $(i) 1 - det(J) > 0$, $(ii) 1 - trace(J) + det(J) > 0$ and $(iii) 0 < a_{11} < 1$, $0 < a_{22} < 1$.

Theorem 4.1 Suppose that conditions of Theorem 1.1 hold. Then the fixed point E^* of the system (13) is locally asymptotically stable.

Proof At the interior fixed point E^*, the variational matrix reads as

$$J(N^*, P^*) = \begin{pmatrix} a_{11}^* & a_{12}^* \\ a_{21}^* & a_{22}^* \end{pmatrix},$$

where

$$\begin{cases} a_{11}^* = 1 + \frac{N^* h(1-2N^*-\alpha P^*)}{G}, \\ a_{12}^* = \frac{N^* h(\alpha - \alpha N^* - 1)}{G}, \\ a_{21}^* = \frac{\beta\delta h P^*}{H}, \\ a_{22}^* = 1 - \frac{\beta h P^*}{H} \end{cases} \tag{17}$$

with $G = \{1 + h + h(N^* + \alpha P^*)\}(N^* + \alpha P^*)$ and $H = (1 + \beta\delta h)N^*$.

Using $P^* = \delta N^*$ in (17), we have

$$
\begin{cases}
a_{11}^* = 1 + \frac{N^* h (1 - 2N^* - \alpha \delta N^*)}{G}, \\
a_{12}^* = \frac{N^* h (\alpha - \alpha N^* - 1)}{G}, \\
a_{21}^* = \frac{\beta \delta^2 h N^*}{H}, \\
a_{22}^* = 1 - \frac{\beta \delta h N^*}{H}.
\end{cases}
\tag{18}
$$

One can compute that $1 - det(J) = -\frac{(N^*)^2 h \{(1 - \beta \delta - \alpha \beta \delta^2) - (2 + \alpha \delta) N^*\}}{GH} + \frac{\beta \delta h^2 (N^*)^2 \{N^* (1 + \alpha \delta) + N^* (1 + \alpha \delta)^2 + \frac{\alpha \delta^2}{1 + \alpha \delta}\}}{GH} > 0$, provided $-(1 - \beta \delta - \alpha \beta \delta^2) + (2 + \alpha \delta) N^* > 0$, i.e. $\delta(2 + \alpha \delta) < (1 + \alpha \delta)^2 (1 + \beta \delta)$. Note that $trace(J) = \frac{(N^*)^2}{GH} [(1 + \alpha \delta) \{2 + h(2 + \beta \delta + 2N^* \alpha \delta)\} + h(1 + N^* \alpha \delta) + h^2 \beta \delta \{\frac{2\delta}{1 + \alpha \delta} + \alpha \delta (1 + N^* + N^* \alpha \delta) + N^*\}] > 0$ and $1 - trace(J) + det(J) = \frac{\beta \delta h^2 (N^*)^2 (1 + \alpha \delta - \delta)}{GH} > 0$, following the existing condition of E^*. Therefore, the positive fixed point E^* is locally asymptotically stable provided conditions of Theorem 1.1 hold. Hence the theorem is proven.

4.3 The Euler Forward Method

By Euler forward method, we transform the continuous model (6) in the following discrete model:

$$
\frac{N_{n+1} - N_n}{h} = N_n \left[1 - N_n - \frac{P_n}{N_n + \alpha P_n} \right],
\tag{19}
$$

$$
\frac{P_{n+1} - P_n}{h} = \beta P_n \left[\delta - \frac{P_n}{P_n} \right],
$$

where $h > 0$ is the step size. Rearranging the above equations, we have

$$
N_{n+1} = N_n + h N_n \left[1 - N_n - \frac{P_n}{N_n + \alpha P_n} \right],
\tag{20}
$$

$$
P_{n+1} = P_n + h \beta P_n \left[\delta - \frac{P_n}{N_n} \right].
$$

It is to be noticed that the system (20) with positive initial values is not unconditionally positive due to the presence of negative terms. The system may therefore exhibit spurious behaviors and numerical instabilities [5].

4.3.1 Existence and Stability of Fixed Points

At the fixed point, we substitute $N_{n+1} = N_n = N$ and $P_{n+1} = P_n = P$. One can easily compute that (20) has the same interior fixed points as in the previous case. The fixed point $E_1 = (1, 0)$ always exist and the fixed point $E^* = (N^*, P^*)$ exists if $1 + \alpha\delta > \delta$, where $N^* = \frac{1 + \alpha\delta - \delta}{1 + \alpha\delta}$, $P^* = \delta N^*$. We are interested for interior equilibrium only.

The variational matrix of the system (20) at any arbitrary fixed point (N, P) is given by

$$J(x, y) = \begin{pmatrix} a_{11} & a_{12} \\ a_{21} & a_{22} \end{pmatrix},$$

where
$$
\begin{cases}
a_{11} = 1 + h\left[1 - N_n - \frac{P_n}{N_n + \alpha P_n}\right] + hN_n\left[-1 + \frac{P_n}{(N_n + \alpha P_n)^2}\right], \\
a_{12} = -h\left(\frac{N_n}{N_n + \alpha P_n}\right)^2, \\
a_{21} = h\beta\left(\frac{P_n}{N_n}\right)^2, \\
a_{22} = 1 + h\left[\beta\delta - \frac{\beta P_n}{N_n} - \beta\frac{P_n}{N_n}\right].
\end{cases}
$$

Theorem 4.2 *Suppose that the conditions of Theorem 1.1 hold. The interior fixed point E^* of the system (20) is then locally asymptotically stable if $h < min[\frac{G}{H}, \frac{2(1 + \alpha\delta)^2}{G}]$, where $G = (1 + \alpha\delta)^2(1 + \beta\delta) - \delta(2 + \alpha\delta)$, $H = \beta\delta(1 + \alpha\delta - \delta)(1 + \alpha\delta)$.*

Proof At the interior equilibrium point E^*, the Jacobian matrix is evaluated as

$$J(N^*, P^*) = \begin{pmatrix} a_{11} & a_{12} \\ a_{21} & a_{22} \end{pmatrix},$$

where $a_{11} = 1 - hN^*[1 - \frac{P^*}{(N^* + \alpha P^*)^2}]$, $a_{12} = -h(\frac{N^*}{N^* + \alpha P^*})^2$, $a_{21} = h\beta(\frac{P^*}{N^*})^2$, $a_{22} = 1 - h\beta\frac{P^*}{N^*}$. Note that $1 - trace(J) + det(J) = h^2\beta P^*$ is always positive, following the existence conditions of E^*. Thus, condition (ii) of Lemma 4.1 is satisfied. One can compute that $det(J) = 1 - hN^*[\frac{G}{H} - h]$. Here H is positive following the existence condition of E^* and $G > 0$ if $(1 + \alpha\delta)^2(1 + \beta\delta) > \delta(2 + \alpha\delta)$. Thus condition (i) of Lemma 4.1 is satisfied if $h > \frac{G}{H}$. Simple computations give $1 + trace(J) + det(J) = 2(2 - h\frac{G}{(1 + \alpha\delta)^2}) + h^2 H$. This expression will be positive if $0 < h < \frac{2(1 + \alpha\delta)^2}{G}$. Therefore, coexistence equilibrium point E^* exists and becomes stable if $1 + \alpha\delta > \delta$, $\delta(2 + \alpha\delta) < (1 + \alpha\delta)^2(1 + \beta\delta)$ and $h < min[\frac{G}{H}, \frac{2(1 + \alpha\delta)^2}{G}]$. Hence the theorem. $\qquad\square$

Remark 4.1 Note that if $h > \frac{G}{H}$ then E^* is unstable even when the other two conditions are satisfied.

5 Numerical Simulations

In this section, we present some numerical simulations to validate our analytical results of the NSFD discrete system (13) and the Euler system (20) with their continuous counterpart (6). For this experiment, we consider the parameters set as in Celik [18]: $\alpha = 0.7$, $\beta = 0.9$, $\delta = 0.6$. The step size is kept fixed at $h = 0.1$ in all simulations, if not stated otherwise. We consider the initial value $I_1 = (0.2, 0.2)$ as in Celik [18] for all simulations. For the above parameter set, the interior fixed point is evaluated as $E^* = (N^*, P^*) = (0.5775, 0.3465)$. We first reproduce the phase plane diagrams (Fig. 3) of the continuous system (6), the NSFD discrete system (13) and the Euler discrete system (20) by using ODE45 of the software Matlab 7.11. Following the analytical results stated in Sect. 3, the phase plane diagrams show that the equilibrium E^* is stable for all three cases.

To compare step-size dependency of the Euler model and NSFD model, we have plotted the bifurcation diagrams of prey population of the systems (20) and (13)

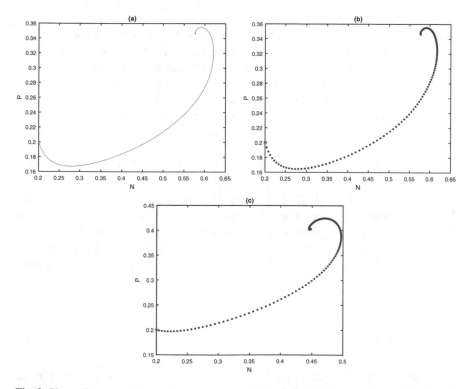

Fig. 3 Phase diagrams of the continuous system (6) (**a**), the NSFD discrete system (13) (**b**) and the Euler system (20) (**c**). These figures show that solution in each case converges to the stable coexistence equilibrium E^* for the parameters $\alpha = 0.7$, $\beta = 0.9$, $\delta = 0.6$. Here $G = (1 + \alpha\delta)^2(1 + \beta\delta) - \delta(2 + \alpha\delta) = 1.6533$ and $h = 0.1 < min\{\frac{G}{H}, \frac{2(1+\alpha\delta)^2}{G}\} = min\{2.6293, 2.4393\}$

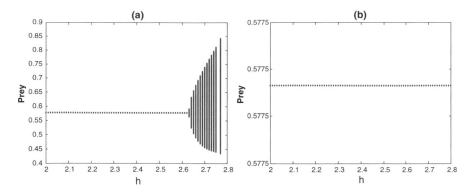

Fig. 4 Bifurcation diagrams of prey population of Euler forward model (20) (**a**) and NSFD
model (13) (**b**) with step-size h as the bifurcation parameter. All the parameters and initial value
are same as in Fig. 3. The first figure shows that the prey population is stable for small step-size h
and unstable for higher value of h. The second figure shows that the prey population is stable for
all step-size h

considering step-size h as the bifurcation parameter (Fig. 4) for the same parameter
values as in Fig. 3. Figure 4a shows that behavior of the Euler model depends on
the step-size. If step-size is small, system population is stable and the dynamics
resembles with the continuous system (6). As the step-size is increased, system
population becomes unstable and therefore the dynamics is inconsistent with the
continuous system. However, the second figure (Fig. 4b) shows that the NSFD
model (13) remains stable for all h, indicating that the dynamics is independent
of step-size.

In particular, we plot (Fig. 5) time series behavior of the NSFD system (13) and
Euler discrete system (20) for $h = 2 (< min\{\frac{G}{H}, \frac{2(1+\alpha\delta)^2}{G}\} = min\{2.6293, 2.4393\})$
and for $h = 2.67 (> min\{2.6293, 2.4393\})$. The first two Fig. 5a, b show that
both populations are stable when the step-size is $h = 2$. Figure 5c shows that
populations of NSFD system (13) remain stable for all step-size, indicating its
dynamic consistency with the continuous system, but Fig. 5d shows that populations
of Euler system (20) oscillate for $h = 2.67$, indicating its dynamic inconsistency
with its continuous counterpart.

6 Summary

Nonstandard finite difference (NSFD) scheme has gained lot of attention in the
last few years mostly for two reasons. First, it generally does not show spurious
behavior as compared to other standard finite difference methods and second,
dynamics of the NSFD model does not depend on the step-size. NSFD scheme also
reduces the computational cost of traditional finite-difference schemes. In this work,

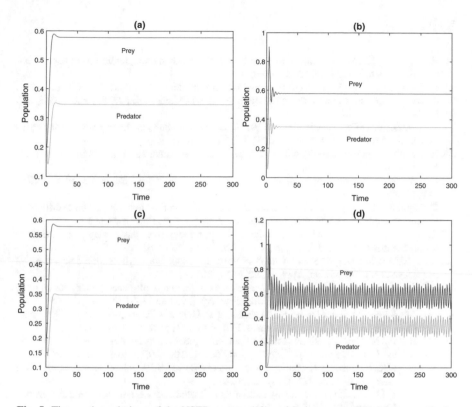

Fig. 5 Time series solutions of the NSFD system (13) and Euler system (20) for two particular values of step-size (h). Here $h = 2$ for (**a** and **b**) and $h = 2.67$ for (**c** and **d**). Other parameters are in Fig. 4

we have studied two discrete systems constructed by NSFD scheme and forward Euler scheme of a well-studied two-dimensional Holling-Tanner type predator–prey system with ratio-dependent functional response. We have shown that dynamics of the discrete system formulated by NSFD scheme are same as that of the continuous system. It preserves the local stability of the fixed point and the positivity of the solutions of the continuous system for any step size. Simulation experiments show that NSFD system always converge to the correct steady-state solutions for any arbitrary large value of the step size (h) in accordance with the theoretical results. However, the discrete model formulated by forward Euler method does not show dynamic consistency with its continuous counterpart. Rather it shows scheme-dependent instability when step-size restriction is violated.

Acknowledgements Nandadulal Bairagi's work is supported by DST, India; Ref. No. SB/EMEQ-046/2018.

References

1. R. Anguelov, J.M.S. Lubuma, Nonstandard fnite difference method, by non-local approximation. S.A. Math. Soc. **31**, 143–543 (2000)
2. R.E. Mickens, Exact solutions to a finite-difference model of a nonlinear reaction-advection equation: implications for numerical analysis. Numer. Methods Partial Differ. Equ. **5**, 313–325 (1989)
3. R.E. Mickens, Difference equation models of differential equations having zero local truncation errors. North-Holland Math. Stud. **92**, 445–449 (1984)
4. R.E. Mickens, Difference equation models of differential equations. Math. Comp. Model. **11**, 528–530 (1988)
5. R.E. Mickens, *Nonstandard Finite Difference Models of Differential Equations* (World Scientific, Singapore, 1994)
6. R. Anguelov, J. Lubuma, Contributions to the mathematics of the nonstandard finite difference method and applications. Numer. Methods Partial Differ. Equ. **17**, 518–543 (2001)
7. R.E. Mickens, Dynamic consistency: a fundamental principle for constructing NSFD schemes for differential equations. J. Differ. Equ. Appl. **11**, 645–653 (2005)
8. Y. Li, Bifurcation analysis of a non-standard finite difference scheme for a time-delayed model of asset prices. J. Differ. Equ. Appl. **19**, 507–519 (2013)
9. R.J. Spiteri, M.C. MacLachlan, An efficient non-standard finite difference scheme for an ionic model of cardiac action potentials. J. Differ. Equ. Appl. **9**, 1069–1081 (2003)
10. S.M. Moghadas, M.E. Alexander, B.D. Corbett, A.B. Gumel, A Positivity-preserving Mickens-type discretization of an epidemic model. J. Differ. Equ. Appl. **9**, 1037–1051 (2003)
11. M. Biswas, N. Bairagi, Discretization of an eco-epidemiological model and its dynamic consistency. J. Differ. Equ. Appl. (2017). https://doi.org/10.1080/10236198.2017.1304544
12. M. Sekiguchi, E. Ishiwata, Global dynamics of a discretized SIRS epi-demic model with time delay. J. Math. Anal. Appl. **371**, 195–202 (2010)
13. L.I. Roege, G. Lahodny, Dynamically consistent discrete Lotka-Volterra competition systems. J. Differ. Equ. Appl. **19**, 191–200 (2013)
14. G. Gabbriellini, Nonstandard finite difference scheme for mutualistic interaction description. Int. J. Differ. Equ. **9**, 147–161 (2012)
15. M. Biswas, N. Bairagi, A predator-prey model with Beddington-DeAngelis functional response: a non-standard finite-difference method. J. Differ. Equ. Appl. (2017). https://doi.org/10.1080/10236198.2017.1304544
16. R.E. Mickens, Numerical study of a non-standard finite difference scheme for the Van Der Pol equation. J. Sound Vib. **250**, 955–963 (2002)
17. A. Mohsen, A simple solution of the Bratu problem. Comput. Math. Appl. **67**, 26–33 (2014)
18. C. Celik, Stability and hopf bifurcation in a delay ratio dependent Holling-Tanner type model. App. Maths. Comput. **255**, 228–237 (2015)
19. D.T. Dimitrov, H.V. Kojouharov, Nonstandard finite-difference schemes for geneal two-dimensional autonomus dynamical systems. Appl. Maths. Lett. **18**, 769–774 (2005)
20. R. Anguelov, J.M.-S. Lubuma, Nonstandard finite difference method by nonlocal approximation. Math. Comput. Simul. **61**, 465–475 (2003)
21. D.T. Dimitrov, H.V. Kojouharov, Positive and elementary stable nonstandard numerical methods with applications to predator-prey models. J. Comput. Appl. Math. **189**, 98–108 (2006)
22. L.-I.W. Roeger, G. Lahodny, Dynamically consistent discrete Lokta-Volterra competition systems. J. Differ. Equ. Appl. **19**, 191–200 (2013)

Frequent Temporal Pattern Mining with Extended Lists

A. Kocheturov and P. M. Pardalos

1 Introduction

In this paper we consider a problem of extracting features from records composed of multivariate time series. It lies on the intersection of knowledge discovery and classification. Our main emphasis is on a faster approach for mining class-specific patterns having temporal resolution which can be used as features for classification purposes.

Mining predictive features is a difficult task. Data usually comes in the form of a collection of records where each record is characterized by a number of numerical time series (e.g., heart rate or blood pressure during a surgical procedure) combined with categorical or numerical attributes like gender and age. Each data record has several outcomes such as a complication or death of the patient within 90 days after surgery. In this paper, we limit ourselves with extracting temporal patterns from multivariate time series only, independent of the attributes available.

Due to several limitations of data-acquisition process, samples of time series may be taken in different time moments among the records and within them. For a given record, some portions or whole time series may be missing as well.

Mining Frequent Patterns was first formulated in [1, 14]. Over the years, several extensions and algorithms have emerged [3, 8, 9, 16]. The framework was successfully applied in a number of medical domains [7, 10, 12, 13] where the importance of temporal relations between patterns was realized.

A. Kocheturov (✉)
Center for Applied Optimization (CAO), University of Florida, Gainesville, FL, USA

P. M. Pardalos
Center for Applied Optimization (CAO), University of Florida, Gainesville, FL, USA

Laboratory of Algorithms and Technologies for Networks Analysis (LATNA), National Research University, Higher School of Economics, Moscow, Russia

© Springer International Publishing AG, part of Springer Nature 2018
R. P. Mondaini (ed.), *Trends in Biomathematics: Modeling, Optimization and Computational Problems*, https://doi.org/10.1007/978-3-319-91092-5_16

In this paper, we continue the work of Batal et al. [4–7] by introducing a new faster algorithm for Frequent Temporal Pattern Mining based on the idea of extended lists which we will explore further. The structure of the manuscript is as follows. In Sect. 2, we introduce essential definitions and concepts. In Sect. 3, the general framework for Mining Frequent Temporal Patterns is described. In Sect. 4, the new algorithm is presented. Computational results are given in Sect. 5. Finally, the paper is concluded in Sect. 6.

2 Definitions

We follow the definitions given in [4–7]. For the sake of clarity, we repeat them here with minor modifications.

The approach starts with reducing dimensionality through converting each time series into a set of temporal abstractions in the form $\langle v_1[s_1, e_1], \ldots, v_k[s_k, e_k]\rangle$, where $v_i \in \Sigma$ is a temporal abstraction that is in effect from start time s_i till end time e_i. Σ is the alphabet or set of possible abstractions. For a given time series, we also require $s_1 \leq e_1 \leq s_2 \cdots \leq s_k \leq e_k$ meaning that an abstraction may not start earlier than any previous one finishes. Based on common logic, we also forbid two consecutive temporal abstractions inside the same time variable to be represented by a single time stamp (e.g., $s_k = e_k = s_{k+1} = e_{k+1}$). Those are rather technical constraints but they are important for constructing temporal patterns.

The alphabet Σ can be defined in several ways. Two examples that we use in the paper are:

(1) **Value Abstractions.** $\Sigma = \{VL, L, N, H, VH\}$ where VL, L, N, H, and VH stand for Very Low, Low, Normal, High and Very High, respectively. Particular ranges for transformation can be, for instance, set up by a field expert or be data-driven and unique for each time series.
(2) **Trend Abstractions.** $\Sigma = \{S, I, D\}$ where S, I, and D stand for Steady, Increasing, and Decreasing, respectively.

If one decides to combine several ways and let the time abstractions overlap, copying the time series and applying one way per copy will solve the issue.

Definition 1 $S = (F, V)$ is a **state** where F is a temporal variable (e.g., heart rate) and $V \in \Sigma$ is an abstraction value (e.g., very low).

Definition 2 $E = (F, V, s, e)$ is a **state interval** where F is a temporal variable, $V \in \Sigma$ is an abstraction value, and s and e are the start and end times, (e.g., heart rate, very high, 0, 10).

In other words, a state interval is a temporal interval for a specific time variable.

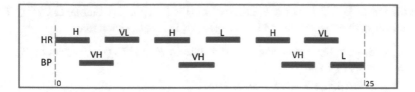

Fig. 1 An artificial record with two time variables heart rate (HR) and blood pressure (BP) after converting them into temporal abstractions

Definition 3 $Z = \langle E_1, \ldots, E_l \rangle$, $E_i.s \leq E_{i+1}.s\ \forall i = 1, \ldots, l-1$ is a **Multivariate State Sequence (MSS)** consisting of l state intervals which are arranged according to non-decreasing order of their start times.

In Fig. 1, the MSS $Z = \langle$ (HR,H,0,3), (BP,VH,2,5), (HR,VL,4,7), (HR,H,8,11), (BP,VH,10,13), (HR,L,12,15), (HR,H,16,19), (BP,VH,18,21), (HR,VL,20,23), (BP,L,22,25) $\rangle$ is depicted.

The next level of abstraction which is called Temporal Pattern will allow us to make a transition from particular values of start and end times and concentrate instead on temporal relationships of the state intervals inside MSS. For this purpose, we utilize Allen's logic [2] but instead of using all 13 possible time relations we take only 2.

For two states E_i and E_j with $E_i.s \leq E_j.s$, we say that E_i finishes **before** E_j starts if $E_i.e < E_j.s$ and denote it as $\mathbf{b(E_i, E_j)}$. Otherwise, we say that E_i **co-occurs** with E_j and denote it as $\mathbf{c(E_i, E_j)}$.

Because we always have uncertainty and errors in the data, other temporal relations of Allen's logic create patterns that represent notions that are very close to each other. Moreover, using them simultaneously makes pattern mining computationally expensive.

Definition 4 $P = (\langle S_1, \ldots, S_k \rangle, R)$ is a k-**Temporal Pattern** (k-TP), or simply Temporal Pattern (TP), of k states $S_1, \ldots, S_k$, where R is an upper-triangular matrix describing pair-wise temporal relationships between the states: $R_{i,j} \in \{b, c\}\ \forall\ 1 \leq i < j \leq k$.

Definition 5 Given an MSS $Z = \langle E_1, E_2, \ldots, E_l \rangle$ and a temporal pattern $P = (\langle S_1, \ldots, S_k \rangle, R)$, we say that Z contains P, denoted as $\mathbf{P} \in \mathbf{Z}$, if there is an injective mapping $\pi : \{1, \ldots, k\} \rightarrow \{1, \ldots, l\}$ ($k \leq l$) that matches every state S in P to a state interval $E_{\pi(i)}$ in Z such that:

(1) $S_i.F = E_{\pi(i)}.F \wedge S_i.V = E_{\pi(i)}.V\ \forall\ 1 \leq i \leq k$,
(2) $\pi(i) < \pi(j)\ \forall\ i < j$,
(3) $R_{i,j}(E_{\pi(i)}, E_{\pi(j)})\ \forall\ i < j$

The first requirement guarantees that the states of P match the corresponding state intervals of Z, while the last two constraints enforce the temporal relations in P and Z to coincide.

Definition 6 k_1-TP P' is a subpattern of k_2-TP P ($k_1 < k_2$), denoted as $P' \subset P$, if there is injective mapping $\pi : \{1, \ldots, k_1\} \rightarrow \{1, \ldots, k_2\}$ such that:

(1) $S'_i = S_{\pi(i)} \; \forall \; 1 \leq i \leq k$ where S'_i is a state in P' and $S_\pi(i)$ is a state in P,
(2) $\pi(i) < \pi(j) \; \forall \; i < j$
(3) $R'_{i,j} = R_{\pi(i),\pi(j)} \; \forall \; 1 \leq i < j \leq k_1$.

It is straightforward to check that:

Corollary 1 *If P' is subpattern of P and $P \in Z$, then Z contains P' as well.*

This corollary, also known as the a priori property, is the main driving force of our approach.

Definition 7 Let $Sub(P)$ denote the set of all subpatterns of P:

$$Sub(P) = \{P' : P' \subset P\}$$

Definition 8 Let $Sub_k(P)$ denote the set of all subpatterns of P of length k:

$$Sub_k(P) = \{P' : P' \text{ is } k - TP \land P' \subset P\}$$

The overall goal is to mine class-specific temporal patterns which appear in a number of MSSs belonging to a certain class. For this purpose, we use the threshold $g \in [0, 1]$ and define the minimum supports, or class-specific thresholds g_i's.

$$g_i = g \times |D_i|.$$

Assume that $D = \{Z_1, \ldots, Z_n\}$ is a data set of n MSSs and $Y = \{y_1, \ldots, y_c\}$ is a set of possible classes, or outcomes. Let D_i denote a set of records from D which belong to class y_i (each record belongs to one class). $Z_j \in D_i$ denotes that record j is in class y_i.

Definition 9 For a given P, threshold g, and class y, we define support of P in class y, denoted as $TP - sup_g{}^y(P)$, as a number of MSSs from D_y having P:

$$TP - sup_g{}^y(P) = |\{Z \in D_y : P \in Z\}|.$$

Definition 10 k-temporal pattern P is a k-Frequent Temporal Pattern (k-FTP) in D, if:

$$\exists i : TP - sup_g{}^{y_i}(P) \geq g_i.$$

In other words, P is an FTP, if the number of MSSs having it is not less than the corresponding class-specific threshold for at least one class.

Corollary 2 $(P' \subset P) \wedge (P' \text{ is } \textbf{not } FTP) \implies P \text{ is } \textbf{not } FTP.$

The proof is a straightforward consequence of Corollary 1 and Definition 10.

3 Mining Frequent Temporal Patterns

To find all FTPs, a breadth-search procedure is applied. First, all FTPs of length 1 are found. Then a list of candidate TPs of length 2 is generated. After that each candidate TP is checked for being an FTP and a list of FTPs of length 2 is formed. The procedure is repeated until all FTPs are found or another stopping criterion is met, e.g., length is no more than a predefined threshold $k_{\max}$.

Input: D, 1-FTPs
Output: FTPs
$FTPs \leftarrow 1\text{-}FTPs$;
$k \leftarrow 1$;
while $|k\text{-}FTPs| > 0$ **and no** other stopping criteria are met **do**
$\qquad (k+1)\text{-}FTPs \leftarrow \emptyset$;
$\qquad (k+1)\text{-}candidates \leftarrow CreateCandidates(k\text{-}FTPs, 1\text{-}FTPs)$;
$\qquad$ **forall the** $TP \in (k+1)\text{-}candidates$ **do**
$\qquad\qquad$ **if** TP is FTP in D **then**
$\qquad\qquad\qquad (k+1)\text{-}FTPs \leftarrow (k+1)\text{-}FTPs \cup \{TP\}$;
$\qquad\qquad$ **end**
$\qquad$ **end**
$\qquad FTPs \leftarrow FTPs \cup (k+1)\text{-}FTPs$;
$\qquad k \leftarrow k+1$;
end

Algorithm 1: The general framework

An important remark is that stopping criteria must not contradict with the a priori property, because some FTPs may be missing, otherwise.

The "create candidates" function takes all FTPs of lengths k and 1 and return candidates of length $k+1$. We use the so-called backward extension of temporal patterns.

Definition 11 P_{new} is a **backward extension** of $P = (\langle S_1, \ldots, S_k \rangle, R)$ with state S_0. It is constructed in the following manner:

(1) $S_i^{new} = S_{i-1} \; \forall i = 1, \ldots, k+1$,
(2) $R_{i+1,j+1}^{new} = R_{i,j} \; \forall 1 \le i < j \le k$,
(3) $R_{1,j}^{new}$ can be either b or c $\forall j = 2, \ldots, k+1$.

We call P a suffix, or parent of P^{new}. We write:

$$P = parent(P^{new}), \text{ or } P^{new}.parent = P$$

Thus, this process can create 2^k possible new $(k + 1)$-patterns for each pair (P_0, P_1) depending on how the temporal relations are chosen. Indeed, less than $k + 1$ possible patterns are coherent (feasible):

Theorem 3.1 *There are at most $k+1$ coherent candidates that result from extending a single k-FTP backward with a new state [7].*

Let us illustrate it on an example. If we extend a 3-FTP backward with a state, the temporal relations of the first state of the resulting candidate TP can be only $\{b, b, b\}$, $\{c, b, b\}$, $\{c, c, b\}$, and $\{c, c, c\}$. It may not be of the form $\{b, c, b\}$, because it means that the first state co-occurs with the third state but is before the second one which is impossible in any MSS since the state intervals are ordered according to their start times. Moreover, if let say the third state's temporal variable is the same as that one of the first state, then only $\{b, b, b\}$, $\{c, b, b\}$ are possible because of the restriction that the state intervals of the same temporal variable cannot co-occur.

These two properties allow to reduce the search space by elimination of impossible TPs.

The most computationally expensive part of this framework is checking if a candidate TP is indeed frequent. Thus, further careful elimination of infrequent TPs at the step of creating candidates is very important.

Based on the a priori property, $(k + 1)$-TP is frequent only if all of its subpatterns are frequent. Indeed, we need to check only if subpatterns of length k are frequent due to transitivity.

Also a simple idea of assigning to each FTP a list of record identifiers which contain it:

$$P.ids = \{i : P \in Z_i\}$$

reduces the search space drastically [7]. It is based on the vertical data format [17, 18]. Again due to the a priori property, a candidate TP of length $k + 1$ will appear only in records where all its subpatterns appear as well. And again we need to check only its k-subpatterns because the record id list of a subpattern of smaller lengths includes the list for at least one k-subpattern (for which it is its subpattern). Such a list is called the list of **potential records**:

$$P.p\_ids = \bigcap_{P' \in Sub(P)} P'.ids = \bigcap_{P' \in Sub_k(P)} P'.ids$$

If for all classes number of the potential records is smaller than the corresponding minimal support values, then this pattern is not frequent, and it can be discarded.

4   Extended Lists

For given record and FTP we keep track of positions (indices of the state intervals in the record) where the first state of the pattern appears inside the record. We say that the pattern starts at those positions.

Assume we have this information for all FTPs of all length k. A coherent candidate $(k+1)$-temporal pattern P constructed from k-FTP and state S has exactly $k+1$ subpatterns of length k ($|Sub_k(P)| = k+1$). From $Sub_k(P)$, exactly k patterns start with state S and they all are in

$$Sub_k(P) \setminus parent(P).$$

It is easy to see that P cannot start at a position i inside Z if at least one k-subpatterns doesn't start at the same position.

For example, assume that we want to find if temporal pattern (Fig. 2)

$$P = (\langle (HR, H), (BP, VH), (HR, L) \rangle, R)$$

with $R_{1,2} = c$, $R_{1,3} = b$, $R_{2,3} = c$ is inside MSS Z (Fig. 1).

Pattern P has two subpatterns P_{s1} (Fig. 3) and P_{s2} (Fig. 4) which have the same first state:

$$P_{s1} = (\langle (HR, H), (BP, VH) \rangle, R_{1,2} = c),$$

$$P_{s2} = (\langle (HR, H), (HR, L) \rangle, R_{1,2} = b).$$

P_{s1} starts at positions 3 and 6 in Z, while P_{s2} starts at positions 0 and 3. Thus, P may potentially start only at position 3 where both the subpatterns start.

Definition 12 Let $P.ppos[i]$ denote all **potential positions** at which P may appear inside Z_i.

Fig. 2 3-TP

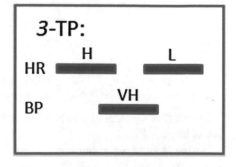

Fig. 3 First 2-FTP
subpattern

Fig. 4 Second 2-FTP
subpattern

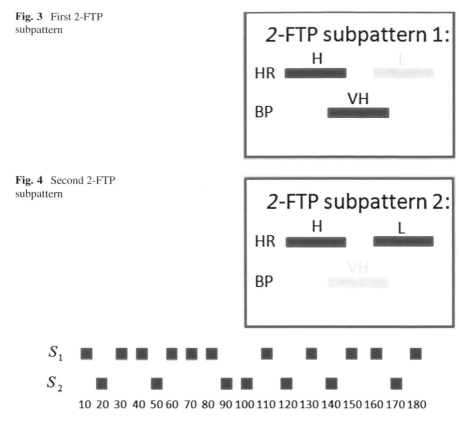

Fig. 5 Locations of S_1 and S_2 in Z

Fig. 6 Possible starting positions of P_1 and starting positions of P_2

Thus,

$$P.p\_pos[i] = \bigcap_{P' \in X} P'.pos[i], \text{ where } X = Sub_k(P) \setminus parent(P).$$

Furthermore, potential positions can be used very efficiently while checking if a
record has a TP.

To see this, let us assume that there is a candidate $(k + 1)$-TP P_1 with states
$S_1, \ldots, S_{k+1}$ and its parent P_2 starting with state S_2 is FTP as well as all other
subpatterns of P_1. Assume also that we know that states S_1 and S_2 appear in MSS

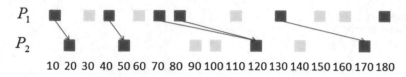

Fig. 7 Links between positions of the patterns before checking

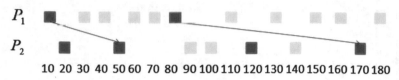

Fig. 8 Links between positions of the patterns after checking

Z as presented in Fig. 5 (all other states are located at different positions and are not depicted here). Then, we know that P_2 starts only at positions as at the bottom of Fig. 6 (all locations of S_2 where P_2 doesn't start are shaded). After intersecting all starting positions of patterns from $Sub(P_1) \setminus P_2$, we have potential starting position of P_1 as at the top of Fig. 6. Then we link all possible positions of P_1 with closest larger positions of P_2 as in Fig. 7. Thus, one can see that the set of possible combinations of S_1 and S_2 locations is significantly smaller than if we had to check all initial locations of the states: now for states S_1 and S_2, we need to check only pairs $(10, 20), (10, 50), \dots, (10, 170), (40, 50), \dots, (130, 170)$.

Then, after checking some combinations have gone away and we have only a few combinations of real positions of P_1 in Z with links to the positions of P_2, e.g., it could be as in Fig. 8.

Definition 13 For given **frequent** temporal pattern P and record Z_i, **extended list** denoted as **P.ex_list[i]** specifies the starting positions of P inside Z_i with links onto the starting positions of $parent(P)$.

Definition 14 For given temporal pattern P and record Z_i, **possible extended list** denoted as **P.p_ex_list[i]** specifies the **possible** starting positions of P inside Z_i with links onto the starting positions of $parent(P)$.

Now we are ready to present our version of the algorithm called **"Frequent Temporal Pattern Mining with Extended Lists"**:

Input: D, *1-FTPs*
Output: *FTPs*
$FTPs \leftarrow 1\text{-}FTPs$;
$k \leftarrow 1$;
while $|k\text{-}FTPs| > 0$ **and no** *other stopping criteria are met* **do**
    $(k + 1)\text{-}FTPs \leftarrow \emptyset$;
    $(k + 1)\text{-}candidates \leftarrow CreateCoherentCandidates(k\text{-}FTPs, 1\text{-}FTPs)$;
    forall the $P \in (k + 1)\text{-}candidates$ **do**
        if $\exists P' \in Sub_k(P)$ *such that* P' *isn't frequent* **then**
        | continue
        end
        $P.p\_ids \leftarrow \cap_{P' \in Sub_k(P)} P'.ids$;
        if *not* **PotentiallyFrequent**$(P.p\_ids)$ **then**
        | continue
        end
        forall the $id \in P.p\_ids$ **do**
            $P.p\_pos[id] = \cap_{P' \in Sub_k(P) \setminus parent(P)} P'.pos[id]$;
            if $P.p\_pos[id] = \emptyset$ **then**
            | $P.p\_ids \leftarrow P.p\_ids \setminus id$
            end
            else
                $P.p\_ex\_list[id] \leftarrow CreateLinks(P.p\_pos[id], P.parent.pos[id])$;
                if $P.p\_ex\_list[id] = \emptyset$ **then**
                | $P.p\_ids \leftarrow P.p\_ids \setminus id$
                end
            end
        end
        if *not* **PotentiallyFrequent**$(P.p\_ids)$ **then**
        | continue
        end
        $P.ids \leftarrow P.p\_ids$;
        forall the $id \in P.ids$ **do**
            $P.ex\_list[id] \leftarrow FindPositionsAndLinks(P.p\_ex\_list[id])$;
            if $P.ex\_list[id] = \emptyset$ **then**
            | $P.ids \leftarrow P.ids \setminus id$
            end
        end
        if P *is FTP in D* **then**
        | $(k + 1)\text{-}FTPs \leftarrow (k + 1)\text{-}FTPs \cup \{P\}$;
        end
    end
    $FTPs \leftarrow FTPs \cup (k + 1)\text{-}FTPs$;
    $k \leftarrow k + 1$;
end

Algorithm 2: Frequent temporal pattern mining with extended lists

5 Computational Results and Simulation

To check the performance of our approach we tested it on two data sets. First data set **rand_DS** consists of $n = 5000$ records of 2 classes (2500 for each class) where each record has two time series with $m = 30$ randomly sampled points from uniform distribution from [0, 1] interval.

The second data set **AKI_DS** consists of $n = 5200$ medical records of time series taken during different surgical procedures [11, 15]. Two time variables were chosen for examination: Blood Pressure and Heart Rate. Each record has an outcome associated with it: 1 if Acute Kidney Injury (AKI) was diagnosed after the surgery (2700 records), and 0, otherwise (2500 records).

In both data sets, the value abstractions were used to convert time series from time domain to abstraction domain. The support threshold of 0.8 was used to mine frequent patterns.

The results are depicted below:

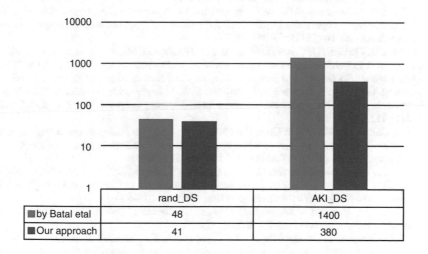

	rand_DS	AKI_DS
■ by Batal etal	48	1400
■ Our approach	41	380

6 Conclusions

In this paper we presented a new faster algorithm for mining Frequent Temporal Patterns. It outperformed the existing approach by an order of magnitude faster on real-life data. Even on random data, where each pattern appears almost everywhere, the structure of extended lists exhibited more efficiency.

The approach can be generalized and applied on other domains where the notion of pattern is defined in other ways.

Acknowledgements Research was supported by RSF grant 14-41-00039.

References

1. R. Agrawal, R. Srikant, Mining sequential patterns, in *Proceedings of the Eleventh International Conference on Data Engineering* (1995), pp. 3–14
2. J.F. Allen, Towards a general theory of action and time. Artif. Intell. **23**(2), 123–154 (1984)
3. J. Ayres, J. Flannick, J. Gehrke, T. Yiu, Sequential pattern mining using a bitmap representation, in *Proceedings of the Eighth ACM SIGKDD International Conference on Knowledge Discovery and Data Mining* (ACM, New York, 2002), pp. 429–435
4. I. Batal, L. Sacchi, R. Bellazzi, M. Hauskrecht, Multivariate time series classification with temporal abstractions, *Proceedings of the 22nd International Artificial Intelligence Research Society Conference (FLAIRS - 22)* (2009), pp. 344–349
5. I. Batal, H. Valizadegan, G.F. Cooper, M. Hauskrecht, A pattern mining approach for classifying multivariate temporal data, in *2011 IEEE International Conference on Bioinformatics and Biomedicine (BIBM)* (2011), pp. 358–365
6. I. Batal, D. Fradkin, J. Harrison, F. Moerchen, M. Hauskrecht, Mining recent temporal patterns for event detection in multivariate time series data, in *Proceedings of the 18th ACM SIGKDD International Conference on Knowledge Discovery and Data Mining* (2012), pp. 280–288
7. I. Batal, G.F. Cooper, D. Fradkin, J. Harrison Jr, F. Moerchen, M. Hauskrecht, An efficient pattern mining approach for event detection in multivariate temporal data. Knowl. Inf. Syst. **46**(1), 115–150 (2016)
8. D.-Y. Chiu, Y.-H. Wu, A.L.P. Chen, An efficient algorithm for mining frequent sequences by a new strategy without support counting, in *Proceedings of the 20th International Conference on Data Engineering* (IEEE, Piscataway, 2004), pp. 375–386
9. J. Han, J. Pei, B. Mortazavi-Asl, H. Pinto, Q. Chen, U. Dayal, M.C. Hsu, Prefixspan: mining sequential patterns efficiently by prefix-projected pattern growth, in *Proceedings of the 17th International Conference on Data Engineering*, 2001, pp. 215–224
10. M. Hauskrecht, S. Visweswaran, G.F. Cooper, G. Clermont, Data-driven identification of unusual clinical actions in the ICU, in *AMIA* (2013)
11. D. Korenkevych, T. Ozrazgat-Baslanti, P. Thottakkara, C.E. Hobson, P. Pardalos, P. Momcilovic, A. Bihorac, The pattern of longitudinal change in serum creatinine and 90-day mortality after major surgery. Ann. Surg. **263**(6), 1219–1227 (2016)
12. R. Moskovitch, Y. Shahar, Classification-driven temporal discretization of multivariate time series. Data Min. Knowl. Disc. **29**(4), 871–913 (2015)
13. L. Sacchi, C. Larizza, C. Combi, R. Bellazzi, Data mining with temporal abstractions: learning rules from time series. Data Min. Knowl. Disc. **15**(2), 217–247 (2007)
14. R. Srikant, R. Agrawal, Mining sequential patterns: generalizations and performance improvements, in *Advances in Database Technology—EDBT'96* (1996), pp. 1–17
15. P. Thottakkara, T. Ozrazgat-Baslanti, B.B. Hupf, P. Rashidi, P. Pardalos, P. Momcilovic, A. Bihorac, Application of machine learning techniques to high-dimensional clinical data to forecast postoperative complications. PLoS One **11**(5), e0155705 (2016)
16. J. Wang, J. Han, Bide: efficient mining of frequent closed sequences, in *Proceedings of the 20th International Conference on Data Engineering* (IEEE, Piscataway, 2004), pp. 79–90
17. M.J. Zaki, Scalable algorithms for association mining. IEEE Trans. Knowl. Data Eng. **12**(3), 372–390 (2000)
18. M.J. Zaki, Spade: an efficient algorithm for mining frequent sequences. Mach. Learn. **42**(1), 31–60 (2001)

Unravelling the Sensitivity of Two Motif Structures Under Random Perturbation

Suvankar Halder, Samrat Chatterjee, and Nandadulal Bairagi

1 Introduction

There are studies demonstrating multiple outputs produced at the level of changes in gene expression and cellular activities [4, 17, 18, 20, 21, 26, 27]. These are clear evidence on sensitivity of intracellular signalling systems to variations in input stimuli which may potentially be induced through either extracellular noise or intracellular stochastic perturbations of the intracellular components (e.g. mutations, alterations in protein turn-over rates, etc.). However, for these inappropriate and non-specific responses the system has its own safe guards, which still needs further investigation.

To probe the issue of maintenance of response-specificity, we selected a condition wherein cells were subjected to varying levels of stochastic perturbations and evaluated the consequent steady state value, rather than the kinetic features, of the output attained [4, 9, 15, 22]. It is the change in steady state levels that have been shown to govern the outcome in complex biological processes such as adaptability, immune memory, development, and cell differentiation [5, 22, 24]. So, in a large signalling network we seek for steady state input-output (I/O) relation in the presence of random perturbation to capture the cell mechanism that prevents any damage due to inappropriate signalling.

The dynamical properties of signalling networks were defined by motifs that are embedded [9, 19, 29]. The analysis of a network through organisation of components

S. Halder · S. Chatterjee (✉)
Drug Discovery Research Centre, Translational Health Science and Technology Institute, NCR Biotech Science Cluster, Faridabad, India
e-mail: samrat.chatterjee@thsti.res.in

N. Bairagi
Centre for Mathematical Biology and Ecology, Department of Mathematics, Jadavpur University, Kolkata, India

© Springer International Publishing AG, part of Springer Nature 2018
R. P. Mondaini (ed.), *Trends in Biomathematics: Modeling, Optimization and Computational Problems*, https://doi.org/10.1007/978-3-319-91092-5_17

and modules into motifs describes the regulatory features of the network. There are different motif structures identified till date, which collectively constitute the building blocks of biological networks [1, 6, 23]. A very frequently observed motif is the feedback loop. Negative feedback loop can give rise to adaptation and desensitisation, while positive feedback loop can lead to emergent network properties such as ultrasensitivity and bistability [2, 5, 14]. In fact, the pattern of motif organisation defines the information processing capabilities of the signalling network, influencing the specificity and plasticity of I/O relationships [1, 22].

One of the aims of the present study is to capture the I/O relation in the structure and how they are influenced by parameter variation specially under random perturbation. The I/O relation becomes more important when a structure showed bistability (simultaneous existence of two stable equilibrium points) as they play a very vital role in cell signalling and functioning. Depending on the initial value of nodes, the output signal can attain any of the two bistable values for the same parameter set value. It is key for understanding basic phenomena of cellular functioning, such as decision-making processes in cell cycle progression, cellular differentiation [16] and apoptosis [13]. It also plays an important role in diseases like cancer and Prion disease [28]. In cancer it is mainly involved in loss of cellular homeostasis associated with beginning of the disease. Bistability can be generated by a positive feedback loop with an ultrasensitive regulatory step. Positive feedback loops is an important regulatory motif in cellular signal transduction [10].

In the present study, we took two particular motif structures (see Fig. 1) that were showing bistability and analysed signal-noise relation between them. Schematic diagrams of two nodes signalling motifs are given in Fig. 1, where node A receives the input signal which then influences the output of node B. The difference between the first and second structure is in the self-activation and inhibition for node A and node B. In structure 1, node A is a self-activator and node B is a self-inhibitor, while in structure 2 it is opposite. k_A and k_B are the self-regulators for the node A and node B, respectively. There is one more difference between the two structures. In structure 1, node A is activating node B and node B is inhibiting node A, while in the structure 2 it is opposite. Here, k_1 and k_2 are regulatory constants of node A

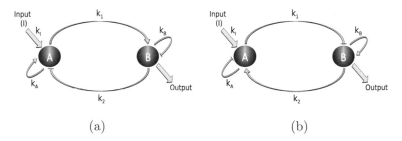

(a)   (b)

Fig. 1 Schematic diagram for two-node motif showing all possible interactions between two nodes. Figure (**a**) depicts the structure 1 and figure (**b**) depicts the structure 2. Details of the structures are given in the text

on B and node B on A, respectively. Here, I is the input signal affecting node A (representing protein A) at a rate k_I. So, the input signal is ultimately the steady state of node A denoted by A^*, say. Finally output signal is the steady state of node B (representing protein B) denoted by B^*, say.

2   Construction of the Deterministic Models

An ordinary differential equation model is constructed based on the pathway map shown in Fig. 1. The first model (for the structure 1) consists of a coupled-differential equations. The nodes represent the proteins present in a cell. Equations of the model describe the rates of loss and creation of particular labelled forms of proteins (nodes) in the system. Our model is based on ODEs and consists of activated form of node A (denoted by A) and node B (denoted by B). Biologically, the total concentration of protein/node (active and inactive form) within the system is constant and for simplicity assumed to be 1 as taken in previous models [22]. The equations are based on the Michaelis-Menten form of equation as described in previous model [22]. The system of differential equations is as follows:

$$
\begin{aligned}
\frac{dA}{dt} &= \frac{k_I I(1-A)}{k_{mI} + (1-A)} + \frac{k_A A(1-A)}{k_{mA} + (1-A)} - \frac{k_2 B A}{k_{m2} + A}, \\
\frac{dB}{dt} &= \frac{k_1 A(1-B)}{k_{m1} + (1-B)} - \frac{k_B B^2}{k_{mB} + B},
\end{aligned}
\tag{1}
$$

The parameter I is the input function, k_I is the activation rate of input signal on node A and k_{mI} is the corresponding half saturation constant, k_i ($i = 1, 2$) denote the activation or inhibition rates of one node on another node, k_j ($j = A, B$) represents the self-activation or self-inhibition rates of the nodes, k_{mi} ($i = 1, 2, A, B$) are the respective half saturation constants.

By similar arguments as of structure 1 (with same parameter) an ordinary differential equation model is constructed for structure 2, which is as follows:

$$
\begin{aligned}
\frac{dA}{dt} &= \frac{k_I I(1-A)}{k_{mI} + (1-A)} + \frac{k_2 B(1-A)}{k_{m2} + (1-A)} - \frac{k_A A^2}{k_{mA} + A}, \\
\frac{dB}{dt} &= \frac{k_B B(1-B)}{k_{mB} + (1-B)} - \frac{k_1 A B}{k_{m1} + B}.
\end{aligned}
\tag{2}
$$

2.1 Analytical Results

2.1.1 Positive Invariance and Boundedness

Let us put the system of Eq. (1) in a vector form by setting $X = \begin{bmatrix} A \\ B \end{bmatrix} \in R^2$.

$$F(X) = \begin{bmatrix} F_1(X) \\ F_2(X) \end{bmatrix} = \begin{bmatrix} \frac{k_I I(1-A)}{k_{mI}+(1-A)} + \frac{k_A A(1-A)}{k_{mA}+(1-A)} - \frac{k_2 BA}{k_{m2}+A} \\ \frac{k_1 A(1-B)}{k_{m1}+(1-B)} - \frac{k_B B^2}{k_{mB}+B} \end{bmatrix}, \quad (3)$$

where $F : C_+ \to R^2$. Then Eq. (1) becomes

$$\dot{X} = F(X), \quad (4)$$

with $X(0) = X_0 \in R_+^2$. It is easy to check in Eq. (3) that whenever choosing $X(0) \in R_+^2$ such that $X_i = 0$, then $F_i(X)|_{X_i=0} \geq 0$, $(i = 1, 2)$. Due to lemma [25], any solution of Eq. (4) with $X(0) \in R_+^2$, say $X(t) = X(t; X_0)$, is such that $X(t) \in R_+^2$ for all $t > 0$.

Since the total concentration of protein/node within the system is constant and is equal to 1, so the maximum value that A, B can take is 1. Hence by model assumption both are bounded.

Similarly for structure 2, again setting $X = \begin{bmatrix} A \\ B \end{bmatrix} \in R^2$ and

$$\widehat{F}(X) = \begin{bmatrix} \widehat{F}_1(X) \\ \widehat{F}_2(X) \end{bmatrix} = \begin{bmatrix} \frac{k_I I(1-A)}{k_{mI}+(1-A)} + \frac{k_2 B(1-A)}{k_{m2}+(1-A)} - \frac{k_A A^2}{k_{mA}+A} \\ \frac{k_B B(1-B)}{k_{mB}+(1-B)} - \frac{k_1 AB}{k_{m1}+B} \end{bmatrix}, \quad (5)$$

By similar argument as in the case of structure 1, we can prove the positive invariance and boundedness of structure 2.

2.1.2 Equilibrium Points of the System (1) and Their Stability Properties

As here we are interested in studying the I/O relation, so we look for only the interior equilibrium point. The interior equilibrium point is denoted by $E^* \equiv (A^*, B^*)$, where

$$A^* = \frac{k_B B^{*2}[k_{m1} + (1 - B^*)]}{k_1(k_{mB} + B^*)(1 - B^*)} \quad (6)$$

and B^* satisfies the equation

$$\frac{k_I I \left(1 - \left(\frac{k_B B^{*2}[k_{m1}+(1-B^*)]}{k_1(k_{mB}+B^*)(1-B^*)}\right)\right)}{k_{mI} + \left(1 - \left(\frac{k_B B^{*2}[k_{m1}+(1-B^*)]}{k_1(k_{mB}+B^*)(1-B^*)}\right)\right)} - \frac{k_2 B^* \left(\frac{k_B B^{*2}[k_{m1}+(1-B^*)]}{k_1(k_{mB}+B^*)(1-B^*)}\right)}{k_{m2} + \left(\frac{k_B B^{*2}[k_{m1}+(1-B^*)]}{k_1(k_{mB}+B^*)(1-B^*)}\right)}$$

$$+ \frac{k_A \left(\frac{k_B B^{*2}[k_{m1}+(1-B^*)]}{k_1(k_{mB}+B^*)(1-B^*)}\right) \left(1 - \left(\frac{k_B B^{*2}[k_{m1}+(1-B^*)]}{k_1(k_{mB}+B^*)(1-B^*)}\right)\right)}{k_{mA} + \left(1 - \left(\frac{k_B B^{*2}[k_{m1}+(1-B^*)]}{k_1(k_{mB}+B^*)(1-B^*)}\right)\right)} = 0. \tag{7}$$

The corresponding Jacobian matrix (J) evaluated at $E^* = (A^*, B^*)$ is denoted by $J(A^*, B^*)$ and given by

$$J(A^*, B^*) = \begin{bmatrix} a_{11} & -a_{12} \\ a_{21} & a_{22} \end{bmatrix} \tag{8}$$

with

$$a_{11} = \frac{k_A(1 - A^*)}{k_{mA} + (1 - A^*)} - \frac{k_I k_{mI} I}{(k_{mI} + (1 - A^*))^2} - \frac{k_A k_{mA} A^*}{(k_{mA} + (1 - A^*))^2} - \frac{k_2 k_{m2} B^*}{(k_{m2} + A^*)^2},$$

$$a_{12} = \frac{k_2 A^*}{k_{m2} + A^*},$$

$$a_{21} = \frac{k_1(1 - B^*)}{k_{m1} + (1 - B^*)},$$

$$a_{22} = -\frac{k_1 k_{m1} A^*}{(k_{m1} + (1 - B^*))^2} - \frac{k_B B^*(2k_{mB} + B^*)}{(k_{mB} + B^*)^2}.$$

The characteristic equation is given by

$$\lambda^2 - \text{trace}(J)\lambda + \text{determinant}(J) = 0. \tag{9}$$

The interior equilibrium point is stable if

$$\text{trace}(J) < 0$$

$$\text{determinant}(J) > 0. \tag{10}$$

So, the interior equilibrium point is stable if

$$a_{11} + a_{22} < 0, \tag{11}$$

$$a_{11}a_{22} + a_{12}a_{21} > 0. \tag{12}$$

2.1.3 Equilibrium Points of the System (2) and Their Stability Properties

The system (2) has an axial equilibrium point $\widehat{E} \equiv (\widehat{A}, 0)$, where $\widehat{A}$ satisfies the following equation

$$k_A \widehat{A}^3 - \{k_I I + k_A(k_{mI} + 1)\} \widehat{A}^2 + (1 - k_{mA})k_I I \widehat{A} + k_I I k_{mA} = 0 . \tag{13}$$

It has also an interior equilibrium $E^* \equiv (A^*, B^*)$, where

$$A^* = \frac{k_B B^*(1 - B^*)(k_{m1} + B^*)}{[k_{mB} + (1 - B^*)]k_1 B^*} \tag{14}$$

and B^* satisfies the equation

$$\frac{k_I I \left(1 - \left(\frac{k_B B^*(1-B^*)(k_{m1}+B^*)}{[k_{mB}+(1-B^*)]k_1 B^*}\right)\right)}{k_{mI} + \left(1 - \left(\frac{k_B B^*(1-B^*)(k_{m1}+B^*)}{[k_{mB}+(1-B^*)]k_1 B^*}\right)\right)} + \frac{k_2 B^* \left(1 - \left(\frac{k_B B^*(1-B^*)(k_{m1}+B^*)}{[k_{mB}+(1-B^*)]k_1 B^*}\right)\right)}{k_{m2} + \left(1 - \left(\frac{k_B B^*(1-B^*)(k_{m1}+B^*)}{[k_{mB}+(1-B^*)]k_1 B^*}\right)\right)}$$

$$- \frac{k_A \left(\frac{k_B B^*(1-B^*)(k_{m1}+B^*)}{[k_{mB}+(1-B^*)]k_1 B^*}\right)^2}{k_{mA} + \left(\frac{k_B B^*(1-B^*)(k_{m1}+B^*)}{[k_{mB}+(1-B^*)]k_1 B^*}\right)} = 0 \tag{15}$$

Following the Jacobian matrix and stability definition (as given in (9) and (10)), the interior equilibrium point is stable if

$$b_{11} + b_{22} < 0 \tag{16}$$

$$b_{11}b_{22} + b_{12}b_{21} > 0 \tag{17}$$

where

$$b_{11} = -\frac{k_1 k_{mI} I}{[k_{mI} + (1 - A^*)]^2} - \frac{k_2 k_{m2} B^*}{[k_{m2} + (1 - A^*)]^2} - \frac{k_A A^*(2k_{mA} + A^*)}{(k_{mA} + A^*)^2}$$

$$b_{12} = \frac{k_2(1 - A^*)}{k_{m2} + (1 - A^*)}$$

$$b_{21} = \frac{k_1 B^*}{k_{m1} + B^*}$$

$$b_{22} = \frac{k_B(1 - B^*)}{k_{mB} + (1 - B^*)} - \frac{k_B k_{mB} B^*}{[k_{mB} + (1 - B^*)]^2} - \frac{k_1 k_{m1} A^*}{(k_{m1} + B^*)^2}$$

Table 1 Parameters description and the initial values

Parameters	Description	Default values
I	Initial input	0.097
k_I	Activation rate of I on node A	1
k_1	Activation rate of node A on node B	1
k_2	Deactivation rate of node B on node A	1
k_A	Self-activation rate of node A	1
k_B	Self-deactivation rate of node B	1
k_{mI}	Half saturation constant respect to k_I	0.1
k_{m1}	Half saturation constant respect to k_1	0.4
k_{m2}	Half saturation constant respect to k_2	0.1
k_{mA}	Half saturation constant respect to k_A	0.1
k_{mB}	Half saturation constant respect to k_B	0.8

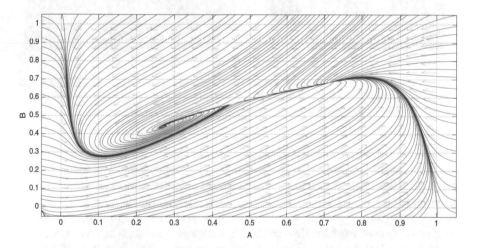

Fig. 2 Phase portrait showing bistability of the system (1). Parameters are as in Table 1

2.2 Simulation Results

2.2.1 Numerical Analysis for the System (1)

We solved the system of differential equations (1) in MATLAB with parameter values as in Table 1 and observed that it shows bistability, see Fig. 2. We, then, varied each parameters ten folds up and down from their base value (given in Table 1). The range of each parameter for which bi- or mono-stability was observed has been plotted in Fig. 3. Bistability was observed around the base value and mono-stability was observed as we moved away from the base value.

Global Sensitivity Analysis The global sensitivity analysis (GSA) helps to identify model parameters that could be particularly important. We used the Partially Ranked

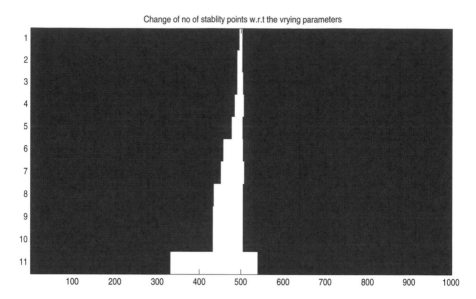

Fig. 3 Parameter ranges for which the system shows mono- or bi-stabilities. Here, the black colour shows the range of each parameter for which the system has only one stable point and the white colour shows the range of each parameter for which the system is bistable

Correlation Coefficients (PRCC) [8] technique for the GSA and their associated p-values to identify the most sensitive parameters. To calculate PRCCs, we used Latin Hypercube Sampling (LHS) method to randomly select vectors of parameter-values used for each run of PRCCs calculations. Over 1000 simulations were performed to calculate PRCCs. In each simulation, system was solved up to 100 time steps, as it was observed from the time series solutions that the system behaves uniformly much before 100 time steps. Figure 4 depicts the sensitivity of each parameter for the variable B. We used a cut-off of ± 0.4 to define the sensitive parameters, i.e., if PRCC value of a particular parameter lies beyond ± 0.4 then that parameter will be called a sensitive parameter. The GSA analysis suggests that the most sensitive parameters in structure 1 are k_1, k_B, k_{m1} and k_{mB}. These parameters are affecting primarily the output signal of Node B and they are associated with node B.

2.2.2 Numerical Analysis for the System (2)

We solved the system of differential equations (2) in MATLAB with parameter values as in Table 2 and observed that it shows bistability, see Fig. 5. We, then, varied each parameters ten folds up and down from their base value (given in Table 2) and divide them in equal partitions and calculated the number of stable equilibrium points. We observed that structure 2 shows bistability for a wider range

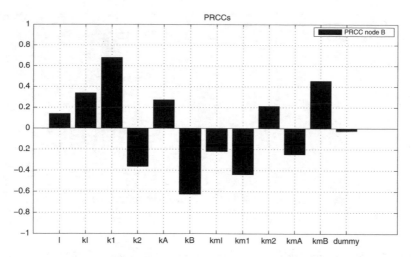

Fig. 4 Global sensitivity analysis (GSA) of model parameters of node B using Latin Hypercube Sampling (LHS) method

Table 2 Parameters description and the initial values

Parameters	Description	Default values
I	Initial input	0.3
k_I	Activation rate of I on node A	0.001
k_1	Deactivation rate of node A on node B	1
k_2	Activation rate of node B on node A	1
k_A	Self-deactivation rate of node A	0.001
k_B	Self-activation rate of node B	1
k_{mI}	Half saturation constant respect to k_I	0.1
k_{m1}	Half saturation constant respect to k_1	0.7
k_{m2}	Half saturation constant respect to k_2	0.1
k_{mA}	Half saturation constant respect to k_A	0.1
k_{mB}	Half saturation constant respect to k_B	0.1

when compare to structure 1, see Fig. 3. In case of structure 2, we also observed a region for non-existence of stable equilibrium point, which was not there in structure 1 (Fig. 6).

Global Sensitivity Analysis

The global sensitivity analysis (GSA) of system (2) was done using Partially Ranked Correlation Coefficients (PRCC) [8] technique similar to the system (1). Figure 7 depicts the sensitivity of each parameter for each variable. We used a cut-off of ± 0.4 to define the sensitive parameters, i.e., if PRCC value of a particular parameter lies beyond ± 0.4 then that parameter will be called a sensitive parameter. The GSA analysis suggests that the most sensitive parameters in structure 2 are k_1, k_B, k_{m1} and k_{mB}. So, here also most of the sensitive parameter effecting the output signal are of Node B.

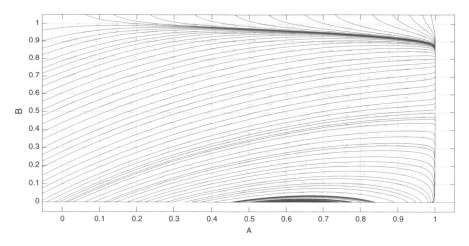

Fig. 5 Phase portrait showing bistability of the system (2). Parameters as in Table 2

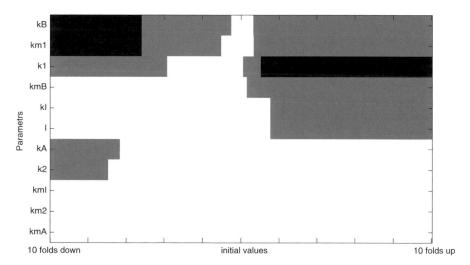

Fig. 6 Parameter ranges for which the number of equilibrium point(s) changes. Here, the black colour shows the range of each parameter for which the system has no stable point, grey colour shows the range of each parameter for which the system has only one stable equilibrium point and the white colour shows the range of each parameter for which the system is bistable

3 Construction of the Stochastic Models

In previous sections we observed that depending on the motif structure and the parameter value, the nature of the output signal may vary from mono-stability to bi-stability. Next we want to see how this rich dynamics behave under random perturbation. This will help us to understand the I/O relationship for the two motifs

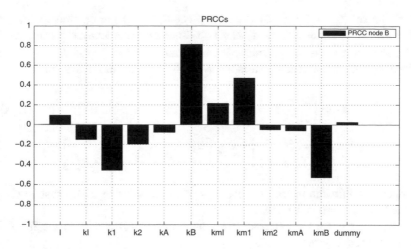

Fig. 7 Global sensitivity analysis (GSA) of model parameters of node B using Latin Hypercube Sampling (LHS) method

in the presence of noise. These random perturbations may arise through mutations and alteration in turnover rates. Random perturbation may also appear due to improper network signaling. To probe this, we incorporated a dispersed stochastic perturbation in our model that could influence any of the components of the motif independent of the signal input. The rationale of using dispersed perturbation was derived from the fact that there exists multiple intrinsic and extrinsic factors, such as cytokines, growth factors, nutrients, environmental stresses, modulation in protein stability and many others, which can potentially influence any of the signaling components through a diverse range of mechanisms [7, 21, 30]. Cumulative effects of such perturbations would exert a heterogeneous influence on the basal state of the signaling network. We considered such random influences as systemic perturbations and incorporated these effects into the model as multiplicative Gaussian white noise [10, 18]. Thus we introduce the stochastic perturbation terms into the equations of both node A and node B. The stochastic perturbations of the state variables around their steady-state values E^* are Gaussian white noise which are proportional to the distances of A, B from their steady-state values A^*, B^*, respectively. So, the deterministic model system (1) results in the following stochastic model system

$$dA = F_1(A, B)dt + \sigma_1(A - A^*)d\xi_t^1 \, ,$$

$$dB = F_2(A, B)dt + \sigma_2(B - B^*)d\xi_t^2 \, . \tag{18}$$

where σ_1 and σ_2 are real constants and known as the intensity of the fluctuations, $\xi_t^i = \xi_i(t)$, $i = 1, 2$ are standard Wiener processes, independent of each other, and F_1, F_2 are defined in Eq. (3). We consider Eq. (18) as an Ito stochastic differential system of the type

$$dX_t = F(t, X_t)dt + G(t, X_t)d\xi_t \tag{19}$$

where the solution $(X_t, t > 0)$ is an Ito process, 'F' is the drift coefficient, 'G' is the diffusion coefficient and ξ_t is a two-dimensional stochastic process having scaler Wiener process components with increments $\Delta \xi_t^j = \xi_j(t + \Delta t) - \xi_j(t)$ are independent Gaussian random variables $N(0, \Delta t)$. In the case of system (18),

$$X_t = \begin{bmatrix} A \\ B \end{bmatrix}, \xi_t = \begin{bmatrix} \xi_t^1 \\ \xi_t^2 \end{bmatrix}, \tag{20}$$

$$F = \begin{bmatrix} F_1(A, B) \\ F_2(A, B) \end{bmatrix}, G = \begin{bmatrix} \sigma_1(A - A^*) & 0 \\ 0 & \sigma_2(B - B^*) \end{bmatrix} \tag{21}$$

Since the diffusion matrix 'G' depends upon the solution of X_t, system (18) is said to have multiplicative noise.

Following above, the deterministic model system (2) results in the following stochastic model system

$$dA = \widehat{F}_1(A, B)dt + \sigma_1(A - A^*)d\xi_t^1 ,$$
$$dB = \widehat{F}_2(A, B)dt + \sigma_2(B - B^*)d\xi_t^2 . \tag{22}$$

where σ_1 and σ_2 are real constants and known as the intensity of the fluctuations, $\xi_t^i = \xi_i(t)$, $i = 1, 2$ are standard Wiener processes, independent of each other, and $\widehat{F}_1, \widehat{F}_2$ are defined in Eq. (5). We consider Eq. (18) as an Ito stochastic differential system of the type

$$dX_t = \widehat{F}(t, X_t)dt + G(t, X_t)d\xi_t \tag{23}$$

with

$$X_t = \begin{bmatrix} A \\ B \end{bmatrix}, \xi_t = \begin{bmatrix} \xi_t^1 \\ \xi_t^2 \end{bmatrix}, \tag{24}$$

$$\widehat{F} = \begin{bmatrix} \widehat{F}_1(A, B) \\ \widehat{F}_2(A, B) \end{bmatrix}, G = \begin{bmatrix} \sigma_1(A - A^*) & 0 \\ 0 & \sigma_2(B - B^*) \end{bmatrix} \tag{25}$$

Since the diffusion matrix 'G' depends upon the solution of X_t, system (22) is said to have multiplicative noise.

3.1 Stochastic Stability of Interior Equilibrium

The stochastic differential system (18) can be centred at its positive equilibrium points $E^*(A^*, B^*)$ by introducing the variables $U_1 = A - A^*, U_2 = B - B^*$. It looks to be a very difficult problem to derive asymptotic stability in mean square sense by Lyapunov functions method working on the complete nonlinear equation (18). For simplicity of mathematical calculations, we deal with the stochastic differential equation obtained by linearising the vector function 'F' in (21) about the positive equilibrium point E^*. The linearised version of (19) around E^* is given by

$$dU(t) = f(U(t))dt + G(U(t))d\xi(t) , \tag{26}$$

where

$$U(t) = \begin{bmatrix} U_1(t) \\ U_2(t) \end{bmatrix} , \tag{27}$$

$$f(U(t)) = \begin{bmatrix} -P_{11}U_1 - P_{12}U_2 \\ P_{21}U_1 - P_{22}U_2 \end{bmatrix} , \tag{28}$$

$$G(U(t)) = \begin{bmatrix} \sigma_1 U_1 & 0 \\ 0 & \sigma_2 U_2 \end{bmatrix} \tag{29}$$

with

$$P_{11} = \frac{k_1 k_{m1} I}{(k_{m1} + (1 - A^*))^2} + \frac{k_2 k_{m2} B^*}{(k_{m2} + A^*)^2},$$
$$+ \frac{k_A k_{mA} A^* - k_A (1 - A^*)(k_{mA} + (1 - A^*))}{(k_{mA} + (1 - A^*))^2} \tag{30}$$

$$P_{12} = \frac{k_2 A^*}{k_{m2} + A^*}, \tag{31}$$

$$P_{13} = \frac{k_1 (1 - B^*)}{k_{m1} + (1 - B^*)}, \tag{32}$$

$$P_{14} = \frac{k_1 k_{m1} A^*}{(k_{m1} + (1 - B^*))^2} + \frac{k_B B^* (2k_{mB} + B^*)}{(k_{mB} + B^*)^2} \tag{33}$$

Note that, in (26) the positive equilibrium E^* corresponds to the trivial solution $(U_1, U_2) = (0, 0)$. Let Ω be the set defined by $\Omega = [(t \geqslant t_0) \times R^2, t_0 \in R^+]$. We define the following theorem [12]

Theorem 3.1 *Suppose there exists a differentiable function* $V(U, t) \in C^2(\Omega)$ *satisfying the inequalities*

$$K_1 |U|^\alpha \leq V(U, t) \leq K_2 |U|^\alpha \tag{34}$$

$$LV(U, t) \leq -K_3 |U|^\alpha, \quad K_i > 0, \quad i = 1, 2, 3, \quad \alpha > 0 . \tag{35}$$

Then the trivial solution of (26) is exponentially α *stable for all time* $t \geq 0$.

Note that, if in (34), (35), $\alpha = 2$, then the trivial solution of (26) is exponentially mean square stable. Furthermore, the trivial solution of (26) is globally asymptotically stable in probability.

Here, following (26),

$$LV(t, U) = \frac{\partial V(t, U(t))}{\partial t} + f^T(U(t))\frac{\partial V(t, U)}{\partial U}$$

$$+ \frac{1}{2}Tr\left[G^T(U(t))\frac{\partial^2 V(t, U)}{\partial U^2}G(U(t))\right] \tag{36}$$

where

$$\frac{\partial V}{\partial U} = \left(\frac{\partial V}{\partial U_1} \frac{\partial V}{\partial U_2}\right)^T, \quad \frac{\partial^2 V(t, U)}{\partial U^2} = \left(\frac{\partial^2 V}{\partial U_j \partial U_i}\right)_{i,j=1,2}$$

and T means transposition.

We can prove the following theorem:

Theorem 3.2 *When the inequality*

$$\frac{k_1 k_{m1} I}{(k_{m1} + (1 - A^*))^2} + \frac{k_A k_{mA} A^*}{(k_{mA} + (1 - A^*))^2} + \frac{k_2 k_{m2} B^*}{(k_{m2} + A^*)^2} > \frac{k_A(1 - A^*)}{k_{mA} + (1 - A^*)} \tag{37}$$

holds true then the zero solutions of the system (18) will be exponentially 2-stable if

$$\sigma_1^2 < 2\left[\frac{k_1 k_{m1} I}{(k_{m1} + (1 - A^*))^2} + \frac{k_A k_{mA} A^*}{(k_{mA} + (1 - A^*))^2}\right]$$

$$+ 2\left[\frac{k_2 k_{m2} B^*}{(k_{m2} + A^*)^2} - \frac{k_A(1 - A^*)}{k_{mA} + (1 - A^*)}\right],$$

$$\sigma_2^2 < 2\left[\frac{k_1 k_{m1} A^*}{(k_{m1} + (1 - B^*))^2} + \frac{k_B B^*(2k_{mB} + B^*)}{(k_{mB} + B^*)^2}\right]. \tag{38}$$

with the positive constants ω_1 and ω_2, where $\omega_1 = \frac{k_{m2} + A^*}{k_2 A^*}$ and $\omega_2 = \frac{k_{m1} + (1 - B^*)}{k_1(1 - B^*)}$.

Proof Let us consider the Lyapunov function

$$V(U(t)) = \frac{1}{2}\left[\omega_1 U_1^2 + \omega_2 U_2^2\right] \tag{39}$$

where ω_i are real positive constants to be chosen later. It is easy to check the inequalities in (34) are true for $\alpha = 2$.

Next, using (29) and (36),

$$LV(U(t)) = \left(-P_{11} + \frac{1}{2}\sigma_1^2\right)\omega_1 U_1^2 + \left(-P_{22} + \frac{1}{2}\sigma_2^2\right)\omega_2 U_2^2 \tag{40}$$

$$+ (P_{21}\omega_2 - P_{12}\omega_1) U_1 U_2$$

Assuming

$$\omega_1 = \frac{k_{m2} + A^*}{k_2 A^*}, \text{ and } \omega_2 = \frac{k_{m1} + (1 - B^*)}{k_1(1 - B^*)},$$

(40) becomes

$$LV(U(t)) = \left(-P_{11} + \frac{1}{2}\sigma_1^2\right)\omega_1 U_1^2 + \left(-P_{22} + \frac{1}{2}\sigma_2^2\right)\omega_2 U_2^2$$

$$= -U^T Q U$$

(41)

where

$$Q = \begin{bmatrix} \left(P_{11} - \frac{1}{2}\sigma_1^2\right)\omega_1 & 0 \\ 0 & \left(P_{22} - \frac{1}{2}\sigma_2^2\right)\omega_2 \end{bmatrix}.$$

The relations (37) and (38) imply that Q is a real symmetric positive definite matrix and therefore all its eigenvalues $\lambda_i(Q)$, $i = 1, 2$ are positive real numbers. Let $\lambda_m = min\{\lambda_i(Q), i = 1, 2\}$, $\lambda_m > 0$. From (41), we get

$$LV(U(t)) \le -\lambda_m |U(t)|^2.$$

If the conditions in Theorem 3.2 hold true, then the zero solutions of the system (18) are exponentially mean square stable.

Hence the proof. □

Thus we observed analytically that under certain threshold on σ_i's the deterministic stable system remains stable under stochastic perturbation, which also agrees with our numerical result, see Fig. 8. But when σ becomes greater than the threshold value given in Theorem 3.2, we observed that the bi-stable points obtained for the system (1) with Table 1 shows scattered dots (Fig. 9). Figure 10 confirms that the system (1) loses its bistability under stochastic perturbation for high noise intensity.

Following similar arguments, one can prove the following theorem for the stochastic differential system (22).

Theorem 3.3 *When the following inequality holds true*

$$\frac{k_B k_{mB} B^*}{(k_{mB} + (1 - B^*))^2} + \frac{k_1 k_{m1} A^*}{(k_{m1} + B^*)^2} > \frac{k_B(1 - B^*)}{k_{mB} + (1 - B^*)}$$

(42)

then the zero solutions of the system (22) will be exponentially 2-stable if

$$\sigma_1^2 < 2\left[\frac{k_I k_{mI} I}{(k_{mI} + (1 - A^*))^2} + \frac{k_2 k_{m2} B^*}{(k_{m2} + (1 - A^*))^2} + \frac{k_A A^*(2k_{ma} + A^*)}{(k_{mA} + A^*)^2}\right]$$

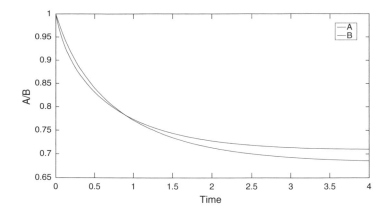

Fig. 8 The figure showing stability of the system (18) under stochastic perturbation for $\sigma_{1,2} = 0.1$

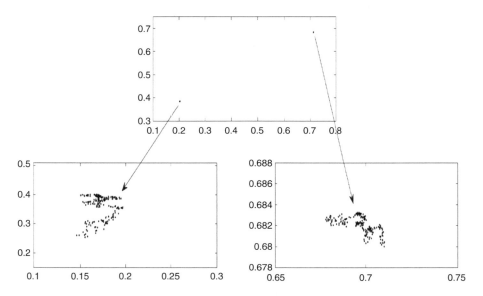

Fig. 9 Phase plane diagram for the system (18). Top figure shows stable nature of E^* for low value of $\sigma_{1,2} = 0.1$ and the bottom figures show the probability clouds for $\sigma_{1,2} = 1.3$, above the threshold value

$$\sigma_2{}^2 < 2\left[\frac{k_B k_{mB} B^*}{(k_{mB} + (1 - B^*))^2} + \frac{k_1 k_{m1} A^*}{(k_{m1} + B^*)^2} - \frac{k_B(1 - B^*)}{k_{mB} + (1 - B^*)}\right]$$

with positive constants ω_1 *and* ω_2 *are* $\omega_1 = \frac{k_{m2} + (1 - A^*)}{k_2(1 - A^*)}$, $\omega_2 = \frac{k_{m1} + B^*}{k_1 B^*}$.

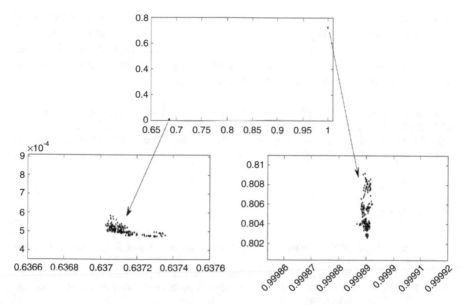

Fig. 10 Phase plane diagram for the system (22). Top figure shows stable nature of E^* for low value of $\sigma_{1,2} = 0.1$ and the bottom figures show the probability clouds for $\sigma_{1,2} = 1.2$, above the threshold value

The behaviour of the system (22) for lower and higher values of sigma than its threshold value (given in Theorem 3.3) is presented in Fig. 10. It shows that the bistable equilibrium points have been replaced by two clods.

4   Discussion

In the stochastic environment of diverse physical and physiological stimuli, the biological system displays remarkable robustness. Some of the attributes that impart robustness to both external and internal perturbations include topological features of the signaling network [3]. The topological features can further be weighted in their magnitude of influence depending on net concentration of the constituent nodes as well as stochastic variations in their level owing to various intrinsic mechanisms [11]. However the contribution of these features towards overall robustness and sensitivity of biological networks are still to be understood.

In the present article, we studied two well-observed motif structures which show bistability, i.e., depending upon the initial conditions the final outcome can take any of the two steady state values. We observed that the range of output signal depends on the structure but the sensitivity of the parameter is independent of the structure. In both the structures, it is the downstream node which is more sensitive in the outcome of output signal. We also observed that under random perturbation with high noise intensity, the systems loses its stability and the bistable points scattered leading to undesirable output signal. This is a small study focusing on only two specific structures, but it shows the importance of the structure and the noise in the signalling mechanism. In future we will extend our study on other structures and on higher dimension with three and possibly four nodes.

Acknowledgements This work is supported by DBT (Govt. of India), Grant no. BT/PR13086/ BRB/10/1380/2015. Samrat Chatterjee thanks to the International Union of Biological Sciences (IUBS) for partial support of living expenses in Moscow, during the 17th BIOMAT International Symposium, October 29–November 04, 2017.

References

1. U. Alon, Network motifs: theory and experimental approaches. Nat. Rev. Genet. **8**, 450–461 (2007)
2. C.P. Bagowski, J.E. Ferrell Jr., Bistability in the JNK cascade. Curr. Biol. **11**, 1176–1182 (2001)
3. Y. Bar-Yam, I.R. Epstein, Response of complex networks to stimuli. Proc. Natl. Acad. Sci. USA **101**, 4341–4345 (2004)
4. U.S. Bhalla, R. Iyengar, Emergent properties of networks of biological signaling pathways. Science **283**, 381–387 (1999)
5. U.S. Bhalla, P.T. Ram, R. Iyengar, MAP kinase phosphatase as a locus of flexibility in a mitogen-activated protein kinase signaling network. Science **297**, 1018–1023 (2002)
6. R.P. Bhattacharyya, A. Remenyi, B.J. Yeh, W.A. Lim, Domains, motifs, and scaffolds: the role of modular interactions in the evolution and wiring of cell signaling circuits. Annu. Rev. Biochem. **75**, 655–680 (2006)
7. M.J. Bissell, D. Radisky, Putting tumours in context. Nat. Rev. Cancer **1**, 46–54 (2001)
8. S.M. Blower, H. Dowlatabadi, Sensitivity and uncertainty analysis of complex models of disease transmission: an HIV model, as an example. Int. Stat. Rev. **62**, 229–243 (1994)
9. O. Brandman, T. Meyer, Feedback loops shape cellular signals in space and time. Science **322**, 390–395 (2008)
10. O. Brandman, J.E. Ferrell Jr., R. Li, T. Meyer, Interlinked fast and slow positive feedback loops drive reliable cell decisions. Science **310**, 496–498 (2005)
11. F.J. Bruggeman, N. Bluthgen, H.V. Westerhoff, Noise management by molecular networks. PLoS Comput. Biol. **5**, e1000506 (2009)
12. M. Carletti, On the stability properties of a stochastic model for phage-bacteria interaction in open marine environment. Math. Biosci. **175**, 117–131 (2002)
13. T. Eissing, H. Conzelmann, E.D. Gilles, F. Allgower, E. Bullinger, P. Scheurich, Bistability analyses of a caspase activation model for receptor-induced apoptosis. J. Biol. Chem. **279**(35), 36892–36897 (2004)
14. J.E. Ferrell Jr., Self-perpetuating states in signal transduction: positive feedback, double-negative feedback and bistability. Curr. Opin. Cell Biol. **14**, 140–148 (2002)
15. J.E. Ferrell, R.R. Bhatt, Mechanistic studies of the dual phosphorylation of mitogen-activated protein kinase. J. Biol. Chem. **272**, 19008–19016 (1997)
16. A. Ghaffarizadeh, N.S. Flann, G.J. Podgorski, Multistable switches and their role in cellular differentiation networks. BMC Bioinf. **15**(Suppl. 7), S7 (2014)
17. L.H. Hartwell, J.J. Hopfield, S. Leibler, A.W. Murray, From molecular to modular cell biology. Nature **402**, C47–52 (1999)
18. T. Helikar, J. Konvalina, J. Heidel, J.A. Rogers, Emergent decision-making in biological signal transduction networks. Proc. Natl. Acad. Sci. USA **105**, 1913–1918 (2008)

19. Q.A. Justman, Z. Serber, J.E. Ferrell Jr., H. El-Samad, K.M. Shokat, Tuning the activation threshold of a kinase network by nested feedback loops. Science **324**, 509–512 (2009)
20. B.N. Kholodenko, Cell-signalling dynamics in time and space. Nat. Rev. Mol. Cell Biol. **7**, 165–176 (2006)
21. D. Kumar, R. Srikanth, H. Ahlfors, R. Lahesmaa, K.V. Rao, Capturing cell-fate decisions from the molecular signatures of a receptor-dependent signaling response. Mol. Syst. Biol. **3**, 150 (2007)
22. W. Ma, A. Trusina, H. El-Samad, W.A. Lim, C. Tang, Defining network topologies that can achieve biochemical adaptation. Cell **138**, 760–773 (2009)
23. S. Mangan, U. Alon, Structure and function of the feed-forward loop network motif. Proc. Natl. Acad. Sci. USA **100**, 11980–11985 (2003)
24. L.A. Martinez, Y. Chen, S.M. Fischer, C.J. Conti, Coordinated changes in cell cycle machinery occur during keratinocyte terminal differentiation. Oncogene **18**, 397–406 (1999)
25. N. Nagumo, Uber die Lage der Integralkurven gewonlicher Dierantialgleichungen. Proc. Phys. Math. Soc. Jpn. **24**, 551 (1942)
26. S.H. Strogatz, Exploring complex networks. Nature **410**, 268–276 (2001)
27. G. Weng, U.S. Bhalla, R. Iyengar, Complexity in biological signaling systems. Science **284**, 92–96 (1999)
28. T. Wilhelm, The smallest chemical reaction system with bistability. BMC Syst. Biol. **3**, 90 (2009)
29. J.W. Williams, X. Cui, A. Levchenko, A.M. Stevens, Robust and sensitive control of a quorum-sensing circuit by two interlocked feedback loops. Mol. Syst. Biol. **4**, 234 (2008)
30. H.C. Yen, Q. Xu, D.M. Chou, Z. Zhao, S.J. Elledge, Global protein stability profiling in mammalian cells. Science **322**, 918–923 (2008)

Modelling the Adaptive Immune Response in HIV Infection with Three Saturated Rates and Therapy

Karam Allali

1 Introduction

HIV stands for the human immunodeficiency virus disease. This pathogen causes the acquired immunodeficiency syndrome (AIDS) which is considered as the last stage of this severe disease. After this stage is reached, the immune system fails to play its essential role which is to protect the whole body [3, 24]. This deficiency is due to the loss of the large majority of CD4$^+$ T cells by the HIV free viruses, reducing their amount to a small number does not exceed 200 cells per μl. In order to study and to better understand this deadly disease many mathematical models describing HIV dynamics were developed (see, for instance, [2, 5, 11, 14, 16, 20], and the references therein). The mathematical model representing the dynamics of HIV with CD4$^+$ T cells and the Cytotoxic T Lymphocytes (CTL) immune response, taking into account two saturated rates that describe viral infection and Cytotoxic T Lymphocytes proliferation is tackled in [20]. The authors study, among others, how the CTL immune response may reduce the HIV viral replication. In a recent work, the role of antibodies in minimizing the load of HIV viruses is studied in [21]. More recently, the same problem was considered by incorporating two kinds of therapies into the model [1], the objective of the first one is to reduce the infected cells number, however the role of the second is to obstruct the free viruses expansion. Indeed, there are two major types of antiretroviral drugs approved for treatment of individuals infected with HIV. These drugs are Reverse Transcriptase Inhibitors (RTIs) and Protease Inhibitors (PIs) [15]. Reverse Transcriptase Inhibitors (RTIs) is one of the chemotherapies which opposes the conversion of RNA of the virus to DNA (reverse transcription), so that the viral population will be minimum and on the other hand

K. Allali (✉)
Laboratory Mathematics and Applications, Department of Mathematics, Faculty of Sciences and Technologies, University Hassan II of Casablanca, Mohammedia, Morocco
e-mail: allali@fstm.ac.ma

© Springer International Publishing AG, part of Springer Nature 2018   265
R. P. Mondaini (ed.), *Trends in Biomathematics: Modeling, Optimization and Computational Problems*, https://doi.org/10.1007/978-3-319-91092-5_18

the CD4+ count remains higher and the host can survive. The Protease Inhibitors (PIs) prevents the production of viruses from the actively infected CD4$^+$ T-cells. In this paper, we extend the latter work [1] by incorporating an antibody proliferation saturated rate into the model and study the dynamics and the local stability of the new derived model. The dynamics of HIV infection with CTL, antibody responses and three saturated rates that we consider is given by the following nonlinear system of differential equations:

$$
\begin{cases}
\dfrac{dT}{dt} = s - dT - \dfrac{(1-\eta)\beta VT}{1+aV} + \rho I, \\[2mm]
\dfrac{dI}{dt} = \dfrac{(1-\eta)\beta VT}{1+aV} - (\delta+\rho)I - pIZ, \\[2mm]
\dfrac{dV}{dt} = (1-\epsilon)N\delta I - \mu V - qVW, \\[2mm]
\dfrac{dW}{dt} = \dfrac{gVW}{1+\gamma V} - hW, \\[2mm]
\dfrac{dZ}{dt} = \dfrac{cIZ}{1+\alpha I} - bZ.
\end{cases}
\tag{1}
$$

With the initial conditions $T(0) = T_0$, $I(0) = I_0$, $V(0) = V_0$, $Z(0) = Z_0$ and $W(0) = W_0$

In this model, T, I, V, Z and W denote the concentration of uninfected cells, infected cells, free virus, CTL cells and antibodies, respectively. Susceptible host cells CD4$^+$ T cells are produced at a rate s, die at a rate dT and become infected by virus at a rate $\dfrac{(1-\eta)\beta VT}{1+aV}$. Infected cells die at a rate δI and are killed by the CTL response at a rate pIZ. Free virus is produced by infected cells at a rate $(1-\epsilon)N\delta I$ and decays in the presence of antibodies at a rate $\mu V + qVW$. CTLs expand in response to viral antigen derived from infected cells at a rate $\dfrac{cIZ}{1+\alpha I}$ and decay in the absence of antigenic stimulation at a rate bZ. Antibodies develop in response to free virus at a rate $\dfrac{gVW}{1+\gamma V}$ and decay at a rate hW. The new parameters to the model η and ϵ stand for the two treatments which measure the efficacy of reverse transcriptase inhibitor and protease inhibitor, respectively.

Note that this model (1) includes a cure rate ρ of the infected cells that reverted to the uninfected state by loss of all cccDNA from their nucleus, these cells are called based gene therapy [10, 12, 17, 25]. The model contains also three saturated rates, the first is the saturated mass action [18, 19] which describe better the rate of viral infection while the second and the third are the saturated functions describing CTL and antibody proliferation when they are reduced by the presence of immune impairment effects caused by HIV infection [9]. The aim of this present work is to study the dynamics of the problem (1) and to highlight the effect of the incorporated third saturated rate and therapy in HIV dynamics.

The rest of the paper is organized as follows. In Sect. 2, we study the positivity and boundedness of solutions. The analysis of the model is described in Sect. 3. Results obtained by numerical simulations are given in Sect. 4 and we conclude in the last section.

2 Positivity and Boundedness of Solutions

For the problems deal with cell population evolution, the cell densities should remain non-negative and bounded. In this section, we will establish the positivity and boundedness of solutions of the model (1). First of all, for biological reasons, the parameters T_0, I_0, V_0, W_0 and Z_0 must be larger than or equal to 0. Hence, we have the following result:

Proposition 2.1 *The solutions of the problem* (1) *exist. Moreover, they are bounded, nonnegative and verify:*

(i) $T_1(t) \leq T_1(0) + \dfrac{s}{\delta_1}$,

(ii) $V(t) \leq V(0) + \dfrac{(1-\epsilon)N\delta}{\mu}\|I\|_\infty$,

(iii) $W(t) \leq W(0) + \dfrac{g}{q}\Big[\max\Big(1; 2 - \dfrac{\mu}{h}\Big) V(0)$

$+ \Big(\dfrac{(1-\epsilon)N\delta}{\mu} + \dfrac{(1-\epsilon)N\delta}{h}\Big)\|I\|_\infty\Big]$,

(iv) $Z(t) \leq Z(0) + \dfrac{c}{p}\Big[\max\Big(1; 2 - \dfrac{d}{b}\Big) T(0) + I(0) + \max\Big(\dfrac{s}{b}; \dfrac{s}{d}\Big)$

$+ \max\Big(0; 1 - \dfrac{\delta}{b}\Big)\|I\|_\infty\Big]$,

where $T_1(t) = T(t) + I(t)$ *and* $\delta_1 = \min(d; \delta)$.

Proof Simple application of Proposition A.1 in [22] shows $(T(t), I(t), V(t), W(t), Z(t)) \in \mathbb{R}_+^5$. By adding the first and second equation in (1), we have $\dot{T}_1 = s - dT - \delta I - pIZ$, thus

$$T_1(t) \leq T_1(0)e^{-\delta_1 t} + \frac{s}{\delta_1}(1 - e^{-\delta_1 t})$$

since $0 \leq e^{-\delta_1 t} \leq 1$ and $1 - e^{-\delta_1 t} \leq 1$, we deduce (i).

From the equation $\dot{V} = (1-\epsilon)N\delta I - \mu V - qVW$, we have

$$V(t) \leq V(0)e^{-\mu t} + (1-\epsilon)N\delta \int_0^t I(\xi)e^{(\xi-t)\mu}d\xi$$

then,

$$V(t) \le V(0) + \frac{(1-\epsilon)N\delta}{\mu} \|I\|_\infty (1 - e^{-\mu t})$$

Since $1 - e^{-\mu t} \le 1$, we have (ii).

The two equations $\dot{V} = (1-\epsilon)N\delta I - \mu V - qVW$ and $\dot{W} = gVW - hW$ imply

$$\dot{W} + hW = \frac{gVW}{1+\gamma V} \le gVW = \frac{g}{q}\left((1-\epsilon)N\delta I - (\dot{V} + \mu V)\right)$$

then,

$$W(t) = W_0 e^{-ht} + \frac{g}{q}\left\{ \int_0^t [(1-\epsilon)N\delta I(\xi) + (h-\mu)V(s)]e^{h(\xi-t)}d\xi - V(t) \right.$$
$$\left. + V(0)e^{-ht} \right\}.$$

If $h - \mu \le 0$, then we have

$$W(t) \le W_0 + \frac{g}{q}\left\{ \frac{(1-\epsilon)N\delta}{h} \|I\|_\infty + V(0) \right\}$$

else, we will have

$$W(t) \le W_0 + \frac{g}{q}\left\{ \left(\frac{(1-\epsilon)N\delta}{h} + \frac{(1-\epsilon)N\delta}{\mu}\right) \|I\|_\infty + \left(2 - \frac{\mu}{h}\right)V(0) \right\}.$$

From the two last inequalities, we deduce (iii).

Finally, from the equation $\dot{Z} = \frac{cIZ}{1+\alpha I} - bZ$ we have

$$\dot{Z} + bZ \le cIZ.$$

Since $cIZ = \frac{c}{p}[s - (\dot{T} + dT) - (\dot{I} + \delta I)]$, we get

$$Z(t) \le \left[\frac{c}{p}\left(T(0) + I(0) - \frac{s}{b}\right) + Z(0)\right]e^{-bt} + \frac{c}{p}\left\{ \frac{s}{b} + \int_0^t [(b-d)T(\xi) \right.$$
$$\left. + (b-\delta)I(\xi)]e^{b(\xi-t)}d\xi - T(t) - I(t) \right\}.$$

Following the same reasoning as in the previous cases for each sign of the $(b-d)$ and $(b-\delta)$, we will deduce (iv). $\square$

3 Analysis of the Model

In this section, we show that there exist a disease free equilibrium point and four infection equilibrium points, we study the stability of these equilibrium points and we will give some numerical simulations.

3.1 Stability of the Disease-Free Equilibrium

System (1) has an infection-free equilibrium $E_f = \left(\frac{s}{d}, 0, 0, 0, 0\right)$, corresponding to the maximal level of healthy CD4$^+$ T-cells. In this case, the disease cannot invade the cell population. By a simple calculation, the basic reproduction number of (1) is given by

$$R_0 = \frac{(1 - \theta)\beta N \delta s}{d\mu(\delta + \rho)}. \tag{2}$$

Here, we put $\theta = \eta + \epsilon - \eta\epsilon$, which represents the combined efficacy of the two drugs. Then $1 - \theta = (1 - \eta)(1 - \epsilon)$ which implies that each drug acts independently.

At any arbitrary point, the Jacobian matrix of the system (1) is given by

$$J = \begin{pmatrix} -d - \dfrac{(1-\eta)\beta V}{1+aV} & \rho & -\dfrac{(1-\eta)\beta T}{(1+aV)^2} & 0 & 0 \\ \dfrac{(1-\eta)\beta V}{1+aV} & -(\delta+\rho) - pZ & \dfrac{(1-\eta)\beta T}{(1+aV)^2} & 0 & -pI \\ 0 & (1-\epsilon)N\delta & -\mu - qW & -qV & 0 \\ 0 & 0 & \dfrac{gW}{(1+\gamma V)^2} & \dfrac{gV}{1+\gamma V} - h & 0 \\ 0 & \dfrac{cZ}{(1+\alpha I)^2} & 0 & 0 & \dfrac{cI}{1+\alpha I} - b \end{pmatrix} \tag{3}$$

Proposition 3.1

(1) The disease-free equilibrium, E_f, is locally asymptotically stable for $R_0 < 1$.
(2) The disease-free equilibrium, E_f, is unstable for $R_0 > 1$.

Proof At the disease-free equilibrium, E_f, the Jacobian matrix is given as follows:

$$J_{E_f} = \begin{pmatrix} -d & \rho & -\dfrac{(1-\eta)\beta s}{d} & 0 & 0 \\ 0 & -(\delta+\rho) & \dfrac{(1-\eta)\beta s}{d} & 0 & 0 \\ 0 & (1-\epsilon)N\delta & -\mu & 0 & 0 \\ 0 & 0 & 0 & -h & 0 \\ 0 & 0 & 0 & 0 & -b \end{pmatrix} \tag{4}$$

The characteristic polynomial of J_{E_f} is

$$P_{E_f}(\xi) = (\xi + d)(\xi + b)(\xi + h)[\xi^2 + (\delta + \rho + \mu)\xi + (\delta + \rho)\mu(1 - R_0)],$$

then the eigenvalues of the matrix J_{E_f} are

$$\xi_1 = -d,$$

$$\xi_2 = -b,$$

$$\xi_3 = -h,$$

$$\xi_4 = \frac{-(\delta + \rho + \mu) - \sqrt{(\delta + \rho + \mu)^2 - 4(\delta + \rho)\mu(1 - R_0)}}{2},$$

$$\xi_5 = \frac{-(\delta + \rho + \mu) + \sqrt{(\delta + \rho + \mu)^2 - 4(\delta + \rho)\mu(1 - R_0)}}{2},$$

it is clear that ξ_1, ξ_2, ξ_3 and ξ_4 are negative. Moreover, ξ_5 is negative when $R_0 < 1$, which means that E_f is locally asymptotically stable.   □

3.2 Infection Steady States

In this section, we focus on the existence and stability of the infection steady states. All these steady states exist when the basic reproduction number exceeds the unity and the disease invasion is always possible. In fact, it is easily verified that the system (1) has four of them:
$E_1 = (T_1, I_1, V_1, 0, 0)$, where

$$T_1 = \frac{s}{d}\left[\frac{a(1 - \epsilon)Ns + \mu}{a(1 - \epsilon)Ns + \mu R_0}\right], \qquad I_1 = \frac{s}{\delta}\left[\frac{\mu(R_0 - 1)}{a(1 - \epsilon)Ns + \mu R_0}\right],$$

$$V_1 = \frac{(1 - \epsilon)Ns(R_0 - 1)}{a(1 - \epsilon)Ns + \mu R_0},$$

$E_2 = (T_2, I_2, V_2, W_2, 0)$, where

$$V_2 = \frac{h}{g - h\gamma},$$

$$T_2 = \frac{s(\delta + \rho)(aV_2 + 1)}{\beta\delta(1 - \eta)V_2 + d(\delta + \rho)(aV_2 + 1)},$$

$$I_2 = \frac{\beta s(1 - \eta)V_2}{\beta\delta(1 - \eta)V_2 + d(\delta + \rho)(aV_2 + 1)},$$

$$W_2 = \frac{\mu}{q}\left[\frac{\beta s N\delta(1-\theta)}{\mu\beta\delta(1-\eta)V_2 + d\mu(\delta+\rho)(aV_2+1)} - 1\right],$$

$E_3 = (T_3, I_3, V_3, 0, Z_3)$, where

$$I_3 = \frac{b}{c-\alpha b}, \qquad V_3 = \frac{N\delta(1-\epsilon)}{\mu I_3},$$

$$T_3 = \frac{a\mu s V_3^2 + a\rho N\delta(1-\epsilon)V_3 + \mu s V_3 + \rho N\delta(1-\epsilon)}{\mu d(1+aV_3)V_3 + \mu\beta(1-\eta)V_3^2},$$

$$Z_3 = \frac{1}{p}\left(\frac{s}{I_3} - \frac{dT_3}{I_3} - \delta\right),$$

and $E_4 = (T_4, I_4, V_4, W_4, Z_4)$, where

$$I_4 = \frac{b}{c-\alpha b}, \qquad V_4 = \frac{-h}{(-g+h\gamma)},$$

$$T_4 = \frac{(s+\rho I_4)(1+aV_4)}{d(1+aV_4)+(1-\eta)\beta V_4},$$

$$W_4 = \frac{1}{q}\left(\frac{(1-\epsilon)N\delta I_4}{V_4} - 1\right), \qquad Z_4 = \frac{1}{p}\left(\frac{s}{I_4} - \frac{dT_4}{I_4} - \delta\right).$$

Here the endemic equilibrium point E_1 represents the equilibrium case in the absence of the adaptive immune response (CTLs and antibody responses). The endemic equilibria points E_2 and E_3 represent the equilibrium case in the presence of only one kind of the adaptive immune response antibody response and CTL response, respectively. While the last endemic equilibrium point E_4 represents the equilibrium case of chronic HIV infection with the presence of both kinds of adaptive immune response CTLs and antibody type. In order to study the local stability of the points E_1, E_2, E_3 and E_4, we first define the following numbers:

$$D_0^W = \frac{(1-\epsilon)gNs}{h\mu}, \qquad \widetilde{D_0^W} = D_0^W \frac{\mu R_0}{a(1-\epsilon)Ns + \mu R_0 + (1-\epsilon)N\gamma s(R_0-1)},$$

$$H_0^W = \frac{1}{\frac{1}{R_0} + \frac{1}{\widetilde{D_0^W}}},$$

$$D_0^Z = \frac{cs}{b\delta}, \qquad \widetilde{D_0^Z} = D_0^Z \frac{\mu\delta R_0}{(a(1-\epsilon)Ns + \mu R_0) + \alpha\mu s(R_0-1)},$$

$$H_0^Z = \frac{1}{\frac{1}{R_0} + \frac{1}{\widetilde{D_0^Z}}}.$$

and

$$H_0^{W,Z} = \frac{ch\beta s(1 - \eta)}{\beta bh(1 - \eta)(\alpha s + \delta) + cd(\delta + \rho)(ah + g - h\gamma)},$$

where D_0^Z represents the CTL immune response reproduction number, D_0^W represents the antibody immune response reproduction number, H_0^W is the half harmonic mean of R_0 and $\widetilde{D_0^W}$ and H_0^Z is the half harmonic mean of R_0 and $\widetilde{D_0^Z}$.

For the first point E_1, we have the following result:

Proposition 3.2

(1) If $R_0 < 1$, then the point E_1 does not exist.
(2) If $R_0 = 1$, then $E_1 = E_f$.
(3) If $R_0 > 1$, then E_1 is locally asymptotically stable for $H_0^W < 1$, and $H_0^Z < 1$; however it is unstable for $H_0^W > 1$ or $H_0^Z > 1$.

Proof It easy to see that if $R_0 < 1$, then the point E_1 does not exist and if $R_0 = 1$ the two points E_1 and E_f coincide. If $R_0 > 1$, the Jacobian matrix at E_1 is given by

$$J_{E_1} = \begin{pmatrix} -d - \dfrac{(1-\eta)\beta V_1}{1 + aV_1} & \rho & -\dfrac{(1-\eta)\beta T_1}{(1 + aV_1)^2} & 0 & 0 \\ \dfrac{(1-\eta)\beta V_1}{1 + aV_1} & -(\delta + \rho) & \dfrac{(1-\eta)\beta T_1}{(1 + aV_1)^2} & 0 & -pI_1 \\ 0 & (1 - \epsilon)N\delta & -\mu & -qV_1 & 0 \\ 0 & 0 & 0 & \dfrac{gV_1}{1 + \gamma V_1} - h & 0 \\ 0 & 0 & 0 & 0 & \dfrac{cI_1}{1 + \alpha I_1} - b \end{pmatrix}$$

then, its characteristic equation is

$$\left(\xi + h - \frac{gV_1}{1 + \gamma V_1}\right)\left(\xi + b - \frac{cI_1}{1 + \alpha I_1}\right)(\xi^3 + a_1\xi^2 + a_2\xi + a_3) = 0,$$

where

$$a_1 = d + \delta + \mu + \rho + \frac{(1-\eta)\beta V_1}{1 + aV_1},$$

$$a_2 = (\delta + \mu + \rho)d + (\mu + \delta)\frac{(1-\eta)\beta V_1}{1 + aV_1} + \mu(\delta + \rho) - \frac{(1-\theta)N\delta\beta T_1}{(1 + aV_1)^2},$$

$$a_3 = \mu d(\delta + \rho) + \frac{\mu\delta(1-\eta)\beta V_1}{1 + aV_1} - \frac{(1-\theta)N\delta\beta T_1 d}{(1 + aV_1)^2},$$

Simple calculation leads to $\dfrac{g V_1}{1 + \gamma V_1} - h = \dfrac{h \widetilde{D_0^W} (H_0^W - 1)}{H_0^W}$ and $\dfrac{c I_1}{1 + \alpha I_1} - b = $

$\dfrac{b \widetilde{D_0^Z} (H_0^Z - 1)}{H_0^Z}$. They are two eigenvalues of J_{E_1}. The sign of the eigenvalue

$\dfrac{h \widetilde{D_0^W} (H_0^W - 1)}{H_0^W}$ is negative if $H_0^W < 1$, zero if $H_0^W = 1$ and positive if $H_0^W > 1$.

The sign of the eigenvalue $\dfrac{b \widetilde{D_0^Z} (H_0^Z - 1)}{H_0^Z}$ is negative if $H_0^Z < 1$, zero if $H_0^Z = 1$

and positive if $H_0^Z > 1$. On the other hand, we have $a_1 > 0$ and $a_1 a_2 - a_3 > 0$
(as $R_0 > 1$). From the Routh-Hurwitz Theorem [7], the other eigenvalues of the
above matrix have negative real parts. Consequently, E_1 is unstable when $H_0^W > 1$
or $H_0^Z > 1$ and locally asymptotically stable when $R_0 > 1$, $H_0^W < 1$ and $H_0^Z < 1$.

$\square$

For the second endemic-equilibrium point E_2, we have the following result:

Proposition 3.3

(1) If $\gamma > \dfrac{g}{h}$ or $H_0^W < 1$, then the point E_2 does not exist.

(2) If $H_0^W = 1$ then $E_2 = E_1$.

(3) If $H_0^W > 1$ and $\gamma < \dfrac{g}{h}$ then E_2 is locally asymptotically stable for $H_0^{W,Z} < 1$
and unstable for $H_0^{W,Z} > 1$.

Proof We notice that the condition $H_0^W > 1$ is equivalent to $V_2 < V_1$. It easy to
verify that the point E_2 does not exist if $H_0^W < 1$; moreover, we have $E_2 = E_1$
when $H_0^W = 1$. Now, we assume that $H_0^W > 1$, the Jacobian matrix at E_2 is

$$J_{E_2} = \begin{pmatrix} -d - \dfrac{(1-\eta)\beta V_2}{1 + a V_2} & \rho & -\dfrac{(1-\eta)\beta T_2}{(1 + a V_2)^2} & 0 & 0 \\ \dfrac{(1-\eta)\beta V_2}{1 + a V_2} & -\rho - \delta & \dfrac{(1-\eta)\beta T_2}{(1 + a V_2)^2} & 0 & -p I_2 \\ 0 & (1-\epsilon)N\delta & -\mu - q W_2 & -q V_2 & 0 \\ 0 & 0 & \dfrac{g W_2}{(1 + \gamma V_2)^2} & \dfrac{g V_2}{1 + \gamma V_2} - h & 0 \\ 0 & 0 & 0 & 0 & \dfrac{c I_2}{1 + \alpha I_2} - b \end{pmatrix}$$

The characteristic equation associated with J_{E_2} is given by

$$\left(\dfrac{c I_2}{1 + \alpha I_2} - b - \xi \right) (\xi^4 + b_1 \xi^3 + b_2 \xi^2 + b_3 \xi + b_4) = 0, \tag{5}$$

where

$b_1 = qW_2 + \mu - gV_2Y + d + (1-\eta)\beta V_2X + \delta + \rho + h,$

$b_2 = h\mu + h\delta + h\rho + hd + \mu\delta + \mu\rho + \mu d + \delta d + \rho d + gW_2qV_2Y^2$

$\qquad - gV_2qW_2Y + qW_2(1-\eta)\beta V_2X - gV_2\mu Y - gV_2\delta Y - gV_2\rho Y - gV_2dY$

$\qquad - gV_2^2(1-\eta)\beta XY + hqW_2 + h(1-\eta)\beta V_2X - (1-\epsilon)N\delta(1-\eta)\beta T_2X^2$

$\qquad + \mu(1-\eta)\beta V_2X + qW_2\delta + qW_2\rho + qW_2d + \delta(1-\eta)\beta V_2X,$

$b_3 = hqW_2(1-\eta)\beta V_2X - gV_2qW_2\delta Y - gV_2qW_2\rho Y - gV_2qW_2dY$

$\qquad - gV_2^2qW_2(1-\eta)\beta XY + gV_2(1-\epsilon)N\delta(1-\eta)\beta T_2X^2Y$

$\qquad + qW_2\delta(1-\eta)\beta V_2X + gV_2qW_2\delta Y^2 + gV_2qW_2\rho Y^2 + gV_2qW_2dY^2$

$\qquad + gV_2^2qW_2(1-\eta)\beta XY^2 - h(1-\epsilon)N\delta(1-\eta)\beta T_2X^2$

$\qquad + hX\mu(1-\eta)\beta V_2X^2 + hqW_2\delta + hqW_2\rho + hqW_2d + h\delta(1-\eta)\beta V_2X$

$\qquad - gV_2\mu\delta Y - gV_2\mu\rho Y - gV_2\mu dY - gV_2^2\mu(1-\eta)\beta XY - gV_2\delta dY$

$\qquad - gV_2^2\delta(1-\eta)\beta XY - gV_2\rho dY + \mu\delta(1-\eta)\beta V_2X + qW_2\delta d + qW_2\rho d$

$\qquad - (1-\epsilon)N\delta(1-\eta)\beta T_2dX^2 + h\mu\delta + h\mu\rho + h\mu d + h\delta d + h\rho d + \mu\delta d + \mu\rho d,$

$b_4 = -gW_2qV_2\rho dY + h\mu\rho d - gV_2\mu\delta dY + hqW_2\delta(1-\eta)\beta V_2X - gV_2\mu\rho dY$

$\qquad + gW_2qV_2\delta dY^2 + gW_2qV_2^2\delta(1-\eta)\beta XY^2 - h(1-\epsilon)N\delta(1-\eta)\beta T_2dX^2$

$\qquad + h\mu\delta(1-\eta)\beta V_2X + h\mu\delta d + hqW_2\delta d + gW_2qV_2\rho dY^2 + hqW_2\rho d$

$\qquad - gW_2qV_2\delta dY - gV_2^2\mu\delta(1-\eta)\beta XY + gV_2(1-\epsilon)N\delta(1-\eta)\beta T_2dX^2Y$

$\qquad - gW_2qV_2^2X\delta(1-\eta)\beta X^2Y,$

here $X = 1/(1 + aV_2)$ and $Y = 1/(1 + \gamma V_2)$. Since $\dfrac{cI_2}{1+\alpha I_2} - b$ is an eigenvalue of J_{E_2}, by assuming $\dfrac{cI_2}{1+\alpha I_2} - b = b(H_0^{W,Z} - 1)$, we deduce that the sign of this eigenvalue is negative when $H_0^{W,Z} < 1$, zero when $H_0^{W,Z} = 1$ and positive for $H_0^{W,Z} > 1$. On the other hand, from the Routh-Hurwitz Theorem applied to the fourth order polynomial in the characteristic equation, the other eigenvalues of the above matrix have negative real parts when $H_0^{W,Z} < 1$ (since $b_1b_2 > b_3$ and $b_1b_2b_3 > b_3^4 + b_1^2b_4$). Consequently, E_2 is unstable when $H_0^W > 1$ and $H_0^{W,Z} > 1$ and locally asymptotically stable when $H_0^W > 1$ and $H_0^{W,Z} < 1$.   □

For the third endemic-equilibrium point E_3, we have the following result:

Proposition 3.4

(1) If $\alpha > \dfrac{c}{b}$ or $H_0^Z < 1$, then the point E_3 does not exist and $E_3 = E_2$ when $H_0^Z = 1$.

(2) If $\alpha < \dfrac{c}{b}$, $H_0^Z > 1$ and $gc^2 \left(1 - \dfrac{\alpha b}{c}\right) D_0^W < b^2 g D_0^Z + c^2 \left(1 - \dfrac{\alpha b}{c}\right) h\gamma$, then E_3 is locally asymptotically stable.

(3) If $\alpha < \dfrac{c}{b}$, $H_0^Z > 1$ and $gc^2 \left(1 - \dfrac{\alpha b}{c}\right) D_0^W > b^2 g D_0^Z + c^2 \left(1 - \dfrac{\alpha b}{c}\right) h\gamma$, then E_3 is unstable.

Proof We notice that the condition $\alpha < \dfrac{c}{b}$ and $H_0^Z > 1$ is equivalent to $I_3 < I_1$. It easy to verify that the point E_3 does not exist if $H_0^Z < 1$ or $\alpha > \dfrac{c}{b}$. Moreover, we have $E_3 = E_1$ for $H_0^Z = 1$. We assume now that $\alpha < \dfrac{c}{b}$ and $H_0^Z > 1$; the Jacobian matrix at E_3 is

$$
J_{E_3} = \begin{pmatrix}
-d - \dfrac{(1-\eta)\beta V_3}{C} & \rho & -\dfrac{(1-\eta)\beta T_3}{(1+aV_3)^2} & 0 & 0 \\[2ex]
\dfrac{(1-\eta)\beta V_3}{1+aV_3} & -\rho - \delta - pZ_3 & \dfrac{(1-\eta)\beta T_3}{(1+aV_3)^2} & 0 & -pI_3 \\[2ex]
0 & (1-\epsilon)N\delta & -\mu & -qV_3 & 0 \\[2ex]
0 & 0 & 0 & \dfrac{gV_3}{1+\gamma V_3} - h & 0 \\[2ex]
0 & \dfrac{cZ_3}{(1+\alpha I_3)^2} & 0 & 0 & \dfrac{cI_3}{1+\alpha I_3} - b
\end{pmatrix}
$$

The characteristic equation associated with J_{E_3} is given by

$$
\left(\xi - \dfrac{gV_3}{1+\gamma V_3} + h\right)(\xi^4 + c_1\xi^3 + c_2\xi^2 + c_3\xi + c_4) = 0 \tag{6}
$$

where

$c_1 = -cI_3 B_3 + \rho + d + B_1 + \delta + pZ_3 + \mu + b,$

$c_2 = -I_3 c B_3 d + I_3 c Z_3 B_3^2 p - I_3 c B_3 B_1 - I_3 c B_3 \rho + \delta d + N\delta B_2\epsilon - I_3 c B_3 \delta$
$\quad + \mu B_1 - I_3 c B_3 \mu - I_3 c B_3 pZ_3 + \delta B_1 + \rho d + \mu\rho + \mu pZ_3 + pZ_3 d + pZ_3 B_1$
$\quad + \mu\delta + \mu d - N\delta B_2$

$c_3 = -I_3 c B_3 pZ_3 d + \mu\delta B_1 - I_3 c B_3 pZ_3 B_1 + N\delta B_2\epsilon d + \mu pZ_3 B_1 - I_3 c B_3 \rho d$
$\quad + I_3 c B_3 N\delta B_2 - I_3 c B_3 \delta B_1 + I_3 c Z_3 B_3^2 p\mu - I_3 c B_3 \delta d - I_3 c B_3 \mu\delta$
$\quad + I_3 c Z_3 B_3^2 pd + I_3 c Z_3 B_3^2 pB_1 - I_3 c B_3 N\delta B_2\epsilon - I_3 c B_3 \mu B_1 - I_3 c B_3 \mu d$
$\quad + \mu pZ_3 d + \mu\rho d - I_3 c B_3 \mu pZ_3 + \mu\delta d - I_3 c B_3 \mu\rho - N\delta B_2 d$

$$c_4 = I_3 c Z_3 B_3^2 p\mu B_1 - I_3 c B_3 N\delta B_2 \epsilon d - I_3 c B_3 \mu \delta B_1 + I_3 c Z_3 B_3^2 p\mu d$$
$$- I_3 c B_3 \mu p Z_3 B_1 - I_3 c B_3 \mu \delta d + I_3 c B_3 N\delta B_2 d - I_3 c B_3 \mu p Z_3 d - I_3 c B_3 \mu \rho d,$$

here $B_1 = \dfrac{(1-\eta)\beta V_3}{1+aV_3}$, $B_2 = \dfrac{(1-\eta)\beta T_3}{(1+aV_3)^2}$ and $B_3 = (1+\alpha I_3)^{-1}$.

It is clear that $gV_3 - h = h\left(\dfrac{gc^2\left(1-\dfrac{\alpha b}{c}\right)D_0^W}{b^2 g D_0^Z + c^2\left(1-\dfrac{\alpha b}{c}\right)h\gamma} - 1\right)$ is an eigenvalue

of J_{E_3}. The sign of this eigenvalue is negative if $gc^2\left(1-\dfrac{\alpha b}{c}\right)D_0^W < b^2 g D_0^Z + c^2\left(1-\dfrac{\alpha b}{c}\right)h\gamma$.

Again by Routh-Hurwitz stability criterion, the other eigenvalues of the above matrix have negative real parts when $H_0^Z > 1$. We conclude that E_3 is unstable when $H_0^Z > 1$ and $D_0^W > \left(1-\dfrac{\alpha b}{c}\right)D_0^Z$ and locally asymptotically stable when $D_0^W < \left(1-\dfrac{\alpha b}{c}\right)D_0^Z$ and $H_0^Z > 1$. $\square$

Using the same reasoning as for the previous theorems, we finally have the following result concerning the last endemic-equilibrium point E_4:

Proposition 3.5

(1) *If $\alpha > \dfrac{c}{b}$ or $gc^2\left(1-\dfrac{\alpha b}{c}\right)D_0^W < b^2 g D_0^Z + c^2\left(1-\dfrac{\alpha b}{c}\right)h\gamma$ or $H_0^{W,Z} < 1$,*

   then the point E_4 does not exist. Moreover $E_4 = E_2$ when $H_0^{W,Z} = 1$ and $E_4 = E_3$ when $D_0^W = D_0^Z$

(2) *If $\alpha < \dfrac{c}{b}$, $gc^2\left(1-\dfrac{\alpha b}{c}\right)D_0^W > b^2 g D_0^Z + c^2\left(1-\dfrac{\alpha b}{c}\right)h\gamma$ and $H_0^{W,Z} > 1$,*

   then E_4 is locally asymptotically stable.

4   Numerical Simulations

For our numerical simulations, we have used the Euler finite-difference scheme method in order to discretize the five equations. The parameters of the simulations are inspired from [4, 6, 20, 21], while the other new three parameters to the problem (1); q, g and h are chosen adequately since they may vary with various types of antibodies [23]. Also, we will take into account the initial conditions approaching the clinical data values for HIV infected individuals during symptomatic phase [8]. Here, the chosen initial conditions are:

$$T_0 = 200, \quad I_0 = 80, \quad V_0 = 12{,}000, \quad W_0 = 50, \quad Z_0 = 100.$$

The chosen interval time of our simulation will vary between 100 days to 200 days. This choice is in good agreement with some biological studies since it was observed that some patients became negative for infectious HIV after 3–6 months of therapy [13]. Our numerical simulations are oriented towards the study of the effect of the antibody proliferation saturated rate and the role of therapy in HIV viral dynamics. We remark that this saturated rate has no effect on the dynamics when the numerical simulations are concerned with the stability of the steady states E_f, E_1 or E_3 since all the curves representing with and without the added saturated rate coincide (not shown). We explain this result by the fact that the fourth component of these steady states, which corresponds to antibodies, is zero. Figure 1 shows the evolution of the infection during the first 200 days for $s = 10$, $\beta = 0.000024$, $d = 0.02$, $\delta = 0.5$, $p = 0.001$, $N = 1200$, $\mu = 3$, $\rho = 0.01$, $a = 0.001$, $\alpha = 0.001$, $c = 0.03$, $b = 0.2$, $q = 0.001$, $g = 10^{-4}$, $h = 0.01$, $\eta = 0.55$ and $\epsilon = 0.45$. This case corresponds to the case of the endemic-equilibrium E_2. It is interesting to point out that with the antibody proliferation saturated rate a significant reduction of the antibodies amount is observed. However, no effect is observed on the other problem variables. Figure 2 depicts the evolution of HIV dynamics during the first days of observation. In this figure, we observe clearly that the antibody proliferation rate has a significant effect on the dynamics of HIV. Indeed, with this saturated rate, an increase of viral load is observed. This leads to a decrease of the healthy cells amount and an increase of the infected cells. Finally, Fig. 3 shows the behavior of the disease for the last endemic points E_4. The plots show that with therapy we may control better the infection than without treatments.

5 Conclusion

A mathematical model describing Human Immunodeficiency Virus (HIV) dynamics in the presence of the adaptive immune response, therapy and three saturated rate is studied in this work. The adaptive immunity is represented by the cytotoxic T lymphocytes (CTL) and the antibody immune responses. The three saturated are considered in order to better describe the viral infection, the CTL and the antibodies proliferations. In this work, attention is focused on the role of added proliferation saturated rate and therapy in HIV dynamics and in controlling viral replication. To this end, two types of treatments were incorporated to the model; the aim of the first one is to reduce the number of infected cells, while the objective of the second one is to block the free viruses. The positivity and the boundedness of solutions are proved; which is consistent with biological studies. Moreover, we have studied the stability of the disease-free equilibrium and endemic equilibria. It was established that the disease free steady state is locally asymptotically stable when the basic reproduction number is less than unity ($R_0 < 1$). When this basic reproduction number exceeds unity ($R_0 > 1$), four infection steady states were observed. Their local stability depends, beyond the basic reproduction number R_0, on the CTL immune response

278   K. Allali

Fig. 1 Behavior of the
infection during the time for
$s = 10$, $\beta = 0.000024$,
$d = 0.02$, $\delta = 0.5$,
$p = 0.001$, $N = 1200$,
$\mu = 3$, $\rho = 0.01$, $a = 0.001$,
$\alpha = 0.001$, $c = 0.03$,
$b = 0.2$, $q = 0.001$,
$g = 10^{-4}$, $h = 0.01$,
$\eta = 0.55$ and $\epsilon = 0.45$ which
correspond to the stability of
the endemic-equilibrium E_2

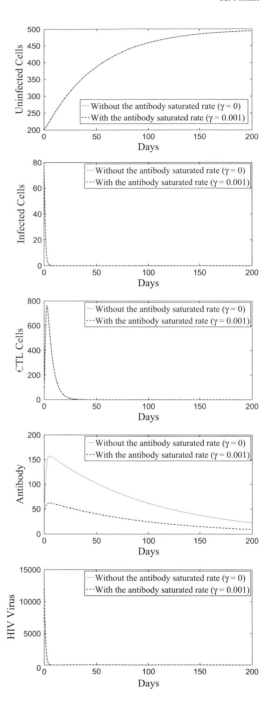

Fig. 2 Behavior of the
infection during the time for
$s = 10$, $\beta = 0.000024$,
$d = 0.02$, $\delta = 0.5$,
$p = 0.001$, $N = 1200$,
$\mu = 3$, $\rho = 0.01$, $a = 0.001$,
$\alpha = 0.001$, $c = 0.03$,
$b = 0.2$, $q = 0.5$, $g = 10^{-4}$,
$h = 0.1$, $\eta = 0.05$ and
$\epsilon = 0.2$, which correspond to
the stability of the
endemic-equilibrium E_4

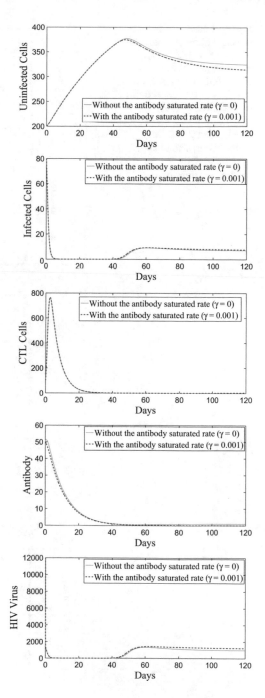

280

K. Allali

Fig. 3 Behavior of the infection during the time for $s = 10$, $\beta = 0.000024$, $d = 0.02$, $\delta = 0.5$, $p = 0.001$, $N = 1200$, $\mu = 3$, $\rho = 0.01$, $a = 0.001$, $\alpha = 0.001$, $c = 0.03$, $b = 0.2$, $q = 0.5$, $g = 10^{-4}$, $h = 0.1$ and $\gamma = 0.001$, which correspond to the stability of the endemic-equilibrium E_4

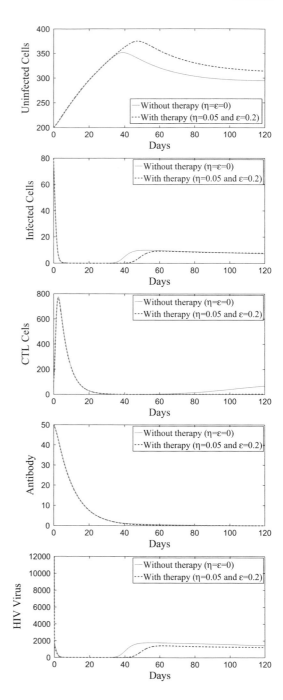

reproduction number D_0^Z and the antibody immune response reproduction number D_0^W. In addition, numerical simulations are performed in order to show the behavior of infection during the first days of therapy. In the presence of therapy, an increase of the uninfected cells and a decrease of the infected cells are observed. Therefore, the therapy plays an essential role in reducing the HIV viral load. Also, in the presence of the antibody proliferation saturated rate a change of HIV dynamics is observed. Hence, this saturated rate must be taken into consideration in HIV viral dynamics.

Acknowledgements Karam Allali thanks to the International Union of Biological Sciences (IUBS) for partial support of living expenses in Moscow, during the 17th BIOMAT International Symposium, October 29–November 04, 2017.

References

1. K. Allali, Y. Tabit, S. Harroudi, On HIV model with adaptive immune response, two saturated rates and therapy. Math. Model. Nat. Phenom. **12**(5), 1–14 (2017)
2. N. Bairagi, D. Adak, Global analysis of HIV-1 dynamics with Hill type infection rate and intracellular delay. Appl. Math. Model. **38**, 5047–5066 (2014)
3. W. Blanttner, R.C. Gallo, H.M. Temin, HIV causes aids. Science **241**, 515–516 (1988)
4. S. Butler, D. Kirschner, S. Lenhart, Optimal control of chemotherapy affecting the infectivity of HIV, in *Advances in Mathematical Population Dynamics: Molecules, Cells, Man*, ed. by O. Arino, D. Axelrod, M. Kimmel, M. Langlais (World Scientific Publishing, Singapore, 1997), pp. 104–120.
5. S. Cobey, P. Wilson, F.A. Matsen, The evolution within us. Philos. Trans. R. Soc. B **370**, 20140235 (2015)
6. R. Fister, S. Lenhart, J.S. McNally, Optimizing chemotherapy in an HIV model. J. Differ. Equ. **1998**, 1–12 (1998)
7. I.S. Gradshteyn, I.M. Ryzhik, Routh Hurwitz theorem, in *Tables of Integrals, Series, and Products*, 6th edn. (Academic Press, San Diego, 2000)
8. K. Hattaf, N. Yousfi, Two optimal treatments of HIV infection model. World J. Model. Simul. **8**, 27–35 (2012)
9. S. Iwami, T. Miura, S. Nakaoka, Y. Takeuchi, Immune impairment in HIV infection: existence of risky and immunodeficiency thresholds. J. Theor. Biol. **260**, 490–501 (2009)
10. R. Kaminski, R. Bella, C. Yin, J. Otte, P. Ferrante, H.E. Gendelman et al., Excision of HIV-1 DNA by gene editing: a proof-of-concept in vivo study. Gene Ther. **23**, 690–695 (2016)
11. D.E. Kirschner, Using mathematics to understand HIV immune dynamics. Not. Am. Math. Soc. **43**, 191–202 (1996)
12. X. Liu, H. Wang, W. Ma, Global stability of an HIV pathogenesis model with cure rate. Nonlinear Anal. Real World Appl. **12**, 2947–2961 (2011)
13. M. Markowitz, M. Saag, W.G. Powderly, A.M. Hurley, A. Hsu, J.M. Valdes et al., A preliminary study of ritonavir, an inhibitor of HIV-1 protease, to treat HIV-1 infection. N. Engl. J. Med. **333**, 1534–1540 (1995)
14. M.A. Nowak, R.M. May, Mathematical biology of HIV infection: antigenic variation and diversity threshold. Math. Biosci. **106**, 1–21 (1991)
15. Panel on Antiretroviral Guidelines for Adults and Adolescents. Guidelines for the use of antiretroviral agents in HIV-1-infected adults and adolescents. Department of Health and Human Services, 10 Jan 2011, pp. 1–166. Available at http://www.aidsinfo.nih.gov/ContentFiles/AdultandAdolescentGL.pdf. Accessed 29 Aug 2011

16. A.S. Perelson, P.W. Nelson, Mathematical analysis of HIV-1 dynamics in vivo. SIAM Rev. **41**, 3–44 (1999)
17. A.S. Perelson, P. Essunger, Y. Cao, M. Vesanen, A. Hurley, K. Saksela, M. Markowitz, D.D. Ho, Decay characteristics of HIV-1-infected compartments during combination therapy. Nature **387**, 188–191 (1997)
18. X. Song, A. Neumann, Global stability and periodic solution of the viral dynamics. J. Math. Anal. Appl. **329**(1), 281–297 (2007)
19. Q. Sun, L. Min, Dynamics analysis and simulation of a modified HIV infection model with a saturated infection rate. Comput. Math. Methods Med. **2014**, Article ID 145162, 14 pp. (2014)
20. Y. Tabit, K. Hattaf, N. Yousfi, Dynamics of an HIV pathogenesis model with CTL immune response and two saturated rates. World J. Model. Simul. **10**, 215–223 (2014)
21. Y. Tabit, A. Meskaf, K. Allali, Mathematical analysis of HIV model with two saturated rates, CTL and antibody responses. World J. Model. Simul. **12**, 137–146 (2016)
22. H.R. Thieme, *Mathematics in Population Biology* (Princeton University Press, Princeton, 2003)
23. X. Wang, W. Wang, An HIV infection model based on a vectored immunoprophylaxis experiment. J. Theor. Biol. **313**, 127–135 (2012)
24. R.A. Weiss, How does HIV cause AIDS? Science **260**, 1273–1279 (1993)
25. X. Zhou, X. Song, X. Shi, A differential equation model of HIV infection of CD4+ T-cells with cure rate. J. Math. Anal. Appl. **342**, 1342–1355 (2008)

Numerical Study on Biological Tissue Freezing Using Dual Phase Lag Bio-Heat Equation

Sushil Kumar and Sonalika Singh

1 Introduction

The heat transfer problems involving melting and freezing process are known as phase change problems and the phase change interface boundary movement depends on the absorption or liberation of latent heat. Phase change heat transfer is a wide area that finds applications in almost all engineering disciplines. Freezing and melting processes are related with high heat transfer rate. The heat transfer involved in phase change is necessary in biomedical applications such as cryopreservation and cryosurgery. Cryosurgery is a technique to treat tumour and can be used inside the body and on the skin. The extreme cold is used for freezing and destroying abnormal cells by introducing liquid nitrogen through cryoprobe into the targeted region. The aim of cryosurgery is to maximize the damage to undesired tissues within the define domain and minimizing the injury to the surrounding healthy tissues [1, 2].

Various models have been proposed to model the heat transport phenomena in blood perfuse tissues, e.g. Pennes Model [3], The Chen and Holmes (CH) Model [4], The Weinbaum, Jiji and Lemons (WJL) Model [5, 6], The Wainbaum and Jiji (WJ) Model [7]. Pennes bio-heat equation is the most widely applied model for temperature distribution in the living biological tissues. This is the earliest model for energy transport in tissues and is represented as

$$\rho c \frac{\partial T}{\partial t} = -\nabla.\mathbf{q} + (\rho c)_b w_b (T_b - T) + Q_m, \tag{1}$$

S. Kumar (✉) · S. Singh
Department of Applied Mathematics and Humanities, S. V. National Institute of Technology, Surat, Gujarat, India
e-mail: sushilk@amhd.svnit.ac.in

© Springer International Publishing AG, part of Springer Nature 2018
R. P. Mondaini (ed.), *Trends in Biomathematics: Modeling, Optimization and Computational Problems*, https://doi.org/10.1007/978-3-319-91092-5_19

where ρ is the density of tissue; k, thermal conductivity; c_b, specific heat of blood; w_b, blood perfusion rate; T, temperature; t, time; T_b, arterial blood temperature and Q_m is the metabolic heat generation in the tissue. In Pennes bioheat equation, the heat conduction in biological tissue is modeled by using Fourier's law

$$\mathbf{q}(t, X) = -k\nabla T(t, X) \tag{2}$$

where $\mathbf{q}$ and $T(t, X)$ represent heat flux and temperature at position $X = (x, y, z)$ and time t, respectively. It assumes that $\mathbf{q}$ and ∇T appear at the same time instant. This implies that thermal signals propagate with an infinite speed [8]. In fact, heat is always found to propagate at finite speed. On the other hand, biological systems are non-homogeneous where heat flux responds to temperature gradient via relaxation behaviour. To solve the paradox occurred in Pennes model, different other models were developed. Cattaneo [9] and Vernotte [10] independently proposed a modified heat flux model as

$$\mathbf{q}(t + \tau_q, X) = -k\nabla T(t, X), \tag{3}$$

where τ_q is the delay between the heat flux vector and the temperature gradient. In Eq. (3), the temperature gradient is established at time t, but the heat flux vector will be established at a later time $t + \tau_q$, at the same point X. The first order Taylor's expansion of Eq. (3) is called as Cattaneo and Vernotte (CV) constitutive relation. Using first order Taylor expansion of Eq. (3) and combining with Eq. (1) one can get the following hyperbolic bioheat equation

$$\tau_q \rho c \frac{\partial^2 T}{\partial t^2} + (\rho c + \tau_q \rho_b c_b w_b)\frac{\partial T}{\partial t} = k\nabla^2 T(t, X) + Q_m + \rho_b c_b w_b(T_b - T) \tag{4}$$

This equation is called thermal wave model of bioheat equation, as it predicts a wave like behaviour of heat transport. Many researchers have studied the heat transfer in tissue using thermal wave bioheat model. Singh and Kumar [11, 12] studied heat transfer during cryosurgery using hyperbolic model. Although a lot of experiments confirmed that CV constitutive relation produces a more accurate prediction than the classical Fourier's law, it still establishes an instantaneous response between the temperature gradient and the energy transport [13–15]. It also establishes that the temperature gradient is always the cause for heat flux while heat flux is always effect in the process of energy transport [13, 14]. Further thermal wave model does not consider the micro-scale response in space, although it considers the micro scale response in time [14, 16, 17].

In-depth study presents that the CV constitutive relation describes only the fast transient effects and not the micro-structural interactions. In order to solve the paradox in Fourier model and to consider the effect of micro structural effect in the fast transient process of heat transport, Tzou [14] proposed a dual phase lag (DPL) model that allows either the temperature gradient to precede heat flux vector or the heat flux vector to precede the temperature gradient, i.e.

$$\mathbf{q}(t + \tau_q, X) = -k\nabla T(t + \tau_T, X). \tag{5}$$

Equation (5) represents that the temperature gradient at a point X of the material at time $t + \tau_T$ corresponds to the heat flux density vector at time $t + \tau_q$ at the same point X. The delay time τ_T is interpreted as being caused by the micro structural interactions and is called the phase-lag of the temperature gradient [13, 14]. The other delay time τ_q is interpreted as the relaxation time due to the fast-transient effects of thermal inertia and is called the phase-lag of the heat flux. Both of the phase-lags are treated as intrinsic thermal or structural properties of the material. The first order Taylor's expansion of Eq. (5) gives the dual phase lag constitutive relation.

$$\mathbf{q} + \tau_q \frac{\partial \mathbf{q}}{\partial t} = -k\nabla \left\{ T(t, X) + \tau_T \frac{\partial T}{\partial t} \right\}. \tag{6}$$

Elimination of $\mathbf{q}$ from energy balance equation (Eq. (1)) and the dual phase lag constitutive relation (Eq. (6)) lead to the following equation

$$\tau_q \rho c \frac{\partial^2 T}{\partial t^2} + (\rho c + \tau_q \rho_b c_b w_b) \frac{\partial T}{\partial t} = k\nabla^2 \left\{ T(t, X) + \tau_T \frac{\partial T}{\partial t} \right\}$$
$$+ Q_m + \rho_b c_b w_b (T_b - T) \tag{7}$$

Equation (7) is the modification of the Pennes bio-heat equation by considering non-Fourier effect and is called as dual-phase lag bio-heat equation. It converts into hyperbolic bio-heat equation if $\tau_T = 0$, and into parabolic bio heat equation for $\tau_T = 0$ and $\tau_q = 0$. Many researchers [18–23] have studied the dual-phase lag bio-heat model without phase change. Liu et al. [18, 19] explain the dual-phase lag bio-heat model during hyperthermia treatment. Majchrzak [20] has described the solution of the dual-phase lag bio-heat model by using the boundary element method. Zhang et al. [23] have studied the dual-phase lag model with non-equilibrium heat transfer in arterial blood, venous blood and biological tissue and also calculated the phase lag of temperature gradient and heat flux in different condition. Zhou et al. [24, 25] have considered the dual-phase lag bio-heat model during the laser heating of living tissues. Singh and Kumar [26] have also studied dual phase change heat transfer model in three layer skin tissue.

In biological tissues, phase change occurs over a wide range and there exist moving boundaries between the two phases thus resulting mathematical models are non-linear. Analytical solution is only possible for one-dimensional, steady state cases [27]. Numerical methods appear to offer a more practical approach for solving these problems. Existing numerical approaches can be divided into two categories: front tracking and non-front tracking [27]. Enthalpy method, a non-front tracking method is easy to implement as fixed grids can be used for computation purpose and the non-linearity at the moving boundary can also be avoided. Finite difference methods are the most popular choice for numerical solution of phase change problems [11, 12, 28–35] though finite element method [36–38], boundary element methods [29] have also been introduced for the phase change problem in biological tissue.

In the present study, the dual-phase lag model is obtained by modifying the classical Pennes bio-heat equation. Pennes bio-heat equation is based on classical Fourier's law of heat conduction while the dual-phase lag bio-heat model is based on dual-phase lag constitutive relation. The dual-phase lag model with nonideal property, blood perfusion and metabolic heat generation has been considered for studying the effect of parameters in the freezing of biological tissue during phase change. The finite difference method is used to solve the enthalpy formulation of the dual-phase lag bio-heat equation during freezing. The effects of both phase lags and blood perfusion on temperature profile and interface positions have been studied. Comparative study of three heat transfer models, i.e. parabolic, hyperbolic and DPL has also been presented here.

2   Mathematical Model

2.1   Governing Equation

One-dimensional dual-phase lag bio-heat equation in frozen region and unfrozen region is given below.

(a) In frozen region: for $0 \leq x \leq s(t)$

$$\tau_q \rho_f c_f \frac{\partial^2 T_f}{\partial t^2} + \rho_f c_f \frac{\partial T_f}{\partial t} = k_f \frac{\partial^2 T_f}{\partial x^2} + \tau_T k_f \frac{\partial^3 T_f}{\partial t \partial x^2}. \tag{8}$$

(b) In unfrozen region: for $s(t) \leq x \leq l$

$$\tau_q \rho_u c_u \frac{\partial^2 T_u}{\partial t^2} + \left(\rho_u c_u + \tau_q \rho_b c_b w_b \right) \frac{\partial T_u}{\partial t} = k_u \left(\frac{\partial^2 T_u}{\partial x^2} + \tau_T \frac{\partial^3 T_u}{\partial t \partial x^2} \right)$$
$$+ Q_m + \rho_b c_b w_b (T_b - T_u). \tag{9}$$

(c) Conditions at phase change interface $x = s(t)$ are

$$\rho_f L \frac{\partial s}{\partial t} + \tau_q \rho_f L \frac{\partial^2 s}{\partial t^2} = k_f \left(\frac{\partial T_f}{\partial x} + \tau_T \frac{\partial^2 T_f}{\partial t \partial x} \right) - k_u \left(\frac{\partial T_u}{\partial x} + \tau_T \frac{\partial^2 T_u}{\partial t \partial x} \right), \tag{10}$$

and

$$T_u(t, s(t)) = T_f(t, s(t)) = T_{ph}, \tag{11}$$

where subscripts f, u and ph denote frozen, unfrozen and phase change, respectively and l denotes the length of tissue.

Major difficulties, those arise in phase change heat transfer of biological tissue is its non-linearity due to variable disconnection between different phase region and unknown position of phase change interfaces. Thus, for the solution purpose we consider the enthalpy formulation of dual-phase lag bio-heat equation for phase change problem associated with freezing.

Using enthalpy $H(T) = \int_{T_0}^{T} c\,dT$, where T_0 is the reference temperature, Eqs. (8)–(11) reduce in single equation as follows:

$$\tau_q \rho \frac{\partial^2 H}{\partial t^2} + \left(\rho + \frac{\tau_q \rho_b c_b w_b}{c}\right) \frac{\partial H}{\partial t} = k\left(\frac{\partial^2 T}{\partial x^2} + \tau_T \frac{\partial^3 T}{\partial t \partial x^2}\right)$$

$$+ Q_m + \rho_b c_b w_b (T_b - T), \tag{12}$$

enthalpy and tissue temperature are related as [39, 40]

$$H = \begin{cases} c_f(T - T_{ms}), & T < T_{ms} \\ c_a(T - T_{ms}) + \dfrac{L}{\Delta T}(T - T_{ms}), & T_{ms} \leq T \leq T_{ml} \\ L + c_a \Delta T + c_u(T - T_{ml}), & T > T_{ml}, \end{cases} \tag{13}$$

where L is the latent heat of freezing, $c_a = \frac{c_f + c_u}{2}$ and $\Delta T = T_{ml} - T_{ms}$.

2.2 Assumptions

The following assumptions have been made to solve the dual phase lag bio-heat transfer model

(i) Heat conduction follows non-Fourier law of heat conduction.
(ii) Latent heat is constant.
(iii) Heat source due to metabolism and blood perfusion is present when tissue is not frozen [28, 29].
(iv) Non-ideal property of tissue is used with liquidus and solidus temperature as -1 and $-8\,°C$, respectively [41, 42].
(v) Thermo-physical properties are different in frozen and unfrozen region.
(vi) One-dimensional model has been considered.

2.3   Initial Condition and Boundary Conditions

The following initial and boundary conditions are used for the mathematical model

(a) at $t = 0$

$$T_i(x, t) = T_0 = 37\,^\circ\mathrm{C} \quad \text{and} \quad \frac{\partial T_i}{\partial t} = 0, \quad i = u, f \tag{14}$$

(b) at $x = 0$

$$T_i(x, t) = T_c, \quad i = u, f \tag{15}$$

(c) at $x = l$

$$\frac{\partial T_i(x, t)}{\partial x} = 0, \quad i = u, f, \tag{16}$$

where T_0 is the body core temperature, $37\,^\circ\mathrm{C}$ and T_c is the cryoprobe temperature, $-196\,^\circ\mathrm{C}$.

3   Numerical Solution

Finite difference approximation has been used to solve the mathematical model. The space length l is divided into N equal parts where $N = \dfrac{l}{\Delta x}$; $x_i = i\Delta x$ and $t_n = n\Delta t$, where i and n are space and time indexes, respectively; Δx and Δt are the increment in space and time, respectively. Introducing forward difference approximation for first order time derivative and central difference approximation for space derivative and second order time derivative into Eq. (12), we get

$$H_i^{n+1} = H_i^n + \left\{ \frac{A_i^n}{A_i^n + B_i^n} \right\} \left(H_i^n - H_i^{n-1} \right) - \left\{ \frac{2E_i^n + 2D_i^n + F_i^n}{A_i^n + B_i^n} \right\} T_i^n$$

$$+ \left\{ \frac{E_i^n + D_i^n}{A_i^n + B_i^n} \right\} \left(T_{i+1}^n + T_{i-1}^n \right) - \left\{ \frac{D_i^n}{A_i^n + B_i^n} \right\} \left(T_{i+1}^{n-1} + T_{i-1}^{n-1} \right.$$

$$\left. - 2T_i^{n-1} \right) + \left\{ \frac{F_i^n}{A_i^n + B_i^n} \right\} T_b + \frac{Q_m}{A_i^n + B_i^n} \tag{17}$$

where

$$A_i^n = \frac{\tau_q \rho_i^n}{(\Delta t)^2}, \quad B_i^n = \left[\frac{\rho_i^n}{(\Delta t)} + \frac{\tau_q \rho_b c_b w_b}{c_i^n (\Delta t)} \right],$$

$$D_i^n = \frac{k_i^n \tau_T}{(\Delta t)(\Delta x)^2}, \quad E_i^n = \frac{k_i^n}{(\Delta x)^2}, \quad F_i^n = \rho_b c_b w_b.$$

The above Eq. (17) can be written as

$$H_i^{n+1} = \left(1 + U_i^n\right) H_i^n - U_i^n H_i^{n-1} - \left(2W_i^n + 2V_i^n + Y_i^n\right) T_i^n + \left(W_i^n + V_i^n\right)$$

$$(T_{i+1}^n + T_{i-1}^n) - V_i^n \left(T_{i+1}^{n-1} - 2T_i^{n-1} + T_{i-1}^{n-1}\right) + Y_i^n T_b + Z_i^n \qquad (18)$$

where

$$U_n^n = \frac{A_i^n}{A_i^n + B_i^n}, \quad V_n^n = \frac{D_i^n}{A_i^n + B_i^n}, \quad W_n^n = \frac{E_i^n}{A_i^n + B_i^n},$$

$$Y_n^n = \frac{F_i^n}{A_i^n + B_i^n}, \quad Z_n^n = \frac{Q_m}{A_i^n + B_i^n}.$$

Equation (18) gives the enthalpy at $(n + 1)$th time step in terms of enthalpy and temperature at nth time level. The time and space increments are adjusted in such a way that they should satisfy the stability criteria,

$$max \frac{(\Delta t) \left\{2k(\Delta t) + 2k\tau_T + \rho_b c_b w_b (\Delta t)(\Delta x)^2\right\}}{(\Delta x)^2 \left\{2c\tau_q \rho + c\rho(\Delta t) + \tau_q \rho_b c_b w_b (\Delta t)\right\}} \leq 1. \qquad (19)$$

After getting the enthalpy at $(n + 1)$th time level, temperature at $(n + 1)$th time level can be obtained by reverting Eq. (13) as follows:

$$T = \begin{cases} \dfrac{H}{c_f} + T_{ms}, & H < 0, \\ \dfrac{H \Delta T}{c_a \Delta T + L} + T_{ms}, & 0 \leq H \leq L + c_a \Delta T, \\ \dfrac{H - L - c_a \Delta T}{c_u} + T_{ml}, & H > L + c_a \Delta T. \end{cases} \qquad (20)$$

Once the new temperature field is obtained from enthalpies the process is repeated. Isotherms at -1 and $-8\,°C$ give the position of upper and lower phase change interfaces, respectively.

4   Results and Discussion

The computer code accuracy has already been validated in our previous study [26]. In the present study, the numerical results are shown for dual phase lag, hyperbolic ($\tau_T = 0$) and parabolic ($\tau_T = 0, \tau_q = 0$) bio-heat transfer with phase change during freezing of biological process. The thermal properties of the biological tissue used are given in Table 1 [30, 35, 41–44]. The initial temperature of the tissue is $T = 37\,°C$. For the numerical solution, the value of the phase-lag of the heat flux are $\tau_q = 0\,s$, $\tau_q = 5\,s$, $\tau_q = 10\,s$, $\tau_q = 15\,s$ and the value of the phase-lag of the temperature gradient are $\tau_T = 0\,s$, $\tau_T = 5\,s$, $\tau_T = 10\,s$ [20, 44–46]. During the freezing process of the biological tissue, the temperature distribution and interface position are important for the prediction of the damage of the infected tissues and minimum damage of the healthy tissues.

The temperature distributions in the tissue at time $t = 600\,s$ with respect to distance x for the DPL, hyperbolic and parabolic models are shown in Fig. 1. It is observed that parabolic model gives lowest temperature in the tissue with comparison to DPL and hyperbolic model, while the highest temperature is obtained for hyperbolic model. This shows that parabolic model gives the fastest heat flow in the media while slowest is for hyperbolic model.

To study the effect of these three models on temperature distribution with respect to time in tissue, the variation of temperature versus time for parabolic, hyperbolic and DPL model at the point $x = 0.01\,m$ is plotted in Fig. 2. Again from Fig. 2, it is clear that decline in temperature is highest for parabolic and lowest for hyperbolic with comparison to DPL model.

The liquidus and solidus interfaces position of the phase change interface during the freezing process for DPL, parabolic and hyperbolic models are shown in

Table 1 Thermal properties of tissue

Parameter	Value
Density of unfrozen tissue (kg/m^3)	1000
Specific heat of unfrozen tissue (J/kg °C)	3600
Thermal conductivity of unfrozen tissue (W/m °C)	0.5
Density of frozen tissue (kg/m^3)	1000
Specific heat of frozen tissue (J/kg °C)	1800
Thermal conductivity of frozen tissue (W/m °C)	2
Density of blood (kg/m^3)	1050
Specific heat of blood (J/kg °C)	3770
Blood perfusion in tissue (ml/s/ml)	0.005
Metabolic heat generation (W/m^3)	4200
Latent heat (J/m^3)	250,000
The upper limit of phase change temperature (°C)	−1
The lower limit of phase change temperature (°C)	−8
Arterial blood temperature (°C)	37

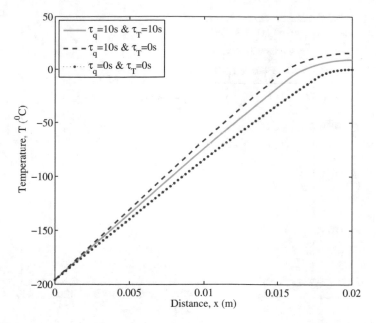

Fig. 1 The temperature profile along the target tissue for parabolic, hyperbolic and DPL model at the time $t = 600$ s

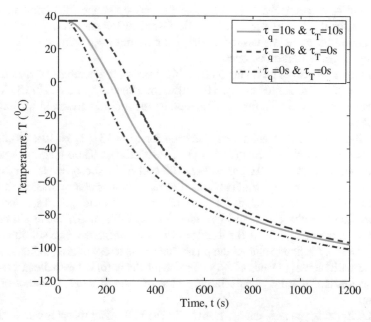

Fig. 2 The variation of temperature versus time for parabolic, hyperbolic and DPL model at the point $x = 0.01$ m

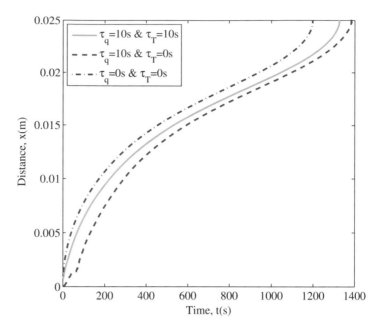

Fig. 3 Liquidus interface position for parabolic, hyperbolic and DPL model

Figs. 3 and 4, respectively. It is clear that the phase change interfaces movement is slowest for the hyperbolic model while fastest is for parabolic model, i.e., the time required for complete tissue freezing is minimum for parabolic model while it is maximum for hyperbolic model.

To study the effect of phase lag on freezing process, temperature profiles in tissue at time $t = 800$ s are plotted in Fig. 5 for different value of τ_q, i.e. 5, 10, 15 s keeping fixed value of $\tau_T = 5$ s. In this case, lowest temperature is observed for minimum value of τ_q.

The liquidus and solidus interface position for $\tau_q = 15$ s; $\tau_q = 10$ s and $\tau_q = 5$ s and $\tau_T = 5$ s are plotted in Figs. 6 and 7, respectively. Time taken to reach the liquidus interface at $x = 0.02$ m is 742, 703 and 661 s for $\tau_q = 15$ s, $\tau_q = 10$ s and $\tau_q = 5$ s, respectively, while solidus interface reaches at $x = 0.02$ m in time is 781.32 s, 741.63 s and 699.33 ms for $\tau_q = 15$ s, $\tau_q = 10$ s and $\tau_q = 5$ s, respectively. It shows that the phase change interfaces for the DPL model move slower with increasing value of τ_q. That is the time required for complete tissue solidification in the DPL model for fixed value of the phase-lag of the temperature gradient $\tau_T = 5$ s increases with increased value of τ_q. τ_q denote the delay time due to the fast transient effect of thermal inertias, thus a larger value of τ_q will result in delay of tissue freezing.

In the freezing process, energy transfer in the biological tissue is due to thermal conduction, blood tissue convection, blood perfusion and metabolic heat generation. These parameters have significant effect of transient temperature profile and position

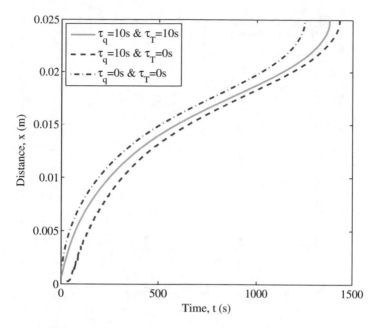

Fig. 4 Solidus interface position for parabolic, hyperbolic and DPL model

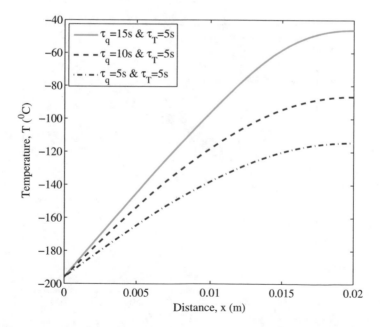

Fig. 5 The temperature profile along the target tissue for DPL model at the time $t = 800$ s

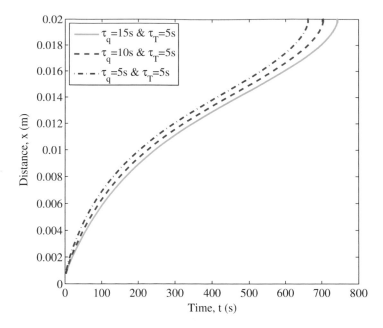

Fig. 6 Liquidus interface position for DPL model

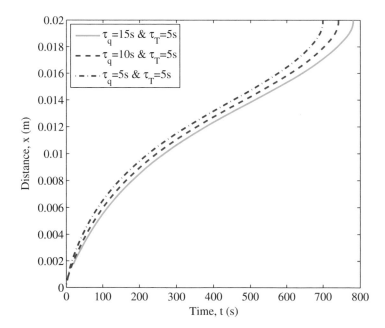

Fig. 7 Solidus interface position for DPL model

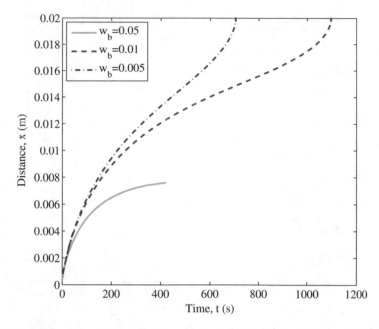

Fig. 8 Liquidus interface position for DPL model at $\tau_q = 10\,\text{s}$ and $\tau_T = 10\,\text{s}$ for different value of blood perfusion

of solidus and liquidus interfaces. The effect of blood perfusion on freezing process using DPL, hyperbolic and parabolic model has been studied in the present study. For DPL model position of liquidus and solidus interfaces for $w_b = 0.05\,\text{ml/s/ml}$, $w_b = 0.01\,\text{ml/s/ml}$ and $w_b = 0.005\,\text{ml/s/ml}$ are plotted in Figs. 8 and 9, respectively. The liquidus interface reaches at distance $x = 0.02\,\text{m}$ in time $t = 1095\,\text{s}$ and $t = 705\,\text{s}$ for $w_b = 0.01\,\text{ml/s/ml}$ and $w_b = 0.005\,\text{ml/s/ml}$, respectively. Similarly solidus interface reaches at distance $x = 0.02\,\text{m}$ in time $t = 1122.4\,\text{s}$ and $t = 742\,\text{s}$ for $w_b = 0.01\,\text{ml/s/ml}$ and $w_b = 0.005\,\text{ml/s/ml}$, respectively.

For hyperbolic model position of liquidus and solidus interfaces for $w_b = 0.05\,\text{ml/s/ml}$, $w_b = 0.01\,\text{ml/s/ml}$ and $w_b = 0.005\,\text{ml/s/ml}$ are plotted in Figs. 10 and 11, respectively. The liquidus interface reaches at distance $x = 0.02\,\text{m}$ in time $t = 1132\,\text{s}$ and $t = 773.15\,\text{s}$ for $w_b = 0.01\,\text{ml/s/ml}$ and $w_b = 0.005\,\text{ml/s/ml}$, respectively. Similarly solidus interface reaches at distance $x = 0.02\,\text{m}$ in time $t = 1163.2\,\text{s}$ and $t = 807.67\,\text{s}$ for $w_b = 0.01\,\text{ml/s/ml}$ and $w_b = 0.005\,\text{ml/s/ml}$, respectively.

For parabolic model position of liquidus and solidus interfaces for $w_b = 0.05\,\text{ml/s/ml}$, $w_b = 0.01\,\text{ml/s/ml}$ and $w_b = 0.005\,\text{ml/s/ml}$ are plotted in Figs. 12 and 13 respectively. The liquidus interface reaches at distance $x = 0.02\,\text{m}$ in time $t = 942.30\,\text{s}$ and $t = 609.15\,\text{s}$ for $w_b = 0.01\,\text{ml/s/ml}$ and $w_b = 0.005\,\text{ml/s/ml}$ respectively. Similarly solidus interface reaches at distance $x = 0.02\,\text{m}$ in time $t = 973\,\text{s}$ and $t = 645.94\,\text{s}$ for $w_b = 0.01\,\text{ml/s/ml}$ and $w_b = 0.005\,\text{ml/s/ml}$ respectively.

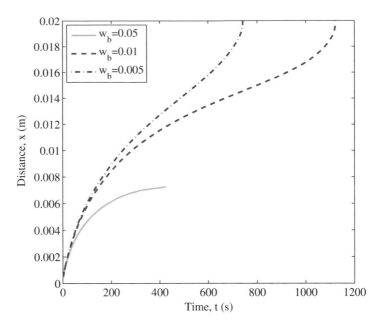

Fig. 9 Solidus interface position for DPL model at $\tau_q = 10$ s and $\tau_T = 10$ s for different value of blood perfusion

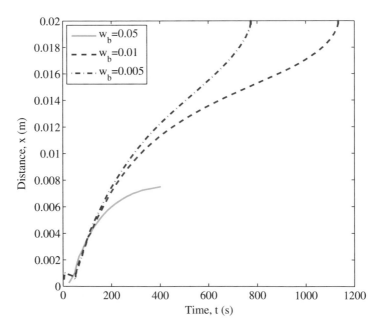

Fig. 10 Liquidus interface position for hyperbolic model at $\tau_q = 10$ s and $\tau_T = 0$ s for different value of blood perfusion

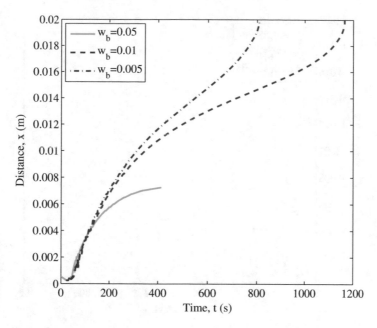

Fig. 11 Solidus interface position for hyperbolic model at $\tau_q = 10$ s and $\tau_T = 0$ s for different value of blood perfusion

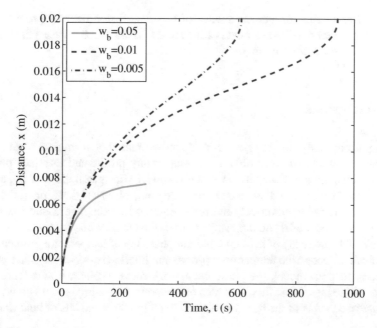

Fig. 12 Liquidus interface position for parabolic model at $\tau_q = 0$ s and $\tau_T = 0$ s for different value of blood perfusion

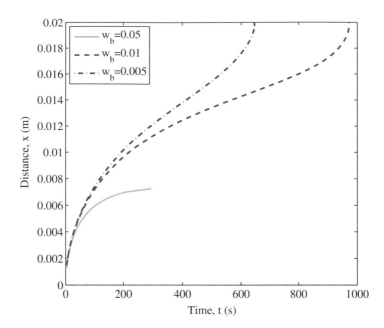

Fig. 13 Solidus interface position for parabolic model at $\tau_q = 0$ s and $\tau_T = 0$ s for different value of blood perfusion

It is observed that for increased value of blood perfusion freezing process slows down. Due to blood perfusion, heat is added to the tissue which opposes the freezing process and hence slows it down.

5 Conclusions

In the present study, the temperature dependent enthalpy formulation and finite difference method is used to obtain the temperature profile and interface position based on DPL model. Comparison of DPL model with parabolic and hyperbolic model of heat transport is also made in the study. It is observed that among the DPL, hyperbolic and parabolic model total time required for complete tissue freezing is least for parabolic model and largest for hyperbolic model while for DPL model it is moderate. The phase lag of heat flux and the phase lag of temperature gradient have a significant effect on the temperature profile and interface position of phase change interfaces. In DPL model, the phase change interface accelerates with decreasing value of phase lag of heat flux. It is also observed that freezing process slows down with increased value of the blood perfusion in all the three models of heat transfer.

Acknowledgements The authors Sushil Kumar and Sonalika Singh are thankful to S. V. National Institute of Technology, Surat, India for providing CPDA grant and Senior Research Fellowship, (SRF) respectively, for the research work presented in this manuscript. Sushil Kumar thanks to the International Union of Biological Sciences (IUBS) for partial support of living expenses in Moscow, during the 17th BIOMAT International Symposium, October 29-November 04, 2017.

List of Abbreviations

T	Temperature (°C)
t	Time (s)
x	Distance $(0 \leq x \leq l)$ (m)
l	Length of tissue $(0 \leq x \leq l)$ (m)
$s(t)$	Position of phase change interface (m)
ρ	Density (kg/m^3)
c	Specific heat (J/kg°C)
k	Thermal conductivity (W/m°C)
w_b	Blood perfusion in tissue (ml/s/ml)
Q_m	Metabolic heat generation (W/m^3)
L	Latent heat (J/m^3)
T_b	Arterial blood temperature (°C)
q	Heat flux
H	Enthalpy (J/kg°C)
$\triangle t$	Time step (s)
$\triangle x$	Space step (m)
τ_q	Heat flux relaxation time (s)
τ_T	Temperature gradient relaxation time (s)

Subscripts

ph	Phase change
f	Frozen
u	Unfrozen
i	Space step
b	Blood
ml	The upper limit of phase change (liquidus)
ms	The lower limit of phase change (solidus)

Superscript

n	Time step

References

1. J.C. Bischof, J. Bastack, B. Rubinsky, ASME J. Biomech. Eng. **114**, 467 (1992)
2. K.J. Chua, S.K. Chou, J.C. Ho, J. Biomech. **40**, 100 (2007)
3. H.H. Pennes, J. Appl. Physiol. **1**, 93 (1948)
4. M.M. Chen, K.R. Holmes, Ann. N. Y. Acad. Sci. **335**, 137 (1980)
5. L.M. Jiji, S. Weinbaum, D.E. Lemons, ASME J. Biomech. Eng. **106**, 331 (1984)
6. S. Weinbaum, L.M. Jiji, D.E. Lemons, ASME J. Biomech. Eng. **106**, 321 (1984)
7. S. Weinbaum, L.M. Jiji, ASME J. Biomech. Eng. **107**, 131 (1985)
8. L. Wang, J. Fan, ASME J. Heat Transfer **133**, 011010-1 (2011)
9. C. Cattaneo, C. R. Acad. Sci. **247**, 431 (1958)
10. P. Vernotte, C. R. Acad. Sci. **246**, 3154 (1958)
11. S. Singh, S. Kumar, Math. Model. Anal. **20**(4), 443 (2015)
12. S. Singh, S. Kumar, J. Mech. Med. Biol. **16**(2), 1650017 (2016)
13. D.Y. Tzou, ASME J. Heat Transfer **117**, 8 (1995)
14. D.Y. Tzou, *Macro-to Microscale Heat Transfer: The Lagging Behavior* (Taylor and Francis, Washington, 1997)
15. L.Q. Wang, Int. J. Heat Mass Transfer **37**, 2627 (1994)
16. D.Y. Tzou, M.N. Ozisik, R.J. Chiffelle, J. Heat Transfer **116**, 1034 (1994)
17. M.N. Ozisik, D.Y. Tzou, ASME J. Heat Transfer **116**, 526 (1994)
18. K.C. Liu, H. Chen, Int. J. Heat Mass Transfer **52**, 1185 (2009)
19. K.C. Liu, H. Chen, Int. J. Therm. Sci. **49**, 1138 (2010)
20. E. Majchrzak, Comput. Model. Eng. Sci. **69**(1), 43 (2010)
21. E. Majchrzak, T. Lukasz, in *19th International Conference on Computer Methods in Mechanics CMM* (2011), p. 337
22. N. Afrin , N. Y. Zhang , J. K. Chen, Int. J. Heat Mass Transfer **54**, 2419 (2011)
23. Y. Zhang, Int. J. Heat Mass Transfer **52**, 4829 (2009)
24. J. Zhou, J.K. Chen, Y. Zhang, Comput. Biol. Med. **39**, 286 (2009)
25. J. Zhou, Y. Zhang, J.K. Chen, Int. J. Therm. Sci. **48**, 1477 (2009)
26. S. Singh, S. Kumar, Int. J. Therm. Sci. **86**, 12 (2014)
27. J. Crank, *Free and Moving Boundary Problems* (University Press, New York, 1984)
28. Z.S. Deng, J. Liu, Numer. Heat Transfer Part A **46**, 487 (2004)
29. Z.S. Deng, J. Liu, Eng. Anal. Bound. Elem. **28**(2), 97 (2004)
30. Y. Rabin, A. Shitzer, J. Biomech. Eng. **120**(1), 32 (1998)
31. S. Kumar, V.K. Katiyar, Int. J. Appl. Math. Mech. **3**(3), 1 (2007)
32. R.I. Andrushkiw, Math. Comput. Model. **13**, 1 (1990)
33. M. Zerroukat, C.R. Chatwin, *Computational Moving Boundary Problem* (Wiley, New York, 1993)
34. J.C. Rewcastle, G.A. Sandison, K. Muldrew, J.C. Saliken, B.J. Donnelly, Med. Phys. **28**, 1125 (2001)
35. S. Kumar, V.K. Katiyar, Int. J. Appl. Mech. **2**(3), 617 (2010)
36. G. Comini, D.S. Giudice, ASME J. Heat Transfer **98**, 543 (1976)
37. A. Weill, A. Shitzer, P.B. Yoseph, J. Biomech. Eng. **115**, 374 (1993)
38. J.Y. Zhang, G.A. Sadison, J.Y. Murthy, L.X. Xu, J. Biomech. Eng. **127**, 279 (2005)
39. C. Bonacina, G. Comini, Int. J. Heat Mass Transfer **16**, 1825 (1973)
40. N.E. Hoffmann, J.C. Bischof, ASME J. Biomech. Eng. **123**, 301 (2001)
41. Y. Rabin, A. Shitzer, J. Heat Transfer **117**, 425 (1995)
42. Y. Rabin, A. Shitzer, ASME J. Heat Biomech. Eng. **119**, 146 (1997)
43. Z.S. Deng, J. Liu, J. Therm. Stresses **26**, 779 (2003)
44. A. Moradi, H. Ahamdikia, J. Eng. Med. **226**(5), 406 (2012)
45. H. Ahamdikia, A. Moradi, R. Fazlali, B. Parsa, J. Mech. Sci. Technol. **26**(6), 1937 (2012)
46. L. Jing, C. Xu, L.X. Xu, IEEE Trans. Biomed. Eng. **46**(4), 420 (1999)

Computational Modeling of Multiple Stenoses in Carotid and Vertebral Arteries

T. Gamilov, S. Simakov, and P. Kopylov

1 Introduction

Cerebrovascular accidents are among the leading reasons for the mortality and disability in the world. Ischemic strokes can be caused by various factors. The most important ones are atherosclerosis of brachiocephalic arteries (BCA), pathological tortuosity of BCA, and occlusions of vertebral arteries. Medical treatment involves drug administration, diet, and/or surgical endovascular interventions (e.g., carotid endarterectomy [1], carotid stenting).

The type of the treatment is selected after the assessment of the stenoses severity (hemodynamic significance/importance). The assessment of the hemodynamic importance is regularly based on the noninvasive Doppler velocity measurements in carotid, vertebral, and cerebral arteries, which are accessible for ultrasound measurements [2]. The final decision is usually made by a surgeon. It is based on the interpretation of the well-known reference values from the literature. This interpretation may be inappropriate for the individual case.

Surgical interventions may be the reason of various postsurgical and perioperative complications, e.g., strokes [3]. Thus, it is important to know the blood flow variations during and after endovascular surgery for the specific patient. Normally, these variations are monitored at several selected points before and after the surgery. Comprehensive analysis of the whole brain circulation is out of current regular practice. It should be mentioned that anatomical features of the circle of Willis may substantially affect the blood flow redistribution after the intervention [4, 5].

T. Gamilov (✉) · S. Simakov
Institute of Numerical Mathematics RAS, Moscow, Russia
e-mail: gamilov@crec.mipt.ru

P. Kopylov
I.M. Sechenov First Moscow State Medical University, Moscow, Russia

© Springer International Publishing AG, part of Springer Nature 2018
R. P. Mondaini (ed.), *Trends in Biomathematics: Modeling, Optimization and Computational Problems*, https://doi.org/10.1007/978-3-319-91092-5_20

In the cases of the multivessel stenotic diseases of the BCA and vertebral arteries, the evaluation of the hemodynamic significance is less obvious due to the complex contribution of each stenosis to the whole hemodynamics and due to the ambiguity of collateral circulation.

Summarizing the above, we conclude that assessment of the stenoses severity in the cases of single and multivessel diseases of the BCA and vertebral arteries and comprehensive analysis of the cerebral circulation during and after the surgical intervention for the specific patient are still not well studied.

The possible solution that may be helpful to surgeons is mathematical modeling of patient-specific cerebral circulation before, during, and after the surgical intervention. It allows to carry out preoperative analysis of possible scenarios, risks, postoperative effect, and negative consequences. The structure of the cerebrovascular network is complex. The simulations based on the straight solution of the Navier–Stokes equations in the network of flexible tubes require substantial computational resources [6]. The 1D models established by cross-sectional averaging are more effective. The 1D modeling of pulsatile blood flow in the brain circulation is a well-developed area [7–13]. It allows simulating a large number of arteries and veins. Some models take into account cerebral autoregulation [10] and baroreflex regulation [8, 9]. Some works make an emphasis on the structure of cerebral veins and investigate associated diseases [7]. In other works, the vein drainage is modeled by 0D compartments or by boundary conditions [10]. Carotid artery stenosis and the impact of surgical interventions were numerically evaluated in the works [12, 13]. The complex shape of the stenosis and pathological tortuosity of BCA can be considered by the multidimensional 1D–3D approach [11].

In this work, the emphasis is made on patient-specific modeling of the blood flow in the BCA, vertebral, and cerebral arteries. The datasets from the randomly picked five patients were analyzed. The 1D structure of the vascular network was extracted from the CT images by the previously developed algorithm [14]. The functional parameters (pulse wave velocity index, hydraulic resistance coefficient, cardiac output) were identified by fitting the computed and measured velocity in the control points before the surgical intervention. A 1D blood flow model was used to predict the changes in blood flow velocities after the carotid endarterectomy [12, 13]. The values were compared to the measured ones. The mean relative error was 6%, and the maximum relative error was 20%. We limit ourselves by the data quality available from the typical hospital CT scans. The structure of the small vessels is poorly visible in the images of such kind. Thus we simulated and compared the two possible situations: A complete circle of Willis and the circle of Willis with missed posterior communicating arteries [5]. The impact of this anatomical feature is illustrated by the case of several stenoses.

An index for the evaluation of the stenosis severity was proposed. It was calculated as a ratio between the blood flow velocity distal to the stenosis and the blood flow velocity in a collateral artery. These velocities can be directly measured by using the noninvasive Doppler ultrasound technique. The values below 0.75 can be associated with the severe stenosis as the reverse blood flow in a closed circle of Willis occurs in these situations.

2 Mathematical Model

2.1 Blood Flow Model

The 1D hemodynamics model used in this work is the model of viscous incompressible fluid in a network of elastic tubes. Networks of arteries are obtained from patients' CT images. In this section, a brief description of the model [15, 16] is presented. Blood flow in each vessel is described by a hyperbolic set of mass and momentum balances:

$$\partial A_k/\partial t + \partial (A_k u_k)/\partial x = 0, \tag{1}$$

$$\partial u_k/\partial t + \partial \left(u_k^2/2 + p_k/\rho\right)/\partial x = f_{fr}(A_k, u_k), \tag{2}$$

where k is the index of the vessel; t is the time; x is the distance along the vessel counted from the vessel junction point; ρ is the blood density (constant); $A_k(t, x)$ is the vessel cross-sectional area; p_k is the blood pressure; $u_k(t, x)$ is the linear velocity averaged over the cross section; f_{fr} is the friction force. The relationship between pressure and cross section is given by the wall-state equation:

$$p_k(A_k) - p_{*k} = \rho_w c_k^2 f(A_k), \tag{3}$$

where ρ_w is vessel wall density (constant); $f(A)$ is a function

$$f(A_k) = \begin{cases} \exp\left(A_k/A_{0k} - 1\right) - 1, & A_k/A_{0k} > 1 \\ \ln A_k/A_{0k}, & A_k/A_{0k} \leqslant 1, \end{cases} \tag{4}$$

p_{*k} is pressure in the tissues surrounding the vessel; A_{0k} is the unstressed cross-sectional area. c_k defines elastic properties of the wall and can be considered as the velocity of small disturbances propagation.

At the entry point of the aorta, the blood flow is assigned

$$u(t, 0) A(t, 0) = Q_H(t). \tag{5}$$

Here function $Q_H(t)$ corresponds to the heart rate value of 1 Hz and stroke volume of 60 ml.

Bifurcation points are divided into two groups: (1) common junctions (between arteries and arteries or veins and veins), and (2) microcirculation junctions (between arteries and veins). At common junctions, continuity of total pressure is postulated

$$p_i\left(A_i\left(t, \tilde{x}_i\right)\right) + \frac{\rho u_i^2\left(t, \tilde{x}_i\right)}{2} = p_j\left(A_j\left(t, \tilde{x}_j\right)\right) + \frac{\rho u_j^2\left(t, \tilde{x}_j\right)}{2}, \tag{6}$$

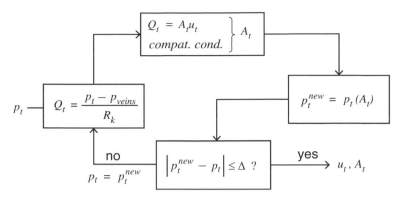

Fig. 1 Iterative process for calculation of cross section and blood flow velocity at terminal points

where i, j are indices of the vessels. $\tilde{x}$ is the coordinate of the boundary point of the vessel. Each terminal artery with an index k was connected to the venous pressure $p_{veins} = 12\,\mathrm{mm\,Hg}$ through a hydraulic resistance R_k

$$p_k\left(A_k\left(t, \tilde{x}_k\right)\right) - p_{veins} = R_k A_k\left(t, \tilde{x}_k\right) u_k\left(t, \tilde{x}_k\right). \tag{7}$$

Parameters R_k are adjusted to simulate pressure drop between arteries and veins. To close the system, we add the mass conservation condition and compatibility conditions of the hyperbolic set (1),(2) (see [17]).

Cross section A_t and velocity u_t at the terminal point are calculated with the help of an iterative process. First, blood pressure p_t is taken from the previous time step. After that blood flow $Q_t = A_t u_t$ is calculated with (7). Compatibility condition is used to calculate cross section A_t and velocity u_t. Finally, blood pressure p_t^{new} is calculated with the wall-state equation (4). If the difference between p_t^{new} and p_t is big enough, p_t^{new} is taken as a new value of p_t and cycle repeats (Fig. 1).

Stenosis is simulated as a separate vessel with decreased diameter. Diameter in the stenosed vessel is calculated based on the degree of stenosis α

$$d_{sten} = d_{non\text{-}sten}(1 - \alpha), \tag{8}$$

where d_{sten} is the diameter in the stenosed vessel, $d_{non\text{-}sten}$ is the diameter in a healthy vessel.

2.2 Reconstruction of Patient-Specific Vessel Structure

The vessel network reconstruction algorithm involves vessel segmentation, thinning-based extraction of centerlines, and graph reconstruction [12, 14]. Input data are 3D DICOM datasets, obtained with contrast-enhanced computed tomography angiography (CTA). Resolution of each 2D transverse slice is 512×512 voxels. Only the quasi-isotopic voxel grids with deviation from cubic grids less than 10% were used; other grids can be resampled into isotopic ones.

Anatomy of the cerebral vessels in the human body is represented by carotid arteries rooting at aorta, vertebral arteries separating from subclavian arteries (that also rooting at aorta), and small arteries in the brain. Carotid and vertebral arteries merged in the circle of Willis that allows blood bypass stenoses in the neck vessels. Some of the segmented arteries (vertebral arteries) are very close to the bones (spine). Separating the bones voxels from the arteries voxels is an important segmentation step. The multiscale MMBE algorithm was used [12] at this stage. It requires two datasets at the input: the CTA contrast-enhanced and not enhanced images.

After the segmentation step, the vessel's centerlines are extracted with a modified version of the thinning method with False Twigs Elimination algorithm [14]. These centerlines are used to produce a graph of the arterial network where each node corresponds to bifurcation or end of the vessel. The length and the mean radius were assigned as parameters to every edge of this graph.

Five patient-specific networks were processed by the above algorithm (Fig. 2). Patient A had an 80% 4 cm stenosis in the left carotid artery (LCA, vessels 8,9); patient B had a 72% 4 cm stenosis in the LCA (vessel 14); patient C had 75% 2 cm stenosis in the LCA (the vessels 8,10); patient D had 75% 3 cm stenosis in the right carotid artery (RCA, vessels 3,4); patient E had 92% 4 cm stenosis in the LCA (vessels 13,15). None of the networks have a fully closed circle of Willis. There are two possible reasons for this. The first is that the contrast agent could not reach the small distant arteries. The second is the anatomical feature of the patient.

Each patient had ultrasound Doppler measurements of blood flow velocities before and after the carotid endarterectomy. The measurements were performed in carotid arteries (common, inner, and outer), subclavian arteries, and vertebral arteries.

3   Results

3.1   Prediction of Blood Flow Velocities After the Carotid Endarterectomy

The measurements before the treatment were used for calibrating the network and setting the parameters c_k and R_k. At the next step, the stenosis was artificially removed and the blood flow velocities for a healthy case were calculated. These velocities were compared to the measured values after the treatment. We neglect the difference in the elasticity of the healthy and repaired vessel wall. Figure 3 shows a comparison between calculated and measured velocities for the considered cases. The mean absolute error is 3 cm/s. The maximum absolute error is 9 cm/s. The mean

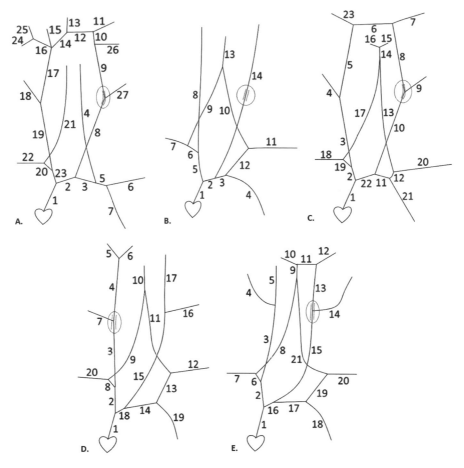

Fig. 2 Patient-specific structures of the head and neck arterial network. Stenoses are designated by dotted lines

relative error is 6%. The maximum relative error is 20%. Figure 4 shows relative errors for the calculated and measured velocity values. It is obvious that the relative error decreases with the increase of the velocity value. This is especially important for the analysis of critical zones with high blood velocity in cerebral circulation during and after the surgical intervention.

Results show that the proposed algorithm of the 1D vessel network reconstruction provides a good basis for the patient-specific simulations. It can be used to predict the results of the stenosis treatment and to calculate the blood flow distribution.

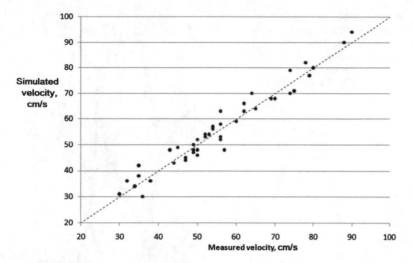

Fig. 3 Velocities after the treatment

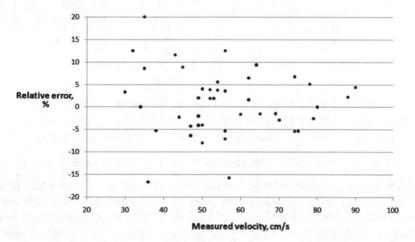

Fig. 4 Relative error between the calculated and measured velocity values after the treatment

3.2 Numerical Simulations of the Collateral Flow for the Complete Circle of Willis

Network C was modified to complete the circle of Willis (Fig. 5). This network was used to compare velocities and pressures in the stenosed LCA (vessel 8) and the collateral RCA (vessel 5). The case of single stenosis (B) is considered.

A series of numerical simulations was performed for stenosis B of the degrees 0%, 50%, 80%, and 95% and the length 2 cm. The ratio $\dfrac{\overline{P}_d}{\overline{P}_a}$ was calculated in every

Fig. 5 The modified network
C (Fig. 2). A, B, C, D denote
the possible stenoses. Dashed
lines denote the virtual
vessels added in order to
complete the circle of Willis

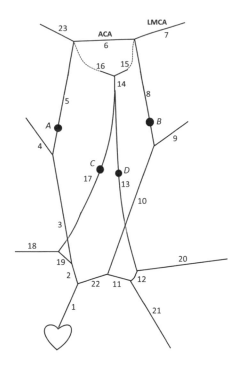

case, where $\overline{P}_d$ is an average pressure distal to the stenosis and $\overline{P}_a$ is an average
aortic pressure. This parameter is analogous to the fractional flow reserve (FFR)
coefficient for the coronary vessels [15]. It is widely used for the coronary stenosis
severity assessment. For the network presented in Fig. 5, the values of the $\dfrac{\overline{P}_d}{\overline{P}_a}$ were
equal to 1.0 with a maximum difference of 1% relative to the initial case (the degree
of 0%). It means that the blood pressure distal to the stenosis remains the same even
in the worst considered case (stenosis of the degree 95%). It can be explained by the
collateral blood flow through the right carotid and right and left vertebral arteries
(the vessels 5, 13, 17 in Fig. 5). In the next series of the numerical simulations,
several stenoses of the degree 95% are arranged in all collateral arteries (A, B, C, D
in Fig. 5). The value of the ratio $\dfrac{\overline{P}_d}{\overline{P}_a}$ is changed to 0.2. This comparison shows that
the ratio $\dfrac{\overline{P}_d}{\overline{P}_a}$ can't be used for the carotid stenosis severity assessment in the case of
single stenosis.

For the stenosis of the degree 50%, the ratio $\dfrac{\overline{P}_d}{\overline{P}_a}$ in the above simulations equals
to 1.0 while the blood flow velocity in the anterior communicating artery (ACA,
vessel 6) is reversed in comparison to the non-stenosed case. It shows that stenosis
of the degree 50% causes the change of the blood flow direction in a circle of Willis

Table 1 Blood flow velocities in collateral carotid arteries and blood flow through ACA

Stenosis	u_{st}, cm/s	u_{coll}, cm/s	$\frac{u_{st}}{u_{coll}}$	Q_{ACA}, ml/s
0%	90	91	0.99	0.21
40%	81	92	0.88	0.08
50%	73	99	0.74	−0.02
80%	10	115	0.09	−0.53
95%	−32	140	−0.22	−0.57

Fig. 6 Configurations of the circle of Willis

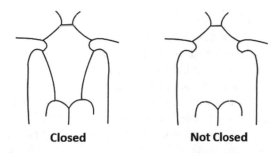

Closed   **Not Closed**

and should be considered as hemodynamically significant. We propose an index of the stenosis severity, which is calculated as the ratio $\frac{u_{st}}{u_{coll}}$, where u_{st} is a systolic velocity in the stenosed carotid artery distal to the stenosis and u_{coll} is a systolic velocity in the non-stenosed collateral carotid artery. The ratio $\frac{u_{st}}{u_{coll}}$ was calculated for the same range of the stenoses B degrees as in the first series above. Results are presented in Table 1. From Table 1, it follows that the change of the blood flow direction in the circle of Willis occurs for the stenosis of the degree 50%, which corresponds to the value $\frac{u_{st}}{u_{coll}} < 0.75$. The relative velocity difference in stenosis B for the cases of the degrees 50% and 0% is less than that of 20%.

3.3 Numerical Simulations of the Blood Flow in the Circle of Willis in the Case of Multiple Stenoses

The network of the vessels presented in Fig. 5 was used for the numerical simulations of the blood flow in the circle of Willis in the case of multiple stenoses in the BCA and vertebral arteries. Stenoses were virtually arranged in the right and left inner carotid arteries (A and B) and right and left vertebral arteries (C and D). Each stenosis is 2 cm long with a degree of 90%. Four configurations of the stenoses arrangement were numerically studied: healthy subject (no stenoses), carotid and vertebral arteries (ABCD), carotid arteries (AB), left carotid and vertebral arteries (BCD). Two cases of the circle of Willis structure were considered (Fig. 6). The first is a complete circle of Willis. The second is the circle of Willis without connections between carotid and vertebral arteries.

Table 2 Blood flow through
left middle cerebral artery
(LMCA, vessel 7 in Fig. 5)
for different senosis
configurations

	LMCA blood flow, ml/s			
Circle of Willis	Healthy	ABCD	AB	BCD
Closed	1.21	0.25	1.15	1.19
Not closed	1.02	0.21	0.20	1.01

Closed circle of Willis—vessels 15 and 16 are
present; Not closed circle of Willis—vessels 15 and
16 are absent

Blood flow through the left middle cerebral artery (LMCA, vessel 7 in Fig. 5)
was calculated in every case. Results are presented in Table 2. In the first case, it
follows that the blood flow through the LMCA remains almost the same if at least
one of the collateral paths is not stenosed. Only the worst case ABCD provides a
substantial decrease of the flow in LMCA. In the second case, the vertebral arteries
are "switched off" from collateral perfusion. It results in the substantial decrease
of the flow in LMCA even in the case of two stenosed carotid arteries AB. This
example emphasizes the importance of the circle of Willis anatomy for the stenosis
severity assessment in the BCA and vertebral arteries.

4 Conclusion

The results of this work show that the proposed vessel network reconstruction
algorithm and 1D model of hemodynamics can be jointly used to predict the results
of stenosis treatment and to calculate the blood flow redistribution during and
after the surgical endovascular intervention. The average absolute error 3 cm/s has
the same order of magnitude as the measurements error by using the ultrasound
Doppler technique. It was demonstrated that the relative error decreases with the
increase of the velocity value. It is especially important for the robust critical zones
identification during and after the surgical intervention. It should be noted that
the value of the blood flow velocity depends on the control point at which the
measurement is performed. This ambiguity can be removed with the development
of a proper clinical protocol for the blood flow velocity measurements.

It was numerically demonstrated that the anatomy of the circle of Willis
substantially affects the result of the BCA and vertebral arteries revascularization.
At least six possible cases of the circle of Willis closing [5] are known which must
be analyzed in the future work. The ratio between the blood flow velocity distal
to the stenosis and the blood flow velocity in a collateral artery was analyzed for
one patient. This value can be directly measured using the ultrasound technique. It
was numerically demonstrated that the values below 0.75 can be associated with
the reverse blood flow in the closed circle of Willis. This index can be used as
one of the possible measures of the stenosis severity in the cerebral arteries. More
patient-specific cases are needed for the detailed analysis of the critical range of the
proposed index. It remains unclear how the changes of hemodynamic parameters
are related to the postoperative complications. More cases should be analyzed to
solve this problem.

Acknowledgements The authors are thankful to scientists from I.M. Sechenov First Moscow State Medical University, particularly to N. Gagarina, E. Fominykh, and A. Dzyundzya for their data and to R. Pryamonosov for assistance in designing the 1D structure of the patient-specific arterial networks. The research was supported by Russian Science Foundation (grant No. 14-31-00024).

References

1. R.W. Hobson, D.G. Weiss, W.S. Fields, J. Goldstone, W.S. Moore, J.B. Towne, C.B. Wright, Efficacy of carotid endarterectomy for asymptomatic carotid stenosis. N. Engl. J. Med. **328**, 221–227 (1993)
2. P. Kopylov, A.A. Bykova, D.Yu. Shchekochikhin, Kh.E. Elmanaa, A.N. Dzyundzya, Yu.V. Vasilevsky, S.S. Simakov, Asymptomatic atherosclerosis of the brachiocephalic arteries: current approaches to diagnosis and treatment. Ter. Arkh. **89**(4), 95–100 (2017). https://doi.org/10.17116/terarkh201789495-100
3. J. Liu, M. Lieb, U. Shah, G.L. Hines, Cerebral hyperperfusion syndrome after carotid intervention: a review. Cardiol. Rev. **20**(2), 84–89 (2012)
4. J. Alastruey, K.H. Parker, J. Peiró, S.M. Byrd, S.J. Sherwin, Modelling the circle of Willis to assess the effects of anatomical variations and occlusions on cerebral flows. J. Biomech. **40**(8), 1794–1805 (2007)
5. G. Zhu, Q. Yuan, J. Yang, J. Yeo, Experimental study of hemodynamics in the circle of Willis. Biomed. Eng. Online **14**, S10 (2015)
6. I. Marshall, S. Zhao, P. Papathanasopoulou, P. Hoskins, Y. Xu, MRI and CFD studies of pulsatile flow in healthy and stenosed carotid bifurcation models. J. Biomech. **37**(5), 679–687 (2004)
7. L.O. Muller, E.F. Toro, A global multiscale mathematical model for the human circulation with emphasis on the venous system. Int. J. Numer. Methods Biomed. Eng. **30**(7), 681–725 (2014)
8. V.B. Koshelev, S.I. Mukhin, T.V. Sokolova, N.V. Sosnin, A.P. Favorski, Mathematical modeling of cardio vascular hemodynamics with account of neuroregulation. Math. Model. **19**(3), 15–28 (2007)
9. A.Y. Bunicheva, M.A. Menyailova, S.I. Mukhin, N.V. Sosnin, A.P. Favorski, Studying the influence of gravitational overloads on the parameters of blood flow in vessels of greater circulation. Math. Models Comput. Simul. **5**(1), 81–91 (2013)
10. J. Alastruey, S.M. Moore, K.H. Parker, T. David, J. Peiro, S.J. Sherwin, Reduced modelling of blood flow in the cerebral circulation: coupling 1-D, 0-D and cerebral auto-regulation models. Int. J. Numer. Methods Fluids **56**(8), 1061–1067 (2008)
11. S.A. Urquiza, P.J. Blanco, M.J. Venere, R.A. Feijoo, Multidimensional modelling for the carotid artery blood flow. Comput. Methods Appl. Mech. Eng. **195**(33–36), 4002–4017 (2016). https://doi.org/10.1016/j.cma.2005.07.014
12. T. Gamilov, R. Pryamonosov, S. Simakov, Modeling of patient-specific cases of atherosclerosis in carotid arteries, in *ECCOMAS Congress 2016—Proceedings of the 7th European Congress on Computational Methods in Applied Sciences and Engineering*, vol. 1 (2016), pp. 81–89
13. D.V. Burenchev, P.Y. Kopylov, A.A. Bykova, T.M. Gamilov, D.G. Gognieva, S.S. Simakov, Y.V. Vasilevsky, Mathematical modelling of circulation in extracranial brachocephalic arteries at pre-operation stage in carotid endarterectomy. Russ. J. Cardiol. **4**, 88–92 (2017, in Russian)
14. A. Danilov, Yu. Ivanov, R. Pryamonosov, Yu. Vassilevski, Methods of graph network reconstruction in personalized medicine. Int. J. Numer. Methods Biomed. Eng. **32**(8), e02754 (2015). https://doi.org/10.1002/cnm.2754
15. T.M. Gamilov, P.Y. Kopylov, R.A. Pryamonosov, S.S. Simakov, Virtual fractional flow reserve assessment in patient-specific coronary networks by 1D hemodynamic model. Russ. J. Numer. Anal. Math. Model. **30**(5), 269–276 (2015)

16. S.S. Simakov, T.M. Gamilov, Y.N. Soe, Computational study of blood flow in lower extremities under intense physical load. Russ. J. Numer. Anal. Math. Model. **28**(5), 485–504 (2013)
17. T. Gamilov, P. Kopylov, S. Simakov, Computational simulations of fractional flow reserve variability. Lecture notes in computational science and engineering, vol. 112 (2016), pp. 499–507

Delay Induced Oscillations in a Dynamical Model for Infectious Disease

A. Kumar and P. K. Srivastava

1 Introduction

Time delays are inherent to natural as well as man-made systems and similar is the case of disease dynamics. There are various interactions in infectious disease models where time delays play an important role. Incorporating these time delays in model systems gives rise to the delay differential equations. Delay differential equation models have been extensively used in literature to explore the dynamics of the infectious diseases considering various delays. Interesting and complex dynamics such as delay induced stability and instability of equilibria, existence of different types of bifurcations, oscillations, etc. have been an important observation of these models [2, 7, 18, 24, 30, 31]. From biological point of view, delay differential equation models are close to real life system in comparison to ordinary differential equation models. However, there is associated mathematical complexity in dealing with models with delays. In the last few years, there has been growing interest among researchers to study the dynamics of infectious diseases considering various time delays in disease progression.

Delays in infectious disease models have been well studied in literature. Cooke et al. proposed an epidemic model in which they considered the effect of maturation delay in the growth of population [4]. They found that when $R_0 > 1$, the disease remains endemic, either approaching to an equilibrium value or oscillating about this value. Greenhalgh et al., in 2005, proposed an SIRS model which accounts for the effect of vaccination and also quantified the impact of delay on waning the vaccine-induced immunity [11]. They established the existence of Hopf bifurcation. Further, in 2005, a delay differential equation model (SIR) was proposed by

A. Kumar · P. K. Srivastava (✉)
Department of Mathematics, Indian Institute of Technology Patna, Patna, India
e-mail: pksri@iitp.ac.in

© Springer International Publishing AG, part of Springer Nature 2018
R. P. Mondaini (ed.), *Trends in Biomathematics: Modeling, Optimization and Computational Problems*, https://doi.org/10.1007/978-3-319-91092-5_21

Kyrychko et al. which accounts for the effect of nonlinear incidence rate along with delay effect in loss of vaccine immunity [20]. They showed that the infected equilibrium is globally stable if the time lag in loss of vaccine immunity crosses a threshold quantity. Wen et al., in 2008, proposed a delayed SIRS model along with temporary immunity [33]. They found the disease free equilibrium is always globally sable whenever the basic reproduction number is less than one. Further, they established the global stability of the infected equilibrium by constructing a Lyapunov function. In 2010, Huang et al. proposed a set of SIR, SIS, SEIR and SEI epidemiological models with time delays and a general incidence rate and studied their global stability properties [14]. Recently, Banerjee et al. have found that the delay may help in stabilizing the system [2].

During the outbreaks, it has been noticed that information about disease induces behavioural changes of the individuals which eventually affects the disease progression [1, 12, 22, 27, 28]. The information about disease is propagated via media which depends on disease prevalence as well as active social and educational programs. In literature two kinds of approaches are found to incorporate behavioural changes due to the information in the infectious disease model. In the first approach, a correction in the incidence rate (force of infection) due to the information was considered [3, 5, 6, 21–23, 32]. Another approach is using subclass of individuals with information or awareness and then model using compartment model technique [9, 10, 15–17, 26]. The effect of human behavioural response on diseases is presented in the book of Manfredi and d'Onofrio [25].

In this paper we propose and analyse a delay differential equation model which also incorporates the effect of information induced behavioural change in the model leaving the susceptible population virtually immune to infection. Thereby moving these virtually immune individuals to recovered class. The waning of immunity is considered and it is assumed that these individuals, after a time lag, rejoin the susceptible class. The model analysis is performed for stability and we observe hopf bifurcation. We numerically validate our analytical results for a fixed set of parameters.

2   Mathematical Model

In this section, we consider the model proposed by Kumar et al. [19] with modified rate equation for the dynamics of information. In Kumar et al. [19], the authors have considered the growth of information as saturated function of infective population. In this study, we consider the growth function as linear function of infective to avoid complex calculations.

Further, we propose the corresponding delay model considering the delay in waning immunity. As in practice, there is always a time lag in loss of immunity and individuals will take some time to become susceptible again after attaining immunity. Hence, we consider a delay effect in waning of immunity and assume that after a time lag $\tau > 0$, recovered individuals will lose the immunity and then will move to susceptible class. In view of this, the corresponding proposed model with delay effect is given by

$$\frac{dS(t)}{dt} = \Lambda - \beta S(t)I(t) - \mu S(t) - u_1 d Z(t) S(t) + \delta_0 R(t - \tau),$$

$$\frac{dI(t)}{dt} = \beta S(t)I(t) - (\mu + \delta + \gamma)I(t),$$

$$\frac{dR(t)}{dt} = \gamma I(t) + u_1 d Z(t) S(t) - \mu R(t) - \delta_0 R(t - \tau),$$

$$\frac{dZ(t)}{dt} = aI(t) - a_0 Z(t),$$

$\qquad\qquad\qquad\qquad\qquad\qquad\qquad\qquad\qquad\qquad\qquad (1)$

with initial conditions $S(\theta) = S_0 \geq 0, I(\theta) = I_0 \geq 0, R(\theta) = R_0 \geq 0$ and $Z(\theta) = Z_0 \geq 0, \theta \in [-\tau, 0]$, where $(S(\theta), I(\theta), R(\theta), Z(\theta)) \in C([-\tau, 0], R_+^4)$, the Banach space of continuous functions mapping the interval $[-\tau, 0]$ into R_+^4 (non-negative cone in R^4). Here, the total variable population $N(t)$ is divided into three subpopulations depending on the state of the disease: $S(t)$-susceptible population, $I(t)$-infective population and $R(t)$-removed population, respectively, at any given time t.

Also, all parameters are taken to be non-negative. The parameter Λ represents growth rate of susceptible population and γ is a constant recovery rate of infected population. The parameter μ represents the natural mortality rate and δ is the disease related death rate. The parameter β is the disease transmission rate from susceptible population to infective population and it is assumed that the interaction follows mass action type contact when population is homogenously mixed. The parameter δ_0 represents the rate of loss of total immunity which includes the loss of natural immunity and the loss of immunity of protective measures. The factor $u_1 d Z(t) S(t)$ represents the behavioural response of susceptible individuals induced by the information about the disease prevalence [19]. Here $u_1 d$ is the corresponding response rate. Parameter d is information interaction rate by which individuals change their behaviour with $0 \leq u_1 \leq 1$ response intensity. Parameter a is the growth rate of information which depends on the infective population, and on active social and educational campaigns. The parameter a_0 denotes natural degradation rate of information.

3   Model Analysis

In this section, first, boundedness of the solutions of delay model system is established. Further, existence and stability of equilibria are discussed. Subsequently, occurrence of Hopf bifurcation due to delay effect is investigated.

3.1 Positivity and Boundedness

We assume that the initial population are so chosen so that all the population components remain positive. Moreover, from the model system (1), we note that the total population $N(t) = S(t) + I(t) + R(t)$ is governed by the following differential equation

$$\frac{dN(t)}{dt} = \Lambda - \mu N(t) - \delta I(t) \leq \Lambda - \mu N(t).$$

This gives that $\limsup_{t \to \infty} N(t) \leq \frac{\Lambda}{\mu}$. Hence all solutions $S(t)$, $I(t)$ and $R(t)$ are bounded by $\frac{\Lambda}{\mu}$. Further $\frac{dZ(t)}{dt} = aI(t - \tau_2) - a_0 Z(t)$, this implies $\limsup_{t \to \infty} Z(t) \leq \frac{a\Lambda}{a_0\mu}$ by using the bound of I. Thus the biologically feasible region of the model system (1) is the following positive invariant set:

$$\Gamma = \left\{ (S(t), I(t), R(t), Z(t)) \in \mathbb{R}_+^4 \mid 0 \leq S(t), I(t), R(t) \leq \frac{\Lambda}{\mu}, Z(t) \leq \frac{a\Lambda}{a_0\mu} \right\}.$$

3.2 Existence of the Equilibrium Points

From Kumar et al. [19], the basic reproduction number (R_0) of the model system (1) is given by

$$R_0 = \frac{\Lambda\beta}{\mu(\mu + \delta + \gamma)}.$$

It is also easy to find that the model system (1) has the following two equilibria:

1. a disease free equilibrium $E_1 = \left(\frac{\Lambda}{\mu}, 0, 0, 0 \right)$ which always exists, and
2. a unique infected equilibrium $E_2 = (S_*, I_*, R_*, Z_*)$, which exists if and only if $R_0 > 1$. Here $S_* = \frac{(\mu+\delta+\gamma)}{\beta}$, $R_* = \frac{I_*}{\mu+\delta_0}\left(\gamma + \frac{du_1 a(\mu+\delta+\gamma)}{a_0\beta} \right)$, $Z_* = \frac{aI_*}{a_0}$ and $I_* = -\frac{C}{B}$, where $B = \frac{\mu(\mu+\delta+\gamma)+\delta_0(\mu+\delta)}{(\mu+\delta_0)} + \frac{\mu du_1 a(\mu+\delta+\gamma)}{\beta a_0(\mu+\delta_0)}$ and $C = \Lambda\left(\frac{1}{R_0} - 1 \right)$.

Further, we shall state the stability result obtained in Kumar et al. [19] for the no delay case, i.e. $\tau = 0$.

Theorem 3.1 ([19]) *For $\tau = 0$,*

 (i) *the disease free equilibrium E_1 of the system (1) is locally asymptotically stable if $R_0 < 1$ and is unstable if $R_0 > 1$,*
(ii) *if $R_0 > 1$ then the unique infected equilibrium E_2 is locally asymptotically stable provided the following conditions are satisfied:*

$$P_1 P_2 > P_3 \quad and \quad P_1(P_2 P_3 - P_1 P_4) > P_3^2.$$

Here, $P_1 = a_0 + 2\mu + \delta_0 + \beta I_* + du_1 Z_*$, $P_2 = a_0(\mu + \delta_0) + (\mu + a_0)(\mu + \beta I_* + du_1 Z_*) + \delta_0(\mu + \beta I_*) + \beta^2 S_* I_*$, $P_3 = \beta I_*((a_0 + \mu)(\mu + \delta + \gamma) + \delta_0(\mu + \delta)) + \mu a(\mu + \beta I_* + du_1 Z_*) + a_0 \delta_0(\mu + \beta I_*) + adu_1 \beta S_* I_*$ and $P_4 = \beta a_0 I_*(\mu(\mu + \delta + \gamma) + \delta_0(\mu + \delta)) + a\mu\beta du_1 S_* I_*$.

3.3 Stability of the Delay Model

In the following, we shall investigate the stability of the equilibrium points of the delay model (1).

3.3.1 Stability of the Disease Free Equilibrium E_1

Theorem 3.2 *For all time delay $\tau \geq 0$, the disease free equilibrium E_1 of the system (1) is locally asymptotically stable if $R_0 < 1$ and is unstable if $R_0 > 1$.*

Proof The Jacobian matrix J at disease free equilibrium E_1 is given by

$$J_{E_1} = \begin{pmatrix} -\mu & -\beta\frac{\Lambda}{\mu} & \delta_0 & -du_1\frac{\Lambda}{\mu} \\ 0 & \beta\frac{\Lambda}{\mu} - (\mu + \delta + \gamma) & 0 & 0 \\ 0 & \gamma & -(\mu + \delta_0) & du_1\frac{\Lambda}{\mu} \\ 0 & a & 0 & -a_0 \end{pmatrix}.$$

The characteristic equation of J_{E_1} is given by

$$(\lambda + \mu)(\lambda + (\mu + \delta_0))(\lambda + a_0)(\lambda + ((\mu + \delta + \gamma)(R_0 - 1))) = 0.$$

If $R_0 < 1$, then all the eigenvalues of J_{E_1} are negative and hence, E_1 is locally asymptotically stable and it is unstable when $R_0 > 1$ for any time delay $\tau \geq 0$. $\square$

3.3.2 Stability of the Infected Equilibrium E_2

Here, we shall establish the local stability of the infected equilibrium E_2. The linearized system corresponding to the delay system (1) around the infected equilibrium E_2 is given as:

$$\frac{dY(t)}{dt} = J_1 Y(t) + J_2 Y(t - \tau). \tag{2}$$

$$\text{Here, } J_1 = \begin{pmatrix} -(\mu + \beta I_* + du_1 Z_*) & -\beta S_* & 0 & -du_1 S_* \\ \beta I_* & 0 & 0 & 0 \\ du_1 Z_* & \gamma & -\mu & du_1 S_* \\ 0 & a & 0 & -a_0 \end{pmatrix}, \quad J_2 = \begin{pmatrix} 0 & 0 & \delta_0 & 0 \\ 0 & 0 & 0 & 0 \\ 0 & 0 & -\delta_0 & 0 \\ 0 & 0 & 0 & 0 \end{pmatrix}$$

and $Y(t) = (S(t), I(t), R(t), Z(t))^T$.

The characteristic equation corresponding to the linearized system (2) is given by

$$D(\lambda, \tau) := det(\lambda I_4 - (J_1 + e^{-\lambda \tau} J_2)) = 0,$$

here, I_4 is the identity matrix of order four. This can be written as

$$D(\lambda, \tau) := \lambda^4 + A_1 \lambda^3 + A_2 \lambda^2 + A_3 \lambda + A_4 + e^{-\lambda \tau}(B_1 \lambda^3 + B_2 \lambda^2 + B_3 \lambda + B_4) = 0, \tag{3}$$

where

$$A_1 = a_0 + 2\mu + \beta I_* + du_1 Z_* > 0$$

$$A_2 = a_0 \mu + (\mu + a_0)(\mu + \beta I_* + du_1 Z_*) + \beta^2 S_* I_* > 0$$

$$A_3 = \beta I_*(a_0 + \mu)(\mu + \delta + \gamma) + \mu a_0(\mu + \beta I_* + du_1 Z_*) + adu_1 \beta S_* I_* > 0$$

$$A_4 = \beta a_0 I_* \mu(\mu + \delta + \gamma) + \mu adu_1 \beta S_* I_* > 0$$

$$B_1 = \delta_0 > 0$$

$$B_2 = \delta_0(a_0 + \mu + \beta I_*) > 0$$

$$B_3 = a_0 \delta_0(\mu + \beta I_*) + \beta I_* \delta_0(\mu + \delta) > 0$$

$$B_4 = \beta a_0 I_* \delta_0(\mu + \delta) > 0.$$

Clearly, the characteristic equation (3) is a transcendental equation in λ and hence it has infinitely many complex roots. For the stability of E_2, we follow the same argument as in Ref. [29]. The sign of real parts of the roots of characteristic equation (3) will determine the stability of infected equilibrium E_2. If all the roots of Eq. (3) have negative real parts, then E_2 will be the locally stable. Whereas existence of a purely imaginary root leads the instability of E_2, i.e. a root crosses the imaginary axis. Determination of sign of roots of Eq. (3) is a cumbersome task due to its transcendental nature. Rouche's Theorem and continuity in τ infer that the sign of roots of Eq. (3) will change if it crosses imaginary axis, i.e. if Eq. (3) has a pair of purely imaginary root.

As Eq. (3) is in transcendental nature, then it would have infinity complex roots. Hence, in presence of τ, the analysis of the sign of roots of Eq. (3) is very complicated. Thus, in order to recognize the purely imaginary root, we put $\lambda = i\omega$ in Eq. (3) and further separate the real and imaginary parts and given as follows:

$$\omega^4 - A_2 \omega^2 + A_4 = (B_2 \omega^2 - B_4) \cos \omega\tau + (B_1 \omega^3 - B_3 \omega) \sin \omega\tau. \tag{4}$$

$$A_3\omega - A_1\omega^3 = (B_1\omega^3 - B_3\omega)\cos\omega\tau - (B_2\omega^2 - B_4)\sin\omega\tau. \tag{5}$$

Now, squaring and adding both the sides of Eqs. (4) and (5), we get

$$\omega^8 + A_{11}\omega^6 + A_{12}\omega^4 + A_{13}\omega^2 + A_{14} = 0. \tag{6}$$

Here, $A_{11} = A_1^2 - 2A_2 - B_1^2$, $A_{12} = A_2^2 + 2A_4 - 2A_1A_3 - B_2^2 + 2B_1B_3$, $A_{13} = A_3^2 - 2A_2A_4 + 2B_2B_4 - B_3^2$ and $A_{14} = A_4^2 - B_4^2$.

Now, we substituting $m = \omega^2$ in Eq. (6), we have

$$\psi(m) = m^4 + A_{11}m^3 + A_{12}m^2 + A_{13}m + A_{14} = 0. \tag{7}$$

Notice that if Routh-Hurwitz criterion is satisfied then Eq. (7) will have all roots with negative real part, i.e. it has no purely imaginary root. Hence, the following result is given as

Theorem 3.3 *The unique infected equilibrium E_2 of the delay system (1) will be locally asymptotically stable for all $\tau > 0$ provided the following conditions hold*

$$A_{11} > 0, A_{13} > 0, A_{14} > 0 \ and \ A_{11}A_{12}A_{13} > A_{13}^2 + A_{11}^2A_{14}.$$

3.4 Existence of Hopf Bifurcation

Here, we show the existence of family of periodic solutions around the infected equilibrium point E_2 via Hopf bifurcation. For this purpose, we consider the delay parameter τ as a bifurcation parameter. For the existence of the Hopf bifurcation there must be a threshold value of the delay τ_0 such that:

(H_1) $\lambda_{1,2}(\tau_0) = \pm i\omega_{10}(\omega_{10} > 0)$ and all other eigenvalues are with negative real parts at $\tau = \tau_0$.

(H_2) $\left[Re\left(\frac{d\lambda_{1,2}}{d\tau}\right)^{-1}\right]\Big|_{\lambda=i\omega_{10}} \neq 0.$

For the (H_1) condition, we require at least one positive root of Eq. (7). In the following, we determine the conditions for the existence of at least one positive root of Eq. (7) using Descartes' rule of signs. Thus we have the following lemma.

Lemma 3.1 *The Eq. (7) has*

(i) at least one positive root (either one or three) if

    (a) $A_{11} > 0, A_{12} < 0, A_{13} > 0, A_{14} < 0.$
    (b) $A_{11} < 0, A_{12} < 0, A_{13} > 0, A_{14} < 0.$
    (c) $A_{11} < 0, A_{12} > 0, A_{13} > 0, A_{14} < 0.$
    (d) $A_{11} < 0, A_{12} > 0, A_{13} < 0, A_{14} < 0.$

(ii) *exactly one positive root if*

   (a) $A_{11} < 0, A_{12} < 0, A_{13} < 0, A_{14} < 0$.
   (b) $A_{11} > 0, A_{12} < 0, A_{13} < 0, A_{14} < 0$.
   (c) $A_{11} > 0, A_{12} > 0, A_{13} < 0, A_{14} < 0$.
   (d) $A_{11} > 0, A_{12} > 0, A_{13} > 0, A_{14} < 0$.

Assuming that Eq. (7) has one positive root satisfying one of the conditions given in Lemma 3.1. Let $m_{10} = \omega_{10}^2$ be a positive root of Eq. (7) then $\pm i\omega_{10}$ is a pair of purely imaginary root of Eq. (3) for the threshold value of the delay τ. In further, we shall determine the threshold value of the delay τ for which the delay system (1) will remain stable. Using Eqs. (4) and (5), we get the following threshold value:

$$\tau_0 = \frac{1}{\omega_{10}}[\arccos(\Upsilon(\omega_{10}))], \tag{8}$$

where $\Upsilon(\omega_{10}) = \frac{(B_2\omega_{10}^2 - B_4)(\omega_{10}^4 - A_2\omega_{10}^2 + A_4 + C_2) + (B_1\omega_{10}^3 - B_3\omega_{10})((A_3 + C_1)\omega_{10} - A_1\omega_{10}^3)}{(B_2\omega_{10}^2 - B_4)^2 + (B_1\omega_{10}^3 - B_3\omega_{10})^2}$.

Now, we further determine the transversality condition (H_2). For this, we differentiate Eq. (3) with respect to τ, we have

$$\left[Re \left(\frac{d\lambda}{d\tau} \right)^{-1} \right]\Bigg|_{\lambda = i\omega_{10}} = \frac{\psi'(m)}{(B_2\omega_{10}^2 - B_4)^2 + (B_1\omega_{10}^3 - B_3\omega_{10})^2}. \tag{9}$$

In the following result, we state the condition for the transversality condition.

Lemma 3.2 *Let $i\omega_{10}$ be a purely imaginary root with $m_{10} = \omega_{10}^2$ such that $\psi(\omega_{10}) = 0$ and $\psi'(\omega_{10}) \neq 0$, then $\left[Re \left(\frac{d\lambda}{d\tau} \right)^{-1} \right]\Big|_{\lambda = i\omega_{10}} \neq 0$ and its sign is the same as $\psi'(\omega_{10}) \neq 0$.*

Thus transversality condition holds. Now we summarise the above discussion in the following result as follows.

Theorem 3.4 *The unique infected equilibrium E_2 is locally asymptotically stable for $\tau < \tau_0$ and is unstable for $\tau > \tau_0$. At $\tau = \tau_0$, a Hopf bifurcation occurs, i.e. a family of periodic solutions bifurcates from the infected equilibrium E_2 as delay parameter τ crosses the threshold value τ_0 [8, 13].*

3.4.1 Numerical Validation

Here, we numerically validate the Hopf bifurcation result obtained above. For this, we consider the set of representative parameters as: $\Lambda = 10, \beta = 0.0325, \mu = 0.04, d = 0.17, \delta = 0.5, \delta_0 = 0.5, a = 0.1, b = 0.1, a_0 = 0.1, \gamma = 0.1, u_1 = 0.9$. The model system has the unique infected equilibrium $E_2 = (19.69, 11.95, 68.91, 11.95)$ along with disease free equilibrium $E_1 = (250, 0, 0, 0)$ when $R_0 = 12.69 > 1$. In this case, note that the coefficients A_{11}, A_{12}, A_{13} and A_{14} satisfy the condition $i(a)$ of Lemma 3.1. Thus Eq. (7) has one positive root (0.0546) and hence Lemma 3.1 follows. We find that the characteristic equation (3) has a pair of purely imaginary root $\pm 0.233i$ with $\omega_{10} = 0.233$. Using (8), the corresponding threshold value of the delay parameter is given as $\tau_0 = 8.73$ and the transversality condition $\left[Re \left(\frac{d\lambda}{d\tau} \right)^{-1} \right]\Big|_{\lambda = i\omega_{10}} = 2.308 > 0$ also holds true. Thus we conclude from Theorem 3.4 that the delay system (1) will be stable for the delay range $\tau \in [0, \tau_0)$ and unstable for $\tau > \tau_0$. At the threshold value of delay $\tau = \tau_0 = 8.73$, periodic oscillations bifurcate near unique infected equilibrium E_2 as τ crosses τ_0.

Further, we solve the delay system (1) using DDE23 in MATLAB to show the stability and instability of the unique infected equilibrium E_2. For this, we first solve the delay system for the delay parameter $\tau = 8.45 < \tau_0$ and other parameters as given above along with initial population size $S(0) = 21, I(0) = 10, R(0) = 65$ and $Z(0) = 11$. The corresponding results are shown in Fig. 1. One can easily see that the solution trajectories approaching to the unique infected equilibrium E_2 showing the asymptotic stability of E_2.

Further from Theorem 3.4, we infer that as τ crosses $\tau_0 = 8.45$, a family of periodic solutions bifurcate around the unique infected equilibrium E_2. For this, we further solve the delay system (1) for $\tau = 9.6$ and the corresponding results are shown in Fig. 2 which clearly show the existence of periodic solutions of populations. This infers that the disease will persist in the oscillatory nature within the population due to the delay effect in waning the immunity of protection. For this set of parameters, we numerically find that if there is a time lag of about 9 days in the waning of immunity related with protection gained from protective measures, then the infected population show oscillations. Thus in this case the disease elimination may be very critical and challenging due to oscillations.

In order to plot the bifurcation diagram to show the occurrence of periodic orbits, we vary $\tau \in [8, 9.75]$ along with other parameters given as above. The corresponding bifurcation diagrams are plotted in Fig. 3. One can easily see that

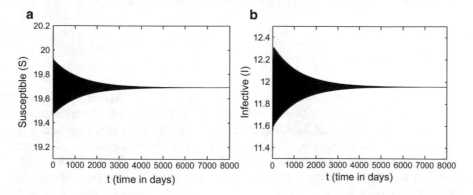

Fig. 1 (**a**) Solution trajectory of susceptible population showing stability for $\tau = 8.45$. (**b**) Solution trajectory of infective population showing stability for $\tau = 8.45$

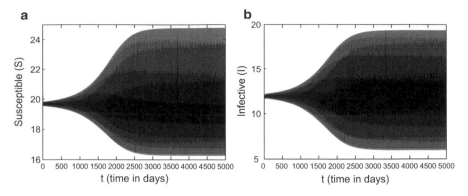

Fig. 2 (**a**) Oscillation in susceptible population for $\tau = 9.6$. (**b**) Oscillation in infective population for $\tau = 9.6$

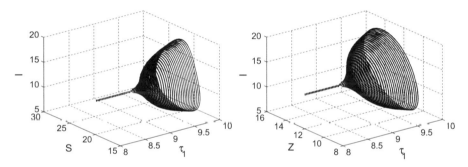

Fig. 3 Plot for the bifurcation diagram showing the occurrence of periodic orbits when τ crosses $\tau_0 = 8.45$ in I-R and Z-R planes

when $\tau \in [8, 8.45)$, the unique infected equilibrium E_2 is stable and when τ passes the threshold value $\tau_0 = 8.45$ periodic orbits arise for $\tau > 8.45$.

4 Conclusion

A nonlinear delay differential equation model SIRS for the disease dynamics is proposed and analysed which accounts for the effect of human behavioural response and the delayed impact of immunity loss. Model analysis is carried out and it is found that the disease free equilibrium is locally stable always for $R_0 < 1$. A unique infected equilibrium is obtained when $R_0 > 1$ which is locally stable when time delay is less than a threshold quantity ($\tau < \tau_0$). Existence of Hopf bifurcation around the unique infected equilibrium is investigated if the time delay crosses a threshold value. Hence, the delay in waning the immunity destabilises the system and causes the occurrence of oscillations. Thus, we finally conclude that the delay effect destabilises the system and induces the oscillatory persistence of the disease within the population. Hence, the delay in waning immunity shows rich and complex dynamics in the model and provides important insight.

Finally, we would like to comment on the fact that if the period of study is long then the survival probability of recovered individuals will come into consideration. This can be done by multiplying the delay term by $e^{-\mu\tau}$ which is survival probability of recovered individuals who have spent time in recovered class and survived the period to return back to susceptible class. This will make mathematical analysis complicated but at the same time will give important insight. We intend to take up the problem in future.

Acknowledgements Anuj Kumar is thankful for the financial support of Council of Scientific and Industrial Research, India (Grant No.: 09/1023(0009)/2012–EMR–I). Prashant Srivastava thanks to the International Union of Biological Sciences (IUBS) for partial support of living expenses in Moscow, during the 17th BIOMAT International Symposium, October 29-November 04, 2017.

References

1. A. Ahituv, V.J. Hotz, T. Philipson, J. Hum. Resour. **31**, 869 (1996)
2. M. Banerjee, Y. Takeuchi, J. Theor. Biol. **412**, 154 (2008)
3. S. Collinson, J.M. Heffernan, BMC Public Health **14**, 376 (2014)
4. K. Cooke, P. Van den Driessche, X. Zou, J. Math. Biol. **39**(4), 332 (1999)
5. J. Cui, Y. Sun, H. Zhu, J. Dyn. Diff. Equa. **20**(1), 31 (2008)
6. A. d'Onofrio, P. Manfredi, J. Theor. Biol. **256**(3), 473 (2009)
7. A. d'Onofrio, P. Manfredi, E. Salinelli, Theor. Popul. Biol. **71**(3), 301 (2007)
8. H.I. Freedman, V. Sree Hari Rao, Bull. Math. Biol. **45**(6), 991 (1983)
9. S. Funk, E. Gilad, V.A.A. Jansen, J. Theor. Biol. **264**(2), 501 (2010)
10. S. Funk, E. Gilad, C. Watkins, V.A.A. Jansen, Proc. Natl. Acad. Sci. U. S. A. **106**(16), 6872 (2009)
11. D. Greenhalgh, Q.J.A. Khan, F.I. Lewis, Nonlinear Anal. Real World Appl. **63**(5), 779 (2005)
12. D. Greenhalgh, S. Rana et al., Appl. Math. Comput. **251**, 539 (2015)
13. J.K. Hale, *Theory of Functional Differential Equations. Applied Mathematical Sciences* (Springer, New York, 1977)
14. G. Huang, Y. Takeuchi, W. Ma, D. Wei, Bull. Math. Biol. **72**(5), 1192 (2010)
15. H. Joshi, S. Lenhart, K. Albright, K. Gipson, Math. Biosci. Eng. **5**(4), 757 (2008)
16. S.M. Kassa, A. Ouhinou, J. Math. Biol. **70**(1–2), 213 (2015)
17. I.Z. Kiss, J. Cassell, M. Recker, P.L. Simon, Math. Biosci. **225**(1), 1 (2010)
18. Y. Kuang, *Delay Differential Equations with Applications in Population Dynamics*. Mathematical in Science and Engineering Series, vol. 191 (Cambridge, 1993)
19. A. Kumar, P.K. Srivastava, Y. Takeuchi, J. Theor. Biol. **414**, 103 (2017)
20. N. Kyrychko, K.B. Blyuss, Nonlinear Anal. Real World Appl. **6**(3), 495 (2005)
21. Y. Li, C. Ma, J. Cui, Rocky Mountain J. Math. **38**(5), 1437 (2008)
22. Y. Liu, J. Cui, Int. J. Biomath. **1**(1), 65 (2008)
23. R. Liu, J. Wu, H. Zhu, Comput. Math. Methods Med. **8**(3), 153 (2007)
24. M. Liu, E. Liz, G. Röst, SIAM J. Appl. Dyn. Syst. **75**(1), 75 (2015)
25. P. Manfredi, A. d'Onofrio, *Modeling the Interplay Between Human Behavior and the Spread of Infectious Diseases* (Springer, New York, 2013)
26. A.K. Misra, A. Sharma, J.B. Shukla, Math. Comput. Model. **53**(3), 1221 (2011)

27. A.K. Misra, A. Sharma, V. Singh, J. Biol. Syst. **19**(2), 389 (2011)
28. T. Philipson, J. Hum. Resour. **31**, 611 (1996)
29. S. Ruan, J. Wei, Dyn. Contin. Discret. Impuls. Syst. Ser. A Math. Anal. **10**, 863 (2003)
30. Y. Song, J. Wei, Chaos, Solitons Fractals **22**(1), 75 (2004)
31. P.K. Srivastava, P. Chandra, Differ. Equ. Dyn. Syst. **16**(1–2), 77 (2008)
32. J.M. Tchuenche, C.T. Bauch, ISRN Biomath., 1–10, Article ID 581274 (2012)
33. L. Wen, X. Yang, Chaos, Solitons Fractals **38**(1), 221 (2008)

Pressure Gradient Influence on Global Lymph Flow

A. S. Mozokhina and S. I. Mukhin

1 Introduction

Modeling of lymph flow through the lymphatic system is an important task which attends great interest in the last years because of different medical problems related to the lymph flow such as lymphedema and distribution of infections and drugs through the organism. Investigation of these problems is based on the models of systemic lymphodynamics, and we propose to use for such modeling quasi-one-dimensional approach, which gives great results in the modeling of systemic blood circulation. One of the aims of this work is investigation of lymph propagation through the complicated topology and determining influence of different driving forces on lymph flow.

The physiological mechanisms of lymph flow are not still clear enough. According to physiological concepts, there are two main forces, which provide lymph flow: pressure gradient and contractions, active and passive, of segments of lymphatic vessels [1]. In the quasi-one-dimensional approach, the pressure gradient is an initial data of the model, but its influence on lymph propagation in the complicated structure of vessels with different sizes and properties should be studied. Contractions are one more driving force, and their mathematical and algorithmic implementation is independent and nontrivial problem.

Lymphatic system contains lymphatic vessels and lymph nodes. Lymphatic vessels have different sizes and properties, they have specific structure, which influence lymph flow. The vessels have valves, which prevent backward flow, and segments of vessels between adjacent pairs of valves can produce active contractions. Such segments are called "lymphangions". The model of systemic lymphodynamics

A. S. Mozokhina (✉) · S. I. Mukhin
Department of Computational Mathematics and Cybernetics, Lomonosov Moscow State University, Moscow, Russia
e-mail: vmmus@cs.msu.ru

© Springer International Publishing AG, part of Springer Nature 2018
R. P. Mondaini (ed.), *Trends in Biomathematics: Modeling, Optimization and Computational Problems*, https://doi.org/10.1007/978-3-319-91092-5_22

requires a mathematically formalized model of lymphatic system as a domain for calculations. Model of the lymphatic system in quasi-one-dimensional approach is a graph with vertices representing vessel bifurcations, and arcs representing vessels and lymph nodes. Creation of such model is a challenge because of complicated and slightly structured topology.

The most well-known model of systemic lymph circulation is the model of Reddy [2], where the lymph flow in the set of lymphangions is considered. This model introduces description of lymph flow in lumped approach in the tree of main lymphatic vessels. There is a simple graph of the lymphatic system, and it is not spatially oriented. Quasi-one-dimensional approximation is used in [3], where lymph flow is considered through the series of lymphangions, without studying systemic circulation. Most of the other models also not consider systemic lymph circulation, and investigate flow through the series of lymphangions, as in [4], or focus on flow mechanisms in one lymphangion [5, 6], using lumped models or multidimensional ones.

In the current work, we introduce a graph of the lymphatic system and quasi-one-dimensional mathematical models for lymph flow through different parts of such graph. This graph was specially designed for joint modeling of lymphatic and cardiovascular systems. On the basis of different physiological data, the typical spatial structure of lymphatic system was segmented with thorough description of its elements: types of vessels, characteristics of vessels, areas of lymph nodes, and their internal topology. Segments of the system are topologically linked to the corresponding areas of cardiovascular system, and the whole graph (with a large set of appropriate physiological data) is physiologically adequate and represents the whole common lymphatic system in the form, suitable for numerical simulations. Topology, parameters of vessels, and nodes can be easily changed in any way according to current task.

In this paper, the results of numerical investigation of the flow under the pressure gradient are shown for both horizontal and vertical positions of the graph. The influence of valves in the biggest (in the sense of the diameter) vessels on the global circulation is also investigated, and the results of such investigation are presented.

2 Lymphatic System Modeling

2.1 Physiology Overview

The lymphatic system complements the cardiovascular system and they form together common vessel system in the human body. About 10 % of blood volume goes to the lymphatic system due to the capillary filtration [7], and then returns to the venous part of the cardiovascular system.

The lymphatic system is not closed. It starts with the initial lymphatics in the interstitial fluid and ends with the biggest lymphatic vessels (the right lymphatic

Fig. 1 Valves in lymphatic vessels in opened (left) and closed (right) states. Arrows show direction of lymph flow

and the thoracic ducts), which open in the upper vena cava (to the right and to the left venous angles, respectively). The lymphatic system is not connected to the heart directly and this leads to low-velocity and low-pressure gradients of lymph flow. Lymph goes into the lymphatic system through the initial lymphatics and flows in one direction: from interstitial space of the periphery to the upper vena cava. This unidirectional movement is provided by the numerous valves in the lymphatic vessels, which restrict backward flow. The lymphatic vessel is divided into segments by valves. These segments are called "lymphangions" (see Fig. 1).

Lymphatic net consists of lymphatic vessels and lymph nodes. The vessels vary in diameter and structure. The biggest vessels in the lymphatic system are trunks and ducts. The diameter of such vessels is about 1.5–2 mm [8]. The length of lymphangions in this case can reach 5 and 10 cm. The velocity of lymph flow in such vessels is about 0.5–1 cm/s. Other vessels (except for capillaries) are not classified as trunks and ducts, but their diameters can reach similar values. The diameter of these vessels varies from 3–5 μm to 1–2 mm [8]. The length of lymphangions can be about 2 mm, so valves in such vessels are located very close to each other. It is assumed that length of the lymphangions correlates with the vessels diameter: the greater the diameter is, the greater the lymphangion length is. Lymphangions of all mentioned vessels can produce active contractions, and elastic properties of such vessels are close to such of veins. The lymphatic capillaries are the entries of lymphatic system. Such vessels have no valves and cannot produce contractions. The diameter of lymphatic capillaries is about 20–200 μm .

There are about 400–500 lymph nodes in the human body. Lymph nodes perform, among others, transport and filtration functions. There are about 3–4 vessels entering the node and 1–2 ones exiting the node [1]. There are valves in regions where vessels connect with the node, so lymph can flow generally in one direction in the node, the same as in lymphatic vessels.

Pressure value in the initial lymphatic vessels is hard to measure, so it is commonly used the value in the interstitial fluid. It is used to consider that the pressure in the upper vena cava is about 10 mm Hg or less [7].

2.2 Graph of the Lymphatic System

On the basis of these anatomical and physiological data, we were able to create a graph of the lymphatic system (see Fig. 2). This graph is anatomically adequate and is spatially consistent with analogical graph of the cardiovascular system [9]. The

Fig. 2 Graph of the lymphatic system: (i) head, (ii) neck, (iii) diaphragm, (iv) elbows, (v) groin; (1) thoracic duct, (2) right lymphatic duct, (3) cisterna chyli, (4) subclavian trunks, (5) lumbar trunks, (6) lymph nodes

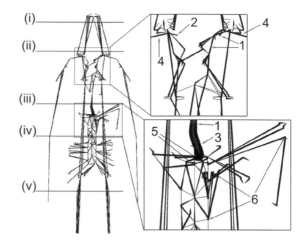

graph contains 543 arcs representing lymphatic vessels and lymph nodes and 478 vertices representing points of vessel bifurcations. One hundred and sixty one arcs of the graph represent lymph nodes.

The graph contains main trunks and ducks, 46 regional groups of lymph nodes according to [1], afferent and efferent vessels for each node, effective representation of lymphatic capillaries, which are the entries of lymphatic system, and two exits: the left and the right venous angles.

All elements of the graph can be divided into four groups. Each group differs from others by the influence on lymph flow and parameters of the vessels forming it. These groups are:

1. Main trunks and ducts, which are characterized by big diameters, long lymphangions, valves preventing backward flow of lymph, and active contractions of lymphangions.
2. All lymphatic vessels except for trunks and ducts (first group), and capillaries (third group). These vessels are characterized by smaller diameters, short lymphangions, valves preventing backward flow of lymph, and active contractions of lymphangions. The main difference from the previous group is that in this group segments between valves are very short (can be about 2 mm).
3. Effective representation of lymphatic capillaries, which have no valves, no lymphangions, and no contractions.
4. Lymph nodes which are complicated structures of the lymphatic system, and modeling of lymph flow through them is a special field of research, e.g., see [10]. In our approximation, lymph nodes are some kind of lymphangions of the vessels of first group: they have big diameters and can produce active contractions.

One more challenge in the creation of the graph of the lymphatic system is a choice of parameters of the vessels and nodes. Available data about diameters of lymphatic vessels is rather poor and concerns mostly the largest vessels. Information about small vessels for obvious reasons is approximate, and their parameters we will define using average meanings.

Table 1 Diameters and cross-section area of the vessels and nodes of the lymphatic system graph

Name	d (cm)	S (cm^2)
Effective vessels	0.020	0.0002
Collectors	0.107	0.0090
Lumbar trunks	0.150	0.0177
Other trunks	0.100	0.0079
Ducts	0.200	0.0314
Cisterna chyli	0.400	0.1257
Lymph nodes	0.200	0.0314

Fig. 3 Binary tree of lymphatic capillaries. It is proposed that on each level of bifurcation vessels have the same pressure gradient Δp_i, length l_i, cross-sections area S_i, and lateral area σ_i, $i = 0, \ldots, n$

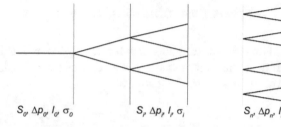

$S_0, \Delta p_0, l_0, \sigma_0$ $S_i, \Delta p_i, l_i, \sigma_i$ $S_n, \Delta p_n, l_n, \sigma_n$

Parameters of the vessels of our graph are presented in Table 1. The diameters for vessels of the first group are taken from the literature. In the current realization, all vessels of the second group are taken with the same diameters.

2.3 Representation of Capillaries

Lymphatic capillaries are very small and numerous, so an effective representation of nets of capillaries is used in the graph.

Let us assume that lymphatic capillaries have the topology of a binary tree (see Fig. 3). We want to substitute this net with one effective element (effective vessel), which contains the parameters the pressure gradient Δp, flux Q, and lateral surface area σ of the net.

Let us assume that there are the following relations between vessel parameters from different levels of bifurcations: $S_i = q S_{i-1}$, $l_i = p l_{i-1}$, $i = 1, \ldots, n$, $q > 0$, $p > 0$. Then, $S_i = q^i S_0$, $l_i = p^i l_0$, $i = 1, \ldots, n$, $q > 0$, $p > 0$ and parameters of the effective vessel, which guarantee the conservation of pressure gradient, flux, and lateral surface area of the net, are defined as follows:

$$\hat{l} = l_0 \left[\sum_{i=0}^{n} \left(\frac{p}{2q^2} \right)^i \left(\sum_{i=0}^{n} \left(2p\sqrt{q} \right)^i \right)^4 \right]^{1/5}, \quad \hat{S} = S_0 \left[\frac{\sum_{i=0}^{n} \left(2p\sqrt{q} \right)^i}{\sum_{i=1}^{n} \left(\frac{p}{2q^2} \right)^i} \right]^{2/5} \tag{1}$$

where $\hat{l}$ is the length of the effective element and $\hat{S}$ is its cross-section area. If we assume that

$$\frac{p}{2q^2} < 1, \quad 2p\sqrt{q} < 1 \tag{2}$$

then series in (1) converge when $n \to \infty$, and Eq. (1) take the following form:

$$\hat{l} = l_0 \left(\frac{2q^2}{(2q^2 - p)(1 - 2p\sqrt{q})^4} \right)^{1/5}, \quad \hat{S} = S_0 \left(\frac{2q^2 - p}{2q^2(1 - 2p\sqrt{q})} \right)^{2/5}. \quad (3)$$

3 Modeling of Lymph Flow

3.1 Mathematical Models of Lymph Flow

Quasi-one-dimensional approach for description of incompressible fluid is applicable if the radial part of velocity is much less than its axial part. In this case, we get the following system of equations:

$$\frac{\partial S}{\partial t} + \frac{\partial u S}{\partial x} = 0, \quad \frac{\partial u}{\partial t} + \frac{\partial}{\partial x}\left(\frac{u^2}{2}\right) + \frac{1}{\rho}\frac{\partial p}{\partial x} = -8\pi\nu\frac{u}{S}, \quad S = S(p), \quad (4)$$

where x is a spatial coordinate, t is time, unknown functions are $u(x, t)$ and $p(x, t)$, velocity and pressure, respectively. $S(x, t)$ is cross-section area, and the third equation defines its dependence on pressure p. ρ is constant density and ν is friction coefficient, which is also constant.

This system can be used for description of lymph flow in the first-order approximation, but there is no respect for specific behavior of lymphatic vessels, namely, for restriction of backward flow of lymph by valves and for contractions of lymphangions. Lymphatic vessels have lymphangions of different length, and this fact leads us to include in the model the valves of two types depending on lymphangion length. If the distance between adjacent pairs of valves is big (about or more than 1 cm, as it is in trunks and ducts), the restriction of backward flow is proposed to be the following condition in the bifurcation point:

$$Q_i = \begin{cases} Q_j, & u_i z > 0 \\ 0, & u_i z < 0 \end{cases}, \quad (5)$$

where Q_i and Q_j are fluxes in i and j vessels, respectively, u_i and u_j are velocities of lymph in the i and j vessels, respectively, z shows which direction the flux is allowed to flow in: from i vessel to j one, or *vice versa*. In other words, if the flux is accepted by valve, condition (5) transforms to flux continuity condition, and if flux is restricted by valve, condition (5) gives flux equating to zero. This valve model is simple enough, gives us appropriate behavior, and is used in many works of modeling lymph flow, e.g., [2, 3, 5], but when valves are too close to each other and when there are too many valves (vessels of second group), this model is unsuitable.

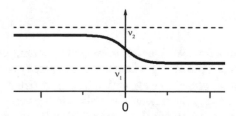

Fig. 4 Sigmoid as an example of monotonic continuous function of viscosity coefficient $v(u)$ in the momentum equation of (4). Sigmoid tends to v_1 when flow is allowed by valves and to v_2 when flow is restricted by valves

If the length of the lymphangions is small enough (about 2 mm, as it is in the vessels of second group), then the valves are proposed to be modeled by some additional force of valve resistance in second equation of (4). This equation takes the following form:

$$\frac{\partial u}{\partial t} + u\frac{\partial u}{\partial x} + \frac{1}{\rho}\frac{\partial p}{\partial x} = -8\pi v\frac{u}{S} + f_{vlv}(u), \tag{6}$$

where f_{vlv} is some function of u. f_{vlv} can be considered as friction force with friction coefficient depending on velocity direction, and in that case the movement equation can be written in the form:

$$\frac{\partial u}{\partial t} + u\frac{\partial u}{\partial x} + \frac{1}{\rho}\frac{\partial p}{\partial x} = -8\pi v(u)\frac{u}{S}, \tag{7}$$

where friction coefficient v is not constant but depends on velocity u. This coefficient can be discontinuous or continuous described by monotonic function with small values in one direction and much more in another (e.g., some kind of sigmoid shown in Fig. 4). This model allows us to deal with huge amount of valves in vessels of second group. The disadvantage of this model is that it restricts backward flow in each point of the vessel, not only in valves. But when the system (4), (7) is solved numerically, the step of spatial net is comparable to lymphangion length, and so the mentioned restriction is negligible in our case. Such approach for valve modeling was used in [11], and similar one in lumped model of chain of contracting elements [4].

Modeling of lymphangion's contractions will be considered in the following work.

3.2 Calculations

The main goal of numeric investigations presented in current work is to find out if the flow is possible under pressure gradient in complicated system of vessels without specific mechanisms of lymph flow regulation both in horizontal and vertical cases. One more goal is to specify the influence of valves in big vessels on lymph flow, in other words, we want to determine if the valves (5) in big vessels are important for our model of lymph flow.

All calculations are performed in Cardio-Vascular Simulation System (CVSS) software [9]. The domain of calculations is the graph of the lymphatic system, described in Sect. 2.2 (see Fig. 2). On each arc, the system (4) with partially linear state equation is stated. In the points of bifurcations, the conditions of mass conservation and pressure equality are stated. In points of bifurcations which have the sense of valves in big vessels, condition (5) instead of mass conservation is stated. There are 25 such valves in the graph. Boundary condition for the system is pressure gradient. Some parameters of the arcs of the graph are shown in Table 1.

First set of calculations we perform in the graph without any valves under the pressure gradient of 5 mm Hg for the horizontal case ($g = 0 \, \text{cm/s}^2$): 5 mm Hg in the interstitial space and 0 mm Hg in the upper vena cava. The results of the calculations in the horizontal case (see Fig. 5a) show that varying of parameters of effective representation of nets of lymphatic capillaries can give physiologically correct output flux in 0.029 ml/s. However, in the vertical case ($g = 1000 \, \text{cm/s}^2$) lymph flow is principally impossible: the lymph flows into the system from each entry point, and there is no output point even if the pressure gradient increases up to 80 mm Hg. So, in the vertical case the specific mechanisms of lymph flow regulation become necessary.

Second set of calculations is performed in the graph with 25 valves in the points of bifurcations. The pressure is 5 mm Hg in the interstitial space and 0 mm Hg in the upper vena cava for the horizontal case, and 60 mm Hg in the interstitial space and 0 mm Hg in the upper vena cava for the vertical case. Calculations in horizontal case give the same results as calculations in the graph without any valves (see Fig. 5b), so in the horizontal case such valves have no influence on lymph flow. In the vertical case, the presence of valves (5) leads to change of pattern of lymph flow. Now, lymph flows in the "right" direction in both main trunks, which open in upper vena cava, so the model has two outputs: the right and the left venous angles, as it must be. However, the pressure gradient required for lymph flow is 60 mm Hg in this case. So, other mechanisms of lymph flow regulations must be implemented to compensate the influence of gravity force.

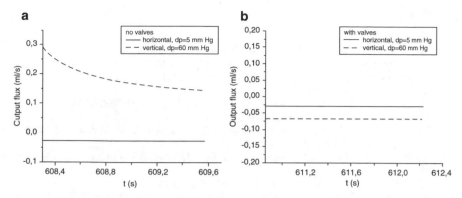

Fig. 5 There are output fluxes upon time as results of the calculations of lymph flow in the graph of the lymphatic system (Fig. 2) under pressure gradient influence: (**a**) calculations in model without valves in horizontal (solid) and vertical (dashed) cases. In the horizontal case, physiologically adequate value 0.029 ml/s of output flux is obtained. Negative values of flux mean that lymph flows out of the system, while positive values mean that lymph flows into the system from the right and the left venous angles. In the vertical case, flux has "wrong" direction and further calculations are not possible. This scenario in the vertical case is true for other pressure gradients (from 5 mm Hg to 80 mm Hg); (**b**) calculations in lymphatic system with valves (5) in horizontal (solid) and vertical (dashed) cases. In the horizontal case, the physiological value 0.029 ml/s of flux is obtained—the same as for model without valves. In the vertical case, lymph flows in the "right" direction

4 Conclusion

The anatomically adequate graph of the lymphatic system was created, and this graph is spatially consistent with analogy graph of the cardiovascular system. Mathematical description of specific mechanisms of lymph flow regulation was offered. Some calculations in the model of first-order approximation were performed, and the results have shown that even without any specific regulatory mechanisms the flow exists in the horizontal case. Calculations also have shown that in the vertical case the valves in big vessels are extremely important: the flow cannot exist without them. The presence of valves gives right pattern of lymph flow, but the required pressure gradient of 60 mm Hg is much more than physiological value for the lymphatic system, so other regulatory mechanisms must be implemented in order to get more accurate model of flow in the lymphatic system.

Acknowledgements Authors would like to thank all interested in the current work, especially Prof. M. V. Abakumov and Prof. V. B. Koshelev for their valuable notes and fruitful discussion on this work.

References

1. E.I. Borzyak, V.Y. Bocharov, M.R. Sapin, Medicine (1993, in Russian)
2. N.P. Reddy, T.A. Krouskop, P.H. Newell Jr., Blood Vessels **12**, 261 (1975)
3. A.J. Macdonald, K.P. Arkill, G.R. Tabor, N.G. McHale, C.P. Winlove, Am. J. Physiol. Heart Circ. Physiol. **295**, H305 (2008)
4. C.D. Bertram, C. Macaskill, J.E. Moore Jr., J. Biomech. Eng. **133**, 011008 (2011)
5. E. Rahbar, J.E. Moore Jr., J. Biomech. **44**, 1001 (2011)
6. C.M. Quick, A.M. Venugopal, A.A. Gashev, D.C. Zawieja, R.H. Stewart, Am. J. Physiol. Regul. Integr. Comp. Physiol. **292**, R1510 (2007)
7. R.F. Schmidt, G. Thews (eds.), *Human Physiology*, 2nd completely revised edn. (Springer, 1987)
8. K.N. Margaris, R.A. Black, J. R. Soc. Interface **9**, 601 (2012)
9. M.V. Abakumov, K.V. Gavrilyuk, N.B. Esikova, A.V. Lukshin, S.I. Mukhin, N.V. Sosnin, V.F. Tishkin, A.P. Favorskii, Differ. Equ. **33**, 895 (1997)
10. G.A. Bocharov, *VIII-th Conference on Mathematical Models and Numerical Methods in Biomathematics* (2016)
11. S. Simakov, T. Gamilov, Y.N. Soe, Rus. J. Numer. Anal. Math. Model. **28**, 485 (2013)

Blood Flows in Vascular Networks: Numerical Results vs Experimental Data

T. K. Dobroserdova, A. A. Cherevko, and E. A. Sakharova

1 Introduction

Methods of blood flow modelling are actively developed nowadays [1]. Such models are demanded and successfully used for many medical applications, in particular, for patient-specific simulations [2, 3]. A number of 1D numerical blood flow models have been developed [4]. The problem of model verification is still actual and significant. It is difficult and often impossible to collect enough data from real humans or animals. Several physical experiments have been designed to provide data for blood flow model verification. Two benchmarks of this sort will be described in this paper. In the first experiment fluid flows in the network of silicone tubes which imitates main arteries of the human body. In the second case the fluid flow in the bifurcation of carotid arteries is modelled. The bifurcation geometry is presented as a cavity in a silicone block. Both experiments provide enough data for simulations. The numerical results will be compared with experimentally measured data.

T. K. Dobroserdova (✉)
Institute of Numerical Mathematics RAS, Moscow, Russia

A. A. Cherevko
Lavrentyev Institute of Hydrodynamics SB RAS, Novosibirsk, Russia
e-mail: cherevko1@ngs.ru

E. A. Sakharova
Higher School of Economics, National Research University, Moscow, Russia

© Springer International Publishing AG, part of Springer Nature 2018
R. P. Mondaini (ed.), *Trends in Biomathematics: Modeling, Optimization and Computational Problems*, https://doi.org/10.1007/978-3-319-91092-5_23

2 Blood Flow Model

Blood is assumed to be viscous incompressible fluid flowing in the vascular network. Every vessel is considered as an elastic tube. Poiseuille's velocity profile is assumed in every cross section. Let S be the area of vessel cross section, $\bar{u}$ is the axial velocity and $\bar{p}$ is the pressure (both averaged over cross section). The model is based on mass and momentum conservation laws. The third equation is the state equation that describes elastic properties of the vessel walls. The model system of equations is the following [5]:

$$
\begin{cases}
\dfrac{\partial S}{\partial t} + \dfrac{\partial (S\bar{u})}{\partial x} = \varphi(t, x, S, \bar{u}) \\[2ex]
\dfrac{\partial \bar{u}}{\partial t} + \dfrac{\partial (\bar{u}^2/2 + \bar{p}/\rho)}{\partial x} = \psi(t, x, S, \bar{u}) \qquad \text{for } x \in [0, l], \\[2ex]
\bar{p} - p_{\text{ext}} = f(S)
\end{cases}
\tag{1}
$$

where l is the length of the vessel, ρ is the fluid density, p_{ext} is the external pressure (in this work $p_{\text{ext}} = 0$), φ is known function of a source or sink of the fluid ($\varphi = 0$ in this work), ψ is known function of external forces, e.g. friction. Function $f(S)$ may be different depending on the vessel wall material properties [6]. One common state equation for tubes with linear elastic properties of wall material is the following:

$$
f(S) = \frac{\sqrt{\pi} h E}{(1 - \sigma^2) S_0} (\sqrt{S} - \sqrt{S_0}),
\tag{2}
$$

where h is the wall thickness, E is Young's modulus, σ is Poisson's ratio, S_0 is the cross section area of a vessel at rest.

The system of Eqs. (1) should be closed by boundary conditions. Hyperbolic type of the system (1) guarantees one condition for every boundary point of the vessel. For the blood flow model in vascular networks we ask for continuity of total pressure and mass conservation at every point of vessel junctions (network node):

$$
p_i + \frac{\rho u_i^2}{2} = p_j + \frac{\rho u_j^2}{2} \quad \forall i, j \in [1, n],
\tag{3}
$$

$$
\sum_{i=1,\dots,n} \epsilon_i S_i u_i = 0,
\tag{4}
$$

where n is the number of connected vessels, $\epsilon_i = 1$ for incoming vessels and $\epsilon_i = -1$ for outgoing vessels. Depending on the task, the condition of the total pressure continuity (3) can be replaced with the condition of pressure continuity

$$
p_i = p_j \quad \forall i, j \in [1, n],
\tag{5}
$$

or with Poiseuille's pressure drop conditions:

$$p_i - p_{node} = R_i \epsilon_i S_i u_i \quad \forall i \in [1, n], \tag{6}$$

where R_i is the hydraulic resistance for the flow between the vessel with the number i and the center of the network node, p_{node} is the pressure in the network node.

The numerical model used in this work is presented in [5, 7, 8] where grid-characteristic method is used for the numerical integration of the 1D model equations. The first order scheme is applied to the characteristic form of (1).

3 Modelling of Fluid Flow in the Vascular Network

The physical experiment is presented in the paper [9]. The system of 37 silicone tapered tubes (Fig. 1) imitates main human arteries. Properties of the silicone are the following: Young's modulus is $E = 1.2\,\text{MPa}$, Poisson's ratio is $\sigma = 0.5$. Water-glycerol mixture has properties similar to blood: density is $\rho = 1050\,\text{kg/m}^3$ and viscosity is $\mu = 2.5\,\text{mPa}$. The fluid is pumped through the silicone network. Terminal vessels are connected to the overflow reservoir via single small-diameter tubes (passive resistance elements).

At the inlet of the network the flow rate measured in vitro is prescribed as the inflow boundary condition. The flow rate corresponds to heart output. Terminal vessels are coupled to single-resistance terminal models. Pressure and flux are measured in middle points of several vessels.

Fig. 1 Scheme of silicone network

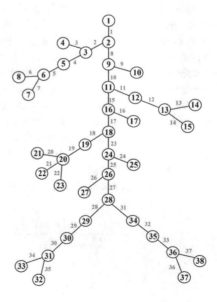

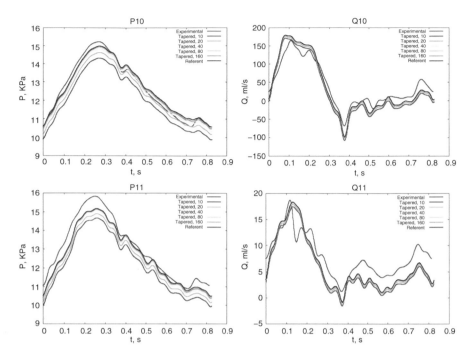

Fig. 2 Convergence of numerical solution in 10 and 11 vessels

The uniform mesh is used in the whole vascular network. Let L be the integer part of the vessel length l cm ($L = [l]$). The series of simulations with mesh consisting of $10L$, $20L$, $40L$, $80L$, $160L$ calculation points in each vessel are presented. Time steps are 1E−5, 5E−6, 2.5E−6, 1.25E−6, 5.125E−7, respectively. The convergence of numerical solutions in the middle points of vessels 10 and 11 is shown in Fig. 2. The red line is experimental (in vitro) data. The black line corresponds to reference data. Let us call the reference data representative curves of pressure and flux calculated in points of experimental measurements by numerical model. Described benchmark is calculated by several 1D blood flow models with different numerical methods. All results are similar and are summarized in [4]. One of such identical numerical solutions in points of experimental measurements is chosen as reference data. Numerical solutions produced in the current series of simulations by our model converge to the reference data. We have calculated the relative $L2$ error norm of numerical flux as follows:

$$\frac{||q(x_i) - q_{ref}(x_i)||_{L_2}}{||q_{ref}(x_i)||_{L_2}} = \frac{\sqrt{\int_T (q(x_i) - q_{ref}(x_i))^2 dt}}{\sqrt{\int_T (q_{ref}(x_i))^2 dt}},$$

where $q_{ref}(x_i)$ and $q(x_i)$ are the reference data and our numerical solution in points x_i, x_i is the point of experimental measurement (the middle point of corresponding vessel), $T \in [14; 16]$. Relative $L2$ error norm of numerical pressure is calculated in the same manner. Errors of numerical pressure and flux are summarized in Tables 1 and 2. Numerical solutions converge to the reference data. In some vessels we can see the first order convergence, in particular, on coarse meshes ($10L$, $20L$, $40L$ meshes). The convergence decays on fine meshes because the error is small and the limit curve (mesh and time steps tend to zero) of numerical solutions does not match exactly the reference curve. We can see that the numerical results on $80L$ and $160L$ meshes are good enough.

Comparisons between experimental (line 1) and numerical (line 2) pressure and flow profiles in eight vessels are shown in Fig. 3. Numerical results (line 2) are shown for mesh with $80L$ points in each vessel. Line 3 corresponds to the reference data [4]. The present 1D blood flow model provides the same results as other 1D blood flow models based on different numerical methods. It is capable to capture the main features of in vitro observed pressure and flow waveforms.

As was stated in [4] and seen in Fig. 3, numerical predictions overestimate the amplitude of the high-frequency oscillations observed in the in vitro pressure and flow waveforms of some vessels (e.g. 7, 14, 20, 34). We presented the same

Table 1 Relative $L2$ error norms for pressure

Vessel	Mesh				
	10L	20L	40L	80L	160L
7	6.3E−2	2.7E−2	1.2E−2	5.9E−3	3.0E−3
10	5.3E−2	2.9E−2	1.5 E−2	8.3E−3	4.5E−3
11	4.2E−2	2.4E−2	1.2E−2	7.1E−3	4.0E−3
14	2.7E−2	1.7E−2	9.7E−3	6.0E−3	4.0E−3
17	3.1E−2	1.9E−2	1.0E−2	5.7E−3	3.3E−3
20	2.4E−2	1.6E−2	9.4E−3	6.3E−3	4.8E−3
29	2.7E−2	1.0E−2	4.9E−2	2.8E−3	2.3E−3
34	4.5E−2	1.8E−2	7.3E−3	3.6E−3	3.2E−3

Table 2 Relative $L2$ error norms for fluid flux

Vessel	Mesh				
	10L	20L	40L	80L	160L
7	7.6E−2	3.6E−2	1.6E−2	9.8E−3	7.4E−3
10	1.2E−1	6.4E−2	3.2E−2	1.9E−3	1.4E−3
11	1.0E−1	5.2E−2	2.7E−2	1.7E−3	1.3E−3
14	4.6E−2	2.8E−2	1.0E−2	8.2E−3	7.4E−3
17	1.6E−1	8.3E−2	4.2E−2	2.7E−2	2.1E−2
20	3.4E−2	2.3E−2	1.6E−2	1.6E−3	1.6E−3
29	8.2E−2	4.7E−2	2.3E−2	2.0E−3	1.9E−3
34	6.3E−2	3.2E−2	1.2E−2	8.2E−3	7.0E−3

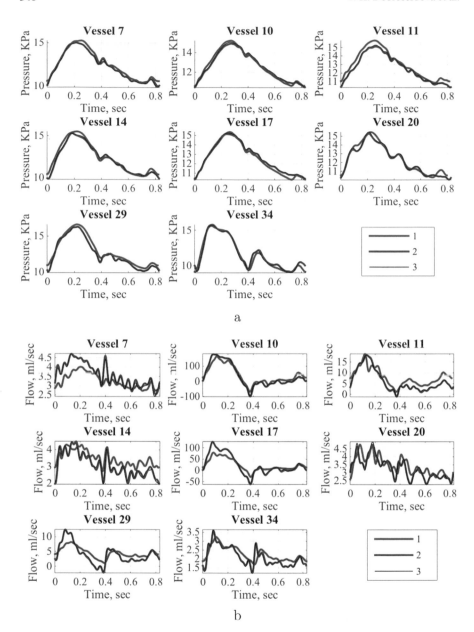

Fig. 3 Comparison of pressure (**a**) and fluid flux (**b**) in several vessels: 1—experimental measurements, 2—numerical results, 3—reference numerical results. All tubes are considered tapered

simulation neglecting the tapered tube shape and assuming the constant (average) diameter along the tube. The numerical results are shown in Fig. 4. Flux and pressure profiles become smoother. Such results better correspond to experimental data.

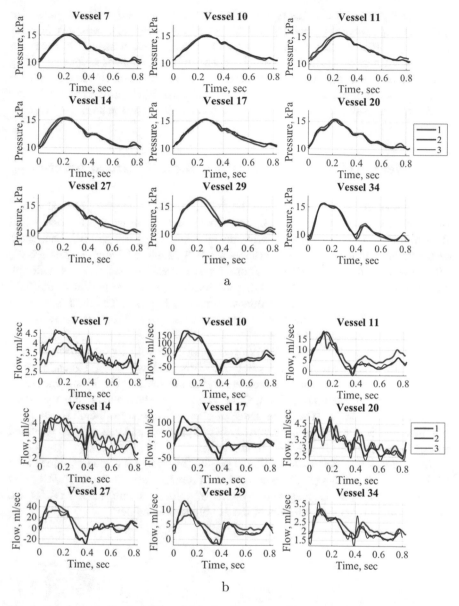

Fig. 4 Comparison of pressure (**a**) and fluid flux (**b**) in several vessels: 1—experimental measurements, 2—numerical results, 3—reference numerical results. All tubes are considered nontapered with constant (average) diameter along the vessel

Tapered vessel shape causes wave reflections from bifurcation and terminal points. Neglecting the tapered form we may underestimate real reflections. At the same time the model still captures all main features of in vitro pressure and flow waveforms, even better for some vessels than the model with tapered vessels.

We also presented a series of numerical experiments with different mesh size and time step. There is no noticeable difference in the numerical solutions even for meshes with $10L$ and $20L$ calculations points at each vessel. It means that the mesh may be considerably coarser for vascular network if the diameter is constant along every vessel. The computational time can be greatly reduced on coarse meshes.

The simulation prescribing the continuity of pressure (5) instead of total pressure (3) in network nodes does not show noticeable difference in numerical results.

4 Modelling of Fluid Flow in Bifurcation of Carotid Arteries

In this numerical experiment we use the data obtained in another physical experiment. A cavity in a silicone block represents geometry of carotid arteries. Fluid flow in this domain is simulated with special engineering device [10]. Diameters of common, external and internal arteries are 8, 4.62 and 5.56 mm, lengths are 90, 60 and 60 mm, respectively. The silicone block is made of Sylgard 184 material, Young's modulus $E = 1.84$ MPa, Poisson's ratio $\sigma = 0.5$. Properties of the fluid are similar to blood (viscosity is 4 mPa s, density is 1 g/cm^3). The fluid is pumped through the cavity. The inflow flux is generated as an average statistical flow rate in the human common carotid artery.

Intravascular guide wire ComboWire is applied for velocity and pressure measurements. It is shown schematically in Fig. 5. This guide wire is used in endovascular operations. It can measure the pressure inside the vessel. Velocity measurement is based on the ultrasound Doppler method. The result is the range of velocity values in vicinity of the device. The dependence of maximal velocity value on time can be obtained in this cross section. Measurements of pressure and velocity are provided in several points: C1, C2, E2, E3, I2, I3 (Fig. 6).

In assumption of Poiseuille's velocity profile the maximal velocity value should be halved to get the velocity averaged over cross section. Therefore Doppler data which corresponds to maximal values of velocity should be halved for the comparison with numerical results of the 1D blood flow model.

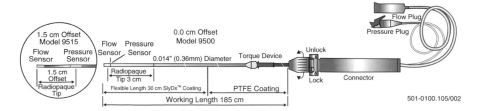

Fig. 5 Scheme of intravascular guide wire ComboWire

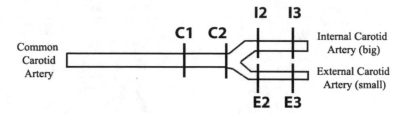

Fig. 6 Scheme of silicone carotid bifurcation

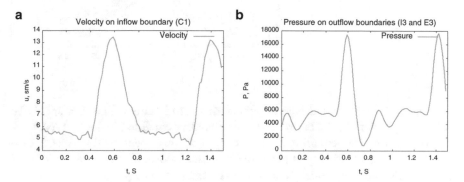

Fig. 7 Measured velocity on inflow boundary (**a**) and pressure on outflow boundaries (**b**)

The computational domain was bounded by the C1, E3, I3 sections. Velocity waveform measured in C1 section and halved is set as the inflow boundary condition (Fig. 7a). Equal pressures measured in E3 and I3 sections are set as the outflow boundary conditions (Fig. 7b). The set of conditions (3) and (4) is used in the bifurcation node.

Numerical results and experimental measurements are compared in sections C2, E2, I2. The pressure drop along the domain is very small comparing with the pressure value. The pressure dependence on time (Fig. 7b) is observed in all studied cross sections. Comparison of calculated velocity profiles and measured data (halved) is shown in Fig. 8. The model reproduces time velocity profile in mother vessel. Nevertheless we can see significant difference in the velocity values in daughter vessels: numerical predictions overestimate experimental data in internal carotid artery and underestimate it in external carotid artery.

We performed the 3D modelling of fluid flow in the bifurcation for better understanding of the flow nature. The 3D model is based on the Navier-Stokes equations. Vessel walls are considered to be fixed. This assumption is possible because the silicone block is rather rigid. The bifurcation geometry is complex: branching vessels are curved. At the moment of maximal velocity vortex appears (Fig. 9) in internal carotid artery which lasts during diastolic part of the cycle. It leads to the flux redistribution in daughter vessels: flux and velocity decrease in

Fig. 8 Comparison of
calculated and measured
average velocities in C2, I2,
E2 cross sections

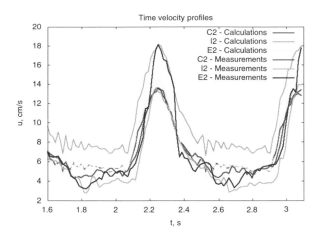

internal carotid artery and increase in external carotid artery. This fact should be taken into account by the bifurcation node boundary condition in the 1D model.

All types of 1D network node boundary conditions mentioned in Sect. 2 have been tested. The continuity of pressure (5) is used instead of the continuity of total pressure (3). The numerical solution has not changed. The pressure drop between C2,I2,E2 cross sections and the point in the middle of bifurcation was calculated by the 3D model of fluid flow. The average hydraulic resistances were obtained from expression (6). The simulation by the 1D blood flow model with (4)–(6) conditions in the bifurcation node was performed. There is no noticeable difference in velocity comparing with the previous numerical results calculated by the 1D model with (3)–(4) conditions. Hydraulic resistance in (6) condition effects on the numerical pressure but not on the flux.

In general Doppler data does not provide enough information for 1D blood flow model because the velocity profile is not necessarily Poiseuille's. In this case the ratio between maximal and mean velocities is also unknown. For this experiment the mean velocities are obtained with MRI scan [10]. Numerical simulations were performed by the 1D blood flow model with MRI mean inflow velocity. Numerical predictions also overestimated experimental MRI measurements in daughter vessels. Therefore the problem of the 1D simulation is the bifurcation node condition and not the measured data.

5 Conclusions

The numerical 1D blood flow model was verified on two benchmarks. First, the fluid flow in the network of silicone tubes was simulated. Numerical pressure and flux correspond to reference data and experimental data very well. Neglecting the vessel tapered form and assuming average constant diameter along every vessel allows us

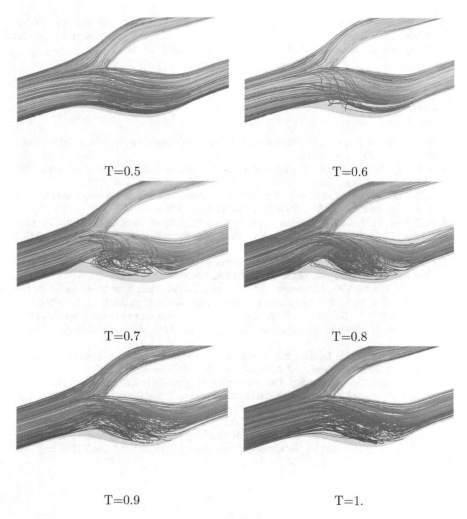

T=0.5   T=0.6

T=0.7   T=0.8

T=0.9   T=1.

Fig. 9 Streamlines of numerical 3D solution at several times: vortex in the bifurcation

to capture all main features of in vitro pressure and flow waveforms. In this case the mesh can be considerably coarser and the simulation can be much faster.

In the second case the fluid flow in bifurcation of carotid arteries was modelled. The fluid flow is rather complex in the bifurcation. The vortex is seen in the 3D numerical solution. Numerical velocities, calculated by the 1D blood flow model, are different from the experimental data in daughter vessels. The improvement of the 1D blood flow model and especially boundary conditions in the network node is demanded for modelling of complex 3D flows.

Acknowledgements This work is supported by the Russian Science Foundation grant 14-31-00024. The authors are grateful to A.Chupakhin, N.Denisenko, A.Yanchenko for measured data used in Sect. 4 (the data obtained with support of RFBR 17-08-01736 grant).

References

1. N. Bessonov, A. Sequeira, S. Simakov et al., Methods of blood flow modelling. Math. Model. Nat. Phenom. **11**(1), 1–25 (2016)
2. A. Danilov, Yu. Ivanov, R. Pryamonosov et al., Methods of graph network reconstruction in personalized medicine. Int. J. Numer. Meth. Biomed. Eng. **32**(8), e02754 (2016)
3. T. Dobroserdova, S. Simakov, T. Gamilov et al., Patient-specific blood flow modelling for medical applications. MATEC Web Conf. **76**(21), 05001 (2016)
4. E. Boileau, P. Nithiarasu, P.J. Blanco, A benchmark study of numerical schemes for one-dimensional arterial blood flow modelling. Int. J. Numer. Meth. Biomed. Eng. **31**(10), e02732 (2015)
5. S. Simakov, A. Kholodov, Computational study of oxygen concentration in human blood under low frequency disturbances. Math. Models Comput. Simul. **1**(2), 283295 (2009)
6. Yu.V. Vassilevski, V.Yu. Salamatova, S.S. Simakov, On the elasticity of blood vessels in one dimensional problems of hemodynamics. Comput. Math. Math. Phys. **55**(9), 1567–1578 (2015)
7. Yu. Vassilevskii, S. Simakov, V. Salamatova et al., Numerical issues of modelling blood flow in networks of vessels with pathologies. Russ. J. Numer. Anal. Math. Model. **26**(6), 605–622 (2011)
8. Y.V. Vassilevski, A.A. Danilov, S.S. Simakov, Patient-specific anatomical models in human physiology. Russ. J. Numer. Anal. Math. Model. **30**(3), 185–201 (2015)
9. J. Alastruey, A. Khir, K. Matthys et al., Pulse wave propagation in a model human arterial network: assessment of 1-D visco-elastic simulations against in vitro measurements. J. Biomech. **44**(12), 2250–2258 (2011)
10. N.S. Denisenko, A.P. Chupakhin, A.K. Khe, A.A. Cherevko et al., Experimental measurements and visualisation of a viscous fluid flow in Y-branching modelling the common carotid artery bifurcation with MR and Doppler ultrasound velocimetry. J. Phys. Conf. Ser. **722** (2016)

Optimization of Combined Antitumor Chemotherapy with Bevacizumab by Means of Mathematical Modeling

M. B. Kuznetsov and A. V. Kolobov

1 Introduction

Pre-existing vascular system limits the rate of tumor growth, mainly due to the fact that tumor cells require much larger nutrient supply than normal cells. This limitation is overcome by tumor angiogenesis, or neovascularization, governed by tumor-induced proangiogenic factors, of which vascular endothelial growth factor, or VEGF, is accepted to be the most important one. Antiangiogenic therapy (AAT) was suggested in 1971 by Folkman [15]. The first antiangiogenic drug, bevacizumab, which irreversibly binds to VEGF, was approved for medical use in 2004 and is widely used nowadays.

Due to overproduction of proangiogenic factors by tumors, their newly formed microvasculatory system is chaotic and the capillaries themselves are dilated, tortuous and their walls are highly permeable [4]. AAT by bevacizumab not only prevents the formation of new capillaries, but also leads to capillaries maturation, i.e., brings them to more physiologically normal state with normalized permeability of walls, both factors resulting in depriving tumor of nutrients and thus decelerating its growth.

Compared to traditional radio- and chemotherapy (CT), AAT has moderate side-effects, but it alone cannot completely eradicate tumor and therefore should have limited efficiency, which has been proved in clinics [13]. Nowadays almost all of

M. B. Kuznetsov (✉)
P.N. Lebedev Physical Institute of the Russian Academy of Sciences, Moscow, Russia
e-mail: postmaster@lebedev.ru

A. V. Kolobov
P.N. Lebedev Physical Institute of the Russian Academy of Sciences, Moscow, Russia

Institute of Numerical Mathematics of the Russian Academy of Sciences, Moscow, Russia
e-mail: director@inm.ras.ru

© Springer International Publishing AG, part of Springer Nature 2018
R. P. Mondaini (ed.), *Trends in Biomathematics: Modeling, Optimization and Computational Problems*, https://doi.org/10.1007/978-3-319-91092-5_24

the approved administration schemes, which include bevacizumab, combine it with various chemotherapy agents [17], aimed at direct killing of actively dividing cells, but also associated with significant side-effects.

Two different types of drugs, administered simultaneously, interact with each other in different ways. Of note, AAT often eventually leads to reduced influx of chemotherapeutic drug in tumor, which was observed experimentally [9, 29]. Therefore, a great challenge concerning CT combined with antiangiogenic agents is the problem of optimal scheduling of drugs administration in order to maximize antitumor effect and minimize side-effects. The topic is being actively investigated now, and mathematical modeling can facilitate it.

However, in contrast to clinical practice, mathematical modeling of combined CT and AAT has begun only recently [3, 19]. The effect of antiangiogenic drugs is described only phenomenologically in existing models. Moreover, all works of this type do not take into account the spatial structure of the tumor, which indicates that they are only suitable for simulation of postoperative (adjuvant) therapy. However, there are a lot of works to rely upon, which model classical CT [10, 14, 33] and AAT with bevacizumab, considering its pharmacokinetics in blood and tissue [8, 26].

In this work we compare the efficiencies of different schemes of palliative combined chemotherapy with bevacizumab. This kind of therapy is administered when surgical intervention is not possible, and drug administration continues until tumor remission or lethal outcome.

2   Model

2.1   Equations

Figure 1 demonstrates the block-scheme of the model under investigation. We consider tumor as heterogeneous colony of malignant cells, introducing variables of normalized density of proliferating cells, $n_1(r, t)$, where r and t are space and time coordinates, and resting cells, $n_2(r, t)$, which are able to move in tissue in accordance with widely accepted principle of migration/proliferation dichotomy of tumor cells [18]. Cells can change their states depending on the concentration of glucose $S(r, t)$, which is selected as the key metabolite. The tissue tumor grows in consists of normal cells with density $h(r, t)$. When dying due to either lack of glucose or CT, tumor and normal cells form necrosis, whose fraction in tissue is denoted as $m(r, t)$. In absence of tumor, the tissue contains preexisting normal capillary network, the bulk density of its surface is $EC(r, t)$. As a result of tumor neovascularization, it expands by addition of angiogenic capillaries, their surface bulk density is $FC(r, t)$. The model also takes into account concentrations of VEGF, $V(r, t)$, bevacizumab $A(r, t)$ and cytotoxic drug. For consideration of pharmacokinetics of the latter we use parameters of cisplatin, which is one of the chemotherapeutic agents, frequently used in combination with bevacizumab [17].

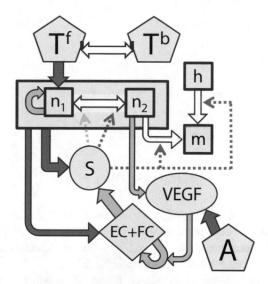

Fig. 1 Block-scheme of the model of tumor progression and combined chemotherapy with bevacizumab: n_1 and n_2 are proliferating and migrating tumor cells, respectively, h is host cells, m is necrosis, S is glucose, VEGF is vascular endothelial growth factor, EC and FC are preexisting and angiogenic microcirculatory networks, respectively, A is bevacizumab, T^f and T^b are free and protein-bound forms of cisplatin, respectively. Gray arrows indicate stimulating relations, black arrows indicate inhibiting relations, white arrows denote cell transitions

Since it possesses high affinity for plasma proteins, we take into account two forms of cisplatin in the model by including variables of concentrations of its free and protein-bound forms, $T^f(r, t)$ and $T^b(r, t)$, respectively. Equations, describing the densities of cell populations and necrosis fraction, are as follows:

$$\frac{\partial n_1}{\partial t} = \overbrace{Bn_1}^{\text{proliferation}} \overbrace{-P_1(S)n_1 + P_2(S)n_2}^{\text{transitions}} \overbrace{-k_T T_n T n_1}^{\text{death by CT}} \overbrace{-\nabla(\mathbf{I}n_1)}^{\text{convection}},$$

$$\frac{\partial n_2}{\partial t} = \overbrace{P_1(S)n_1 - P_2(S)n_2}^{\text{transitions}} \overbrace{-d_n(S)n_2}^{\text{death}} \overbrace{-\nabla(\mathbf{I}n_2)}^{\text{convection}} + \overbrace{D_n n_2}^{\text{migration}},$$

$$\frac{\partial h}{\partial t} = \overbrace{-d_h(S)h}^{\text{death}} \overbrace{-\nabla(\mathbf{I}h)}^{\text{convection}},$$

$$\frac{\partial m}{\partial t} = \overbrace{d_n(S)n_2 + d_h(S)h + k_T T_n T n_1}^{\text{cell death}} \overbrace{-\nabla(\mathbf{I}m)}^{\text{convection}}, \tag{1}$$

where $n_1 + n_2 + m + h = 1$,

$$\nabla \mathbf{I} = Bn_1 + D_n \Delta n_2,$$

$$P_1(S) = k_1 exp(-k_2 S),$$

$$P_2(S) = \frac{1}{2}k_3(1 - \tanh[\epsilon_{tr}(S_{tr} - S)]),$$

$$d_i(S) = \frac{1}{2}d_i^{max}(1 + \tanh[\epsilon_d(S_d - S)]), i = n, h.$$

The forms of functions of transitions from one state to another $P_1(S)$ and $P_2(S)$ are described in detail in our previous work [26]. Herein we consider only cell cycle-specific action of chemotherapeutic agent. The physical meanings of parameters and their values used in this work are discussed in Sect. 2.2.

We consider an incompressible dense tissue, in which space distribution of components is affected by their local kinetics—e.g., dividing cells push out surrounding tissues, providing an increase of tumor size. To account for these effects, we introduce the convective velocity field $\mathbf{I}(r, t)$, which is derived analogically to [25].

As already mentioned, the model uses two variables to describe the vascular network, namely, its preexisting and angiogenic parts, the latter having significantly higher permeability of walls. With tumor progression microcirculatory network locally degrades under the influence of increased local pressure due to cell proliferation and migration [6] and various chemical factors [21], accounted for implicitly via the fact that degradation rate is non-zero inside necrosis, but lower than in presence of tumor cells. Under sufficient amount of VEGF, formation of new angiogenic capillaries takes place along with "dematuration" of preexisting ones, which is introduced in the model to reflect increase in vascular walls permeability due to the action of VEGF and is described by transition from EC to FC. Under small concentrations of VEGF capillaries mature and reverse transition takes place. Of note, angiogenesis rate decreases in presence of cisplatin, which kills dividing endothelial cells as well. The term microvasculature pruning is introduced for description of microcirculatory network tendency to return to constant physiologically reasonable density, this term responsible for its returning to normal state when tumor angiogenesis is stopped by therapeutic intervention. All capillaries move with the convective flows, although slower, than the cells, due to the connectivity of microvasculature. Migration of angiogenic capillaries is also included to reflect stimulation of capillaries motility by VEGF.

Equations describing the dynamics of capillary surface density are as follows:

$$\frac{\partial EC}{\partial t} = \overbrace{-[l(n_1 + n_2) + l_m m]EC}^{\text{degradation}} + \overbrace{\frac{v_{mat} \cdot V^*}{V + V^*}FC}^{\text{maturation}} - \overbrace{\frac{v_{dem} \cdot V}{V + V^*}EC}^{\text{dematuration}}$$

$$\overbrace{-\mu(EC + FC - 1)EC \cdot \Theta(EC + FC - 1)}^{\text{microvessel pruning}} \overbrace{-\nabla(\gamma \mathbf{I} \cdot EC)}^{\text{convection}},$$

$$\frac{\partial FC}{\partial t} = \overbrace{Re^{-k_T^A T}\frac{V}{V + V^*}(EC + FC)\left[1 - \frac{(EC + FC)}{C_{max}}\right]}^{\text{angiogenesis}}$$

$$\overbrace{-[l(n_1 + n_2) + l_m m]FC}^{\text{degradation}} - \overbrace{\frac{v_{mat} \cdot V^*}{V + V^*}FC}^{\text{maturation}} + \overbrace{\frac{v_{dem} \cdot V}{V + V^*}EC}^{\text{dematuration}}$$

$$\overbrace{-\mu(EC + FC - 1)EC \cdot \Theta(EC + FC - 1)}^{\text{microvessel pruning}}$$

$$\overbrace{+D_{FC}\Delta FC}^{\text{migration}} \overbrace{-\nabla(\gamma \mathbf{I} \cdot FC)}^{\text{convection}}. \tag{2}$$

Glucose enters the tissue from the microvasculature, its inflow being governed by the process of transvascular diffusion [28]. The inflow term accounts for difference in permeabilities of walls of different capillaries for glucose. Glucose blood level is considered to be constant. All types of cells consume glucose, consumption rate of proliferating cells being much higher due to specific features of tumor metabolism [20]. Glucose diffuses in tissue, its local diffusion coefficient being known to decrease with the concentration of cells [39], or, which is the same in terms of this model, to increase with concentration of necrosis.

Moreover, since it is known that diffusion coefficient of substances increases in necrosis, DW-MRI being working on this very principle [36], we introduce dependence of diffusion coefficient on necrosis fraction for all the substances in the model, using the simplest linear form of relevant function:

$$D_i(m) = D_i^0(1 + \alpha m), i = S, V, A, T^f, T^b.$$

Thus, change of glucose concentration is defined by the equation:

$$\frac{\partial S}{\partial t} = \overbrace{[P_{S,EC}EC + P_{S,FC}FC](S_{bl} - S)}^{\text{inflow}}$$

$$\overbrace{-[q_{n1}n_1 + q_{n2}n_2 + q_h h]\frac{S}{S + S^*}}^{\text{consumption}} \overbrace{+D_S(m)\Delta S}^{\text{diffusion}}. \tag{3}$$

Proangiogenic factor VEGF is produced by malignant cells. Since metabolic stress significantly upregulates its production [37], its secretion only by resting cells is taken into account. Also we take into account VEGF internalization by endothelial cells, diffusion in tissue, molecular degradation and outflow from tissue, the latter being included in the model in explicit form under the assumption that VEGF concentration in blood is equal to zero.

Bevacizumab is administered intravenously, which is reflected by the ordinary equation for its blood concentration, consisting of administration term, due to which blood concentration of bevacizumab abruptly increases by unity in the moments of its injection, and the term of its blood clearance. Bevacizumab comes from blood into the tissue, where it diffuses and binds irreversibly to VEGF, thus rendering it inactive. Difference in permeabilities of walls of different capillaries is significant for bevacizumab, as its size is comparable to pore sizes of normal body capillaries.

Since bevacizumab is a macromolecule, developed especially for VEGF inhibition, it practically does not interact with tissue elements.

$$\frac{\partial V}{\partial t} = \overbrace{pn_2}^{\text{production}} \overbrace{-\omega V(EC + FC)}^{\text{internalization}} \overbrace{-[P_{V,EC}EC + P_{V,FC}FC]V}^{\text{outflow}}$$

$$\overbrace{-d_V V}^{\text{degradation}} \overbrace{-(k_A A_n)AV}^{\text{neutralization}} \overbrace{+D_V \Delta V}^{\text{diffusion in tissue}},$$

$$\frac{\partial A}{\partial t} = \overbrace{[P_{A,EC}EC + P_{A,FC}FC](A_{bl} - A)}^{\text{inflow}}$$

$$\overbrace{-(k_A V_n)AV}^{\text{binding to VEGF}} \overbrace{+D_A \Delta A}^{\text{diffusion in tissue}},$$

$$\frac{\partial A_{bl}}{\partial t} = \overbrace{F_A^{iv}}^{\text{injection}} \overbrace{-d_A A_{bl}}^{\text{clearance}}. \tag{4}$$

Cisplatin enters the systemic circulation by intravenous administration as well, that is also reflected in the model by abrupt increase by unity. Cisplatin strongly interacts with blood proteins and as a result exists in blood and tissue in two forms—as free small active drug molecules (which concentration is denoted by T^f) and large protein-bound inactive complex (concentration denoted by T^b). Both fractions of drug come from blood into the tissue, where they diffuse and continue to pass from one form to another. The rate on protein binding should depend on the concentration of proteins, which is usually much higher in blood than in normal tissue. However, since we consider tumor tissue, to the time of therapy blood protein should accumulate inside it to the levels comparable to blood ones due to enhanced permeability of walls of angiogenic microvasculature and impaired lymphatic drainage of interstitial fluid, so we consider unchanged rates of binding to blood proteins inside the tissue. We neglect blood clearance of protein-bound complex as well as its binding to tissue elements as being very small compared to the ones of free drug.

$$
\frac{\partial T^f}{\partial t} = \overbrace{[P_{EC,Tf} EC + P_{FC,Tf} FC] \left(T_{bl}^f - T^f \right)}^{\text{inflow}} \overbrace{-k_{on} T^f + k_{off} T^b}^{\text{interactions with proteins}}
$$
$$
\overbrace{-d_T T^f}^{\text{non-specific binding}} \quad \overbrace{+D_{Tf} \Delta T^f}^{\text{diffusion}},
$$

$$
\frac{\partial T^b}{\partial t} = \overbrace{[P_{EC,Tb} EC + P_{EC,Tb} FC] \left(T_{bl}^b - T^b \right)}^{\text{inflow}}
$$
$$
\overbrace{+k_{on} T^f - k_{off} T^b}^{\text{interactions with proteins}} \overbrace{+D_{Tb} \Delta T^b}^{\text{diffusion}},
$$

$$
\frac{\partial T_{bl}^f}{\partial t} = \overbrace{F_T^{iv}}^{\text{injection}} \overbrace{-k_{on} T_{bl}^f + k_{off} T_{bl}^b}^{\text{interactions with proteins}} \overbrace{-d_{T,bl} T^f}^{\text{clearance}},
$$

$$
\frac{\partial T_{bl}^b}{\partial t} = \overbrace{k_{on} T_{bl}^f - k_{off} T_{bl}^b}^{\text{interactions with proteins}}. \tag{5}
$$

2.2 Parameters

The model contains several dozens of parameters, which are taken from various experiments of different nature, where possible, or estimated in order to reflect known features of tumor growth. The basic set of parameters in given in Table 1, where the following normalization parameters are used to obtain their model values: $t_n = 1\,\text{h}$ for time, $L_n = 10^{-2}\,\text{cm}$ for length, $S_n = 1\,\text{mM}$ for glucose concentration. Normalization parameters for VEGF, bevacizumab and cisplatin are used in corresponding terms describing actions of therapies and are chosen to be $V_n = 10^{-11}\,\text{mol/ml}$, $A_n = 1.6 \times 10^{-9}\,\text{mol/ml}$, $T_n = 5 \times 10^{-8}\,\text{mol/ml}$, the latter two values estimated as blood concentration of relevant drug in average man right after its injection. Maximal density of tumor cells is equal to 3×10^8 cells/ml and is taken from [16]. Normal capillary surface density is taken to be $EC_n = 100\,\text{cm}^2/\text{cm}^3$, based on averaged value for human muscle [28].

In estimations of tumor cells proliferation rate and nutrient consumption rates of proliferating cells we rely on the corresponding values obtained at the initial stage of tumor spheroids growth in suspension, but we assume that these values should be proportionally diminished during the growth of relevant tumor in tissue due to such factors as mechanical pressure, increased acidity and production of lactate by tumor cells [32]. Estimation of glucose uptake rate of resting tumor cells is based on

Table 1 Model parameters

Parameter	Description	Value	Model value	Estimations based on
	Tumor cells			
B	Proliferation rate	$0.02\,\mathrm{h}^{-1}$	0.02	[16]
k_1	Maximum rate of transition to rest	$0.4\,\mathrm{h}^{-1}$	0.4	[35]
k_2	Sensitivity of transition to rest to glucose	$19.8\,(\mathrm{mg/ml})^{-1}$	3.6	[35]
k_3	Maximum rate of transition to proliferation	$0.16\,\mathrm{h}^{-1}$	0.16	[26]
ϵ_{tr}	Sensitivity of transition to proliferation to glucose	1.8	1.8	[26]
S_{tr}	Threshold glucose concentration for transition to proliferation	$1.7\,\mathrm{mM}$	1.7	[26]
k_T	Sensitivity to cisplatin		Varies	See text
d_n^{max}	Maximum rate of death	$0.02\,\mathrm{h}^{-1}$	0.02	[22]
ϵ_d	Sensitivity of death rate to glucose	5	5	See text
S_d	Threshold glucose concentration for death		0.55	See text
D_n	Migration coefficient	$10^{-10}\,\mathrm{cm}^2/\mathrm{s}$	0.0036	See text
	Normal cells			
d_h^{max}	Maximum rate of death	$0.03\,\mathrm{h}^{-1}$	0.03	[22]
	Capillaries			
l	Degradation rate in viable region	$1.7 \times 10^{-10}\,(\mathrm{cells/ml})^{-1}\,\mathrm{s}^{-1}$	0.05	See text
l_m	Degradation rate in necrosis	$1 \times 10^{-10}\,(\mathrm{cells/ml})^{-1}\,\mathrm{s}^{-1}$	0.03	See text
v_{mat}	Maturation rate	$0.05\,\mathrm{h}^{-1}$	0.05	See text
V^*	Michaelis constant for angiogenesis rate	$10^{-14}\,\mathrm{mol/ml}$	0.001	See text
v_{dem}	Dematuration rate	$0.05\,\mathrm{h}^{-1}$	0.05	See text
μ	Pruning rate	$10^{-5}\,(\mathrm{cm}^2/\mathrm{cm}^3)^{-1}\,\mathrm{s}^{-1}$	0.001	See text
γ	Network elasticity	0.5	0.5	See text
R	Maximum angiogenesis rate	$0.02\,\mathrm{h}^{-1}$	0.02	See text
C_{max}	Maximum capillary area surface density	$500\,\mathrm{cm}^2/\mathrm{cm}^3$	5	See text
k_T^A	Sensitivity of angiogenesis to cisplatin	$0.35 \times 10^{-8}\,\mathrm{mol/ml}$	10	See text

(continued)

Table 1 (continued)

Parameter	Description	Value	Model value	Estimations based on
D_{FC}	Migration coefficient	10^{-11} cm^2/s	0.00036	See text
	Glucose			
α	Diffusion sensitivity to necrosis fraction	0.3	0.3	[36, 39]
$P_{EC,S}$	Permeability of continuous capillaries' walls	1.1×10^{-5} cm/s	4	[28]
$P_{FC,S}$	Permeability of angiogenic capillaries' walls	2.8×10^{-5} cm/s	10	[28]
S_{bl}	Blood level	5.5 mM	5.5	[1]
q_{n1}	Proliferating tumor cells uptake rate	8.0×10^{-17} mol/cell·s	70	[16]
q_{n2}	Resting tumor cells uptake rate	2.3×10^{-18} mol/cell·s	2	[35]
q_h	Normal cells uptake rate	0.49 mg/min · 100 ml	1.6	[2]
S^*	Michaelis constant for uptake rate	0.04 mM	0.04	[7]
D_S^0	Diffusion coefficient	2.6×10^{-6} cm^2/s	94	[40]
	VEGF			
p	Production rate	2 fg/h cell	1	[23]
ω	Internalization rate	2.8×10^{-4} s^{-1}	1	[31]
$P_{EC,V}$	Permeability of continuous capillaries' walls	6.4×10^{-8} cm/s	0.023	[28]
$P_{FC,V}$	Permeability of angiogenic capillaries' walls	7.8×10^{-7} cm/s	0.28	[28]
d_V	Degradation rate	0.01 h^{-1}	0.01	[24]
k_A	Binding to bevacizumab	5.3×10^5 M^{-1} s^{-1}	1.9×10^{12}	[34]
D_V^0	Diffusion coefficient	5.9×10^{-7} cm^2/s	21.2	[24]
	Avastin			
$P_{EC,A}$	Permeability of continuous capillaries' walls	1.6×10^{-9} cm/s	$6 \cdot 10^{-4}$	[28]
$P_{FC,A}$	Permeability of angiogenic capillaries' walls	1.2×10^{-7} cm/s	0.044	[28]
D_A^0	Diffusion coefficient	4×10^{-7} cm^2/s	14.3	See text
d_A	Blood clearance rate	0.035 day^{-1}	0.0014	[17]
	Cisplatin			
$P_{EC,Tf}$	Permeability of continuous capillaries' walls	8×10^{-6} cm/s	3	[28]
$P_{FC,Tf}$	Permeability of angiogenic capillaries' walls	2.2×10^{-5}	7.9	[28]
k_{on}	Rate of binding to proteins	0.46	0.46	[42]
k_{off}	Rate of unbinding from proteins	0.04	0.04	[42]

(continued)

Table 2 (continued)

Parameter	Description	Value	Model value	Estimations based on
d_T	Rate of non-specific binding with tissue elements		Varies	See text
D_T^0	Diffusion coefficient	2.1×10^{-6} cm^2/s	77.2	See text
$d_{T,bl}$	Clearance rate	$0.13\,\text{h}^{-1}$	0.13	[41]
	Protein-bound cisplatin			
$P_{EC,Tb}$	Permeability of continuous capillaries' walls	2.8×10^{-8} cm/s	0.01	[28]
$P_{FC,Tb}$	Permeability of angiogenic capillaries' walls	4.5×10^{-7} cm/s	0.16	[28]
D_{Tb}^0	Diffusion coefficient	5.8×10^{-7} cm^2/s	20.9	See text

the observation, made in [35], that it should be at least 40 times lower than glucose consumption rate of proliferating cells of corresponding tumor cell line. The values of death rates of normal and tumor cells are assessed on experimental data on cell behavior under extreme nutrient deprivation, the parameters of sensitivity to glucose concentration being chosen so that cell death becomes significant only under its considerable decrease, normal cells being more sensitive to nutrient deprivation. Migration coefficient of tumor cells corresponds to non invasive tumor. We vary the value of tumor cells sensitivity to cisplatin and non-specific binding of cisplatin to investigate relative efficiencies of different schemes of drugs administration under different conditions. Microvasculature parameters are estimated in order for the model microvasculature behavior to adequately approximate general features of structure and dynamics of functional tumor microvasculature [11, 12, 38]. Permeabilities of capillaries for different substances are assessed using Renkin equation, analogically to how it was done in [26]. Diffusion coefficients for bevacizumab and two forms of cisplatin are estimated based on already defined values for glucose and VEGF and molecule radii of all substances in use. The radius of albumin is used for estimations of parameters of protein-bound cisplatin. Since the rate of non-specific binding of cisplatin strongly depends on patient-specific characteristics, tumor localization and various other factors, which introduced a large portion of uncertainty in its clinical estimation, we choose this parameter to vary in simulations presented herein along with tumor cells sensitivity to cisplatin.

2.3 Numerical Solving

The set of Eqs. (1)–(5) was solved in one-dimensional region with size of $L = 2$ cm using plane geometry, since it brings practically no difference to simulation results in comparison with spherically symmetric case, as center of large tumor is occupied with necrosis in our simulations as well as in numerous experimental studies. Initial conditions correspond to normal tissue with a small colony of tumor cells situated on the left border, so $n_1(x, 0) = max(0, 0.25[1 - (x/10)^2])$, $h(x, 0) = 1 - n_1(x, 0)$, $EC(x, 0) = 1$. The initial distribution of glucose $S(x, 0)$ is calculated as its steady-state concentration in normal tissue. The other variables at the initial moment of time are equal to zero. For all variables zero-flux boundary conditions are set on both borders. The convective flow speed is set zero on the left border, free boundary condition is used for it on the right border, which results in the following equation for convective flow speed:

$$I(x, t) = \int_0^x [Bn_1(r, t)]dr + D_n \nabla n_2(x, t).$$

To speed up the calculations, equations for VEGF and glucose are considered in the quasi-stationary approximation due to high rates of their reactions with respect to these rates for other variables and are solved numerically using the tridiagonal matrix algorithm. For other variables, the method of splitting into physical processes is used. Kinetic equations are solved via the fourth-order Runge-Kutta method, and Crank-Nicholson scheme is used for the diffusion equations. Convective equations are solved using the flux-corrected transport algorithm [5] with the use of explicit anti-diffusion stage.

3 Results

Figure 2 demonstrates the distribution of model variables during monochemotherapy under tumor cells sensitivity to cisplatin $k_T = 5 \times 10^9$ and non-specific binding rate of cisplatin $d_T = 1.5$. The scheme of drug administration is one dose every 3 weeks, with total of six injections. Figure 2a relates to the day right before the beginning of the treatment and thus shows structure of untreated tumor and its microenvironment, which adequately represents experimental observations. Few functional capillaries are situated inside the tumor, which main mass consists of necrosis m, while massive angiogenic extension of microvasculature FC is located adjacent to the tumor. Due to elevated capillary density as well as increased permeability of walls of angiogenic microvasculature compared to that of normalized one EC, this region provides enhanced supply of glucose S to the tumor cells, which are concentrated near the tumor boundary, comprising a viable rim of several millimeters in width, with proliferating cells n_1 on its outer side. However, close location to capillaries renders them rather vulnerable to the action of cisplatin, whose inflow in tissue also depends on amount and state of capillaries, and in a more intense way, since considerable amount of drug enters tissue in a protein-bound form, for which relation in permeabilities of capillaries of two types is much bigger than that for small molecules like glucose and free cisplatin (16 versus approximately 2.5 for both).

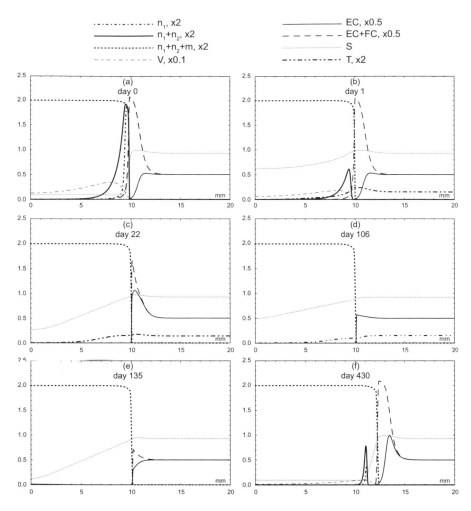

Fig. 2 Profiles of proliferating tumor cells density n_1, total tumor cells density $n_1 + n_2$, fraction of tumor with necrosis $n_1 + n_2 + m$, concentration of VEGF V, normal microcirculatory network surface density EC, total microcirculatory network surface density $EC + FC$, concentrations of glucose S and cisplatin T under tumor cells sensitivity to cisplatin $k_T = 5 \times 10^9$ and non-specific binding of cisplatin $d_T = 1.5$ on the days: (**a**) 0 (the day before the start of chemotherapy), (**b**) 1 (first injection of cisplatin), (**c**) 22 (second injection of cisplatin) (**d**) 106 (sixth and last injection of cisplatin), (**e**) 135 and (**f**) 430

With the first injection of cisplatin T (Fig. 2b), depth of glucose penetration into the tumor increases in result of death of proliferating cells, which are its main consumers. This leads to active transition of resting cells n_2 into proliferating state, in which they are also killed by the action of chemotherapeutic drug. Reduction of number of resting cells leads to decrease in production of VEGF V, which, altogether with direct killing of proliferating endothelial cells by cisplatin, leads to stop of angiogenesis, and subsequent maturation of capillaries, reflected in the model as transition from FC to EC, as well as pruning of capillaries, described by tendency of microvasculature density to return to its normalized state $EC = 1$.

However, not all tumor cells are killed by the first injection of drug, and a small amount of them remains alive on some distance from the tumor border, where sufficient amount of drug does not penetrate. Before the next injection of cisplatin on day 22 (Fig. 2c) they have time to proliferate and consume enough glucose to noticeably lower its concentration inside the tumor, which leads to transition of some cells to resting state which actively produce VEGF anew, due to which during last days before the second drug injection dematuration and growth of capillaries takes place, both factors again enhancing the inflow of cisplatin, and subsequently deepening its penetration, compared to what it would have been in case of normalized microvasculature.

Nevertheless, in result of several drug infusions, after every one of which the majority of tumor cells are killed and the remaining ones settle deeper inside the tumor mass, microvasculature in tissue adjacent to tumor becomes almost fully normalized, thus pointing out implicit slow antiangiogenic action of CT (Fig. 2d). However, if some cells still remain alive inside the necrotic mass, like in this model case, they provide slow tumor remission, which apparently is fostered by renewed angiogenesis resulting in enhanced supply of nutrients, availability of which defines the rate of remission (Fig. 2e, f). Thus, it is clear that administration of angiogenic drug after end of CT will restrain the rate of tumor remission, if the one is to take place. This notion is supported in clinical practice since administration of antiangiogenic drugs as a rule continues after the end of CT [17].

But is it efficient to use AAT during the whole course of CT—the way which is widely accepted in clinical practice? The insights of what happens with tumor during CT, provided above, imply that this question should not have explicit and clear answer, since a large number of processes simultaneously influence the outcome of treatment. From one hand, AAT suppresses the inflow of nutrients to tumor cells, thus limiting their proliferation rate, from the other hand—it as well impairs the delivery of chemotherapeutic agent, which is the one that kills tumor cells directly. It is reasonable to assume that the relative efficiency of combined CT and AAT would depend on specific parameters of case under consideration.

To prove this idea, herein we consider the variation of two parameters of different nature: the first one is tumor cells sensitivity to cisplatin k_T, which is both drug- and tumor-related parameter; the second one is the rate of non-specific binding of chemotherapeutic drug to tissue elements d_T, which is drug- and patient-related parameter. We compare two schemes of drug administration: the one widely used in clinics, where administrations of both drugs take place simultaneously with continuation of AAT after the end of CT, hereafter denoted as scheme A for convenience; and the scheme B, proposed herein and not used in clinics, in which administration of antiangiogenic drug bevacizumab begins together with the last injection of cisplatin. In both cases there are six injections of cisplatin at 3 weeks intervals.

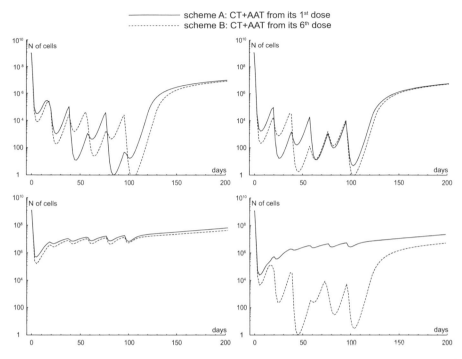

Fig. 3 Total number of tumor cells during combined chemotherapy with bevacizumab under different schemes of drugs administration, tumor cells sensitivity to cisplatin k_T and rate of non-specific binding of cisplatin d_T. CT is chemotherapy, AAT is antiangiogenic therapy. (**a**) $k_T = 3 \times 10^9, d_T = 1$, (**b**) $k_T = 5 \times 10^9, d_T = 1$, (**c**) $k_T = 3 \times 10^9, d_T = 1.5$, (**d**) $k_T = 5 \times 10^9, d_T = 1.5$

Figure 3 illustrates the most typical cases of what the difference between results of two considered treatments may be, demonstrating dependencies of tumor cells number over time. Of note, in every presented case number of cells after the first injection of cisplatin is reduced more markedly in absence of AAT, since, as it has been mentioned above, larger amount of drug enters tumor tissue. However, as Fig. 3a, b demonstrates, in cases of low rate of cisplatin non-specific binding, in long-term run scheme B may temporarily lose its advantage. It happens due to the fact that cisplatin is able to penetrate deep enough inside the tumor to eventually kill almost all tumor cells even under relatively low inflow of cisplatin provided by normalized microvasculature, which also suppresses inflow of glucose, resulting in ambivalent influence of AAT on overall system dynamics. The number of cells exhibits complex quasioscillatory behavior and though the scheme B without initial administration of AAT proves to be slightly more efficient in presented case, in terms of tumor remission time, it is apparent that this result may change provided CT is finished earlier. Nevertheless, the observation that under the same number of tumor cells a single dose of CT administered without AAT turns out to be more efficient, like in the case of fifth injection in Fig. 3b, suggests that the scheme B would still be more profitable in the majority of cases. Of note, the behavior of the system under low value of d_T is difficultly predictable—e.g., using the scheme A results in death of practically all cells at one particular moment under lower value of sensitivity of tumor cell to cisplatin k_T, while a small amount of cells always remains alive under higher k_T.

The situation qualitatively changes under higher value of d_T, when drug penetration inside the tumor becomes compromised and consequently the sensitivity of proliferating tumor cells to cisplatin begins to play crucial role. Under relatively low value of k_T both schemes of treatment turn out to be little effective, as Fig. 3c demonstrates, with only slight advantage of scheme B. However, under bigger value of k_T scheme B reveals its power, as under these values of parameters the ability of cisplatin to penetrate deeper and thus kill more cells proves to be the main factor defining the outcome of treatment. Scheme B alone is able in this case to potentially kill all of the tumor cells, and, if not, to lead to significant delay in tumor remission of approximately 90 days, compared to that of scheme A, since tumor cells have to begin regrowth at larger depth.

4 Discussion

In the presented article, a comparative study of the efficiencies of two protocols of palliative combined chemotherapy with antiangiogenic therapy was conducted. The classical protocol was considered as the basic one, which is the simultaneous administration of chemotherapeutic and antiangiogenic drugs. It was compared with the scheme, proposed by authors, in which the administration of antiangiogenic agent starts together with the last injection of cytotoxic drug. The idea of such a protocol is to take advantage of increased permeability of angiogenic capillary network in the peritumoral region, compared to the mature network which forms in result of the antiangiogenic therapy. Thus, using of this scheme leads to increased penetration of the cytotoxic agent into the tumor.

An important moment is that, like in clinical practice, in the considered model a significant part of the cytotoxic drug enters the tissue in a complex with large blood proteins, such as albumin. The molecular mass of the resulting complex is more than two orders of magnitude greater than the molecular mass of cisplatin itself, and for such heavy molecules the permeabilities of the walls of angiogenic and normalized capillaries in practice differ in dozens of times. This significantly distinguishes the considered model from the one proposed by the authors in the previous work [27], where this binding was not taken into account.

An essential feature of this model is the retention of necrotic tissue after the action of the cytotoxic agent. In clinical practice, this volume can both significantly decrease and remain practically unchanged [30]. Apparently, such differences are determined by the rate of outflow of interstitial fluid through the lymphatic system. In order to take into account the effects associated with the reduction of primary tumor volume in result of therapy, we are going to include in our model the detailed dynamics of interstitial fluid. In the model herein, the proportion of interstitial fluid associated with the necrotic tissue fraction influenced only the diffusion rate of substances within the tumor.

The results are obtained in the limitation of small chemotherapeutic drug. For such drugs, which size is smaller than the size of pores in capillaries walls, transvascular transport occurs via the process of diffusion. For modeling of dynamics of widely used polymer-conjugated macromolecular drugs their transvascular convection must also be taken into account, which would also be possible to do when dynamics of the interstitial fluid will be included in the model. Moreover, it is apparent that variation of other parameters related to tumor, drugs, and patients features will also affect the relative efficiencies of different administration schemes. Their influence will also be studied separately.

It should be noted that the maximum increase in the efficiency of combined CT+AAT by using the proposed protocol instead of the standard one, i.e., the increase in the remission period by 3 months, is not observed for all values of the model parameters. In the cases of strong enough cytotoxic drug with low rate of binding to tissue elements, which infers its ability to freely penetrate throughout the tumor volume, the difference is not so significant. Surely, in practice in every particular case the relative efficiencies of different schemes of drugs administration will depend on colossal number of parameters of different nature. However, the model results presented herein, which are based on the consideration of the most crucial processes taking place during tumor progression, undoubtedly indicate the potential of the proposed protocol of combined palliative therapy and the need for carrying out experimental studies to verify it.

Acknowledgements This work is supported by the Russian Science Foundation under grant 14-31-00024.

References

1. American Diabetes Association, Diabetes care **27**(suppl 1), s11 (2004)
2. P.G.B. Baker, R.F. Mottram, Clin. Sci. **44**(5), 479 (1973)
3. S. Benzekry, G. Chapuisat, J. Ciccolini et al., C.R. Math. **350**(1), 23 (2012)
4. G. Bergers, Nat. Rev. Cancer **3**(6), 401 (2003)
5. J.P. Boris, D.L. Book, J. Comput. Phys. **11**(1), 38 (1973)
6. D.J. Brat, A.A. Castellano-Sanchez, S.B. Hunter, Cancer Res. **64**(3), 920 (2004)
7. J.J. Casciari, S.V. Sotirchos, R.M. Sutherland, Cell Prolif. **25**(1), 1 (1992)
8. J. Chen, B. Liu, J. Yuan et al., Mol. Oncol. **6**(1), 62 (2012)
9. A. Claes, P. Wesseling, J. Jeuken et al., Mol. Cancer Ther. **7**(1), 71 (2008)
10. P.G. Corrie, Medicine **36**(1), 24 (2008)
11. P.V. Dickson, J.B. Hamner, T.L. Sims et al., Clin. Cancer Res. **13**(13), 3942 (2007)
12. R.P. Dings, M. Loren, H. Heun et al., Clin. Cancer Res. **13**(11), 3395 (2007)
13. J. Ebos, R. Kerbel, Nat. Rev. Clin. Oncol. **8**(4), 210 (2011)
14. A.W. El-Kareh, T.W. Secomb, Neoplasia **6**(2), 117 (2004)
15. J. Folkman, N. Engl. J. Med. **285**(21), 1182 (1971)
16. J.P. Freyer, R.M. Sutherland, J. Cell. Physiol. **124**(3), 516 (1985)

17. Genentech Inc., https://www.gene.com/download/pdf/avastin_prescribing.pdf. 13 Apr 2017
18. A. Giese, R. Bjerkvig, M.E. Berens, M. Westphal, J. Clin. Oncol. **21**(8), 1624 (2003)
19. R. Grossman, H. Brastianos, J.O. Blakeley et al., J. Neuro-Oncol. **116**(1), 59–65 (2014)
20. V. Heiden, G. Matthew, L.C. Cantley, C.B. Thompson, Science **324**(5930), 1029 (2009)
21. J. Holash, P.C. Maisonpierre, D. Compton et al., Science **284**(5422), 1994 (1999)
22. K. Izuishi, K. Kato, T. Ogura et al., Cancer Res. **60**(21), 6201 (2000)
23. J.M. Kelm, C.D. Sanchez-Bustamante, E. Ehler et al., J. Biotechnol. **118**(2), 213 (2005)
24. A. Köhn-Luque, W. De Back, Y. Yamaguchi et al., Phys. Biol. **10**(6), 066007 (2013)
25. A.V. Kolobov, A.A. Polezhaev, G.I. Solyanik, Comput. Math. Methods Med. **3**(1), 63 (2000)
26. A.V. Kolobov, V.V. Gubernov, M.B. Kuznetsov, Russ. J. Numer. Anal. Math. Model. **30**(5), 289 (2015)
27. M.B. Kuznetsov, A.V. Kolobov, Russ. J. Numer. Anal. Math. Model. **32**(5) (2017)
28. J.R. Levick, *An Introduction to Cardiovascular Physiology* (Butterworth-Heinemann, Oxford, 2013)
29. J. Ma, S. Pulfer, S. Li et al., Cancer Res. **61**(14), 5491 (2001)
30. J. Ma, S. Yao, X.S. Li et al., Medicine **94**(42) (2015)
31. G.F. Mac, A.S. Popel, Am. J. Physiol. Heart Circ. Physiol. **292**(1), H459 (2007)
32. E. Marx, W. Mueller-Klieser, P. Vaupel, Int. J. Radiat. Oncol. Biol. Phys. **14**(5), 947 (1988)
33. Y. Miyagi, K. Fujiwara, J. Kigawa et al., Gynecol. Oncol. **99**(3), 591 (2005)
34. N. Papadopoulos, J. Martin, Q. Ruan et al., Angiogenesis **15**(2), 171 (2012)
35. O.N. Pyaskovskaya, D.L. Kolesnik, A.V. Kolobov et al., Exp. Oncol. **30**(4), 269 (2008)
36. J.P. Rock, L. Scarpace, D. Hearshen et al., Neurosurgery **54**(5), 1111 (2004)
37. D. Shweiki, M. Neeman, A. Itin, E. Keshet, Proc. Natl. Acad. Sci. **92**(3), 768 (1995)
38. S.K. Stamatelos, E. Kim, A.P. Pathak, A.S. Popel, Microvasc. Res. **91**, 8 (2014)
39. H. Suhaimi, D.B. Das, Biotechnol. Lett. **38**(1), 183 (2016)
40. V.V. Tuchin, A.N. Bashkatov, E.A. Genina, Tech. Phys. Lett. **27**(6), 489 (2001)
41. P.J.M. Van de Vaart, N. Van der Vange, F.A.N. Zoetmulder et al., Eur. J. Cancer **34**(1), 148 (1998)
42. D. Zemlickis, J. Klein, G. Moselhy, G. Koren, Pediatr. Blood Cancer **23**(6), 476 (1994)

Math Model of the Passage of a Diffusible Indicator Throughout Microcirculation Based on a Stochastic Description of Diffusion and Flow

V. V. Kislukhin and E. V. Kislukhina

1 Introduction

In his quest to select from different math models that generate a transport function for a liver's circulation Goresky [1] found out that by the linear transformation of any dilution curves (from RBC-Cr^{51} to the DHO) one get all dilution curves coincided, see Fig. 1. To perform Goresky transformation one takes maximum of RBC curve and maximum of chosen diffusible curve. The transformation coefficient K is equal to ratio $K = $ max(RBC)/max(diffusible indicator). Now all heights of diffusible dilution curve are multiplied by coefficient K. Simultaneously the times of points of diffusible dilution are reduced by the same factor, K. In other words, for diffusible indicator its Y-axis is up K times, and X-axis is shortened K-times. However, for some situations Goresky transform gives poor coincidence. But why the coincidence varies was unclear [2]. Paragraphs should have its first line indented by about 0.25 in. except where the paragraph is preceded by a heading and the abstract should be indented on both sides by 0.25 in. from the main body of the text.

Thus the aim of the manuscript is

(a) To present a mathematical model of the passage of diffusing tracer throughout microcirculation.
(b) To address conditions for dilution curves be coincidental.
(c) To reveal that Goresky transform can be used for estimation of the permeability of a capillary wall.

V. V. Kislukhin (✉)
Medisonic, Moscow, Russia

E. V. Kislukhina
Sklifosovsky Institute for Emergency Medicine, Moscow, Russia

© Springer International Publishing AG, part of Springer Nature 2018
R. P. Mondaini (ed.), *Trends in Biomathematics: Modeling, Optimization and Computational Problems*, https://doi.org/10.1007/978-3-319-91092-5_25

Fig. 1 Original dilution
curves (above) and the same
curves after Goresky
transform

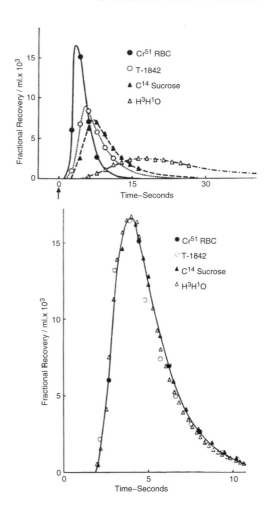

2   Math Equations for First Four Events

The next five events constitute the pass through microcirculation: (1) be in intravascular space, (2) be in extravascular space (for diffusible particles), (3) a microvessel is closed, (4) a microvessel is open, and (5) a particle, being in open microvessel, experiences a variation of velocity.

2.1   Math Equations for First Four Events

Assumed Markovian property leads to exponential distributions for all four first events [3].

List of distributions with characteristic parameters is as follows:

1. Density of distribution to be in intravascular space is

$$f_\delta(t) = \delta \cdot \exp(-\delta t) \tag{1}$$

with $1/\delta$ as the mean time to be in vascular space before entering tissue.
2. Density of distribution to be in extravascular space is

$$f_\gamma(t) = \gamma \cdot \exp(-\gamma t) \tag{2}$$

with $1/\gamma$ as the mean time for a particle to be in the tissue before returning into blood.
3. The time for resuming of flow has the density

$$f_\mu(t) = \mu \cdot \exp(-\mu t) \tag{3}$$

and $1/\mu$ is the mean time for resuming of flow.
4. The time for microvessels to be open has the density

$$f_\beta(t) = \beta \cdot \exp(-\beta t) \tag{4}$$

and $1/\beta$ is the mean time for a microvessel being open.

2.2   The Passage Through Open Microvessels

Let us denote time to pass open microvessels as T. A distribution of T, $G(T)$, can be taken as Gamma distribution. The choice for $G(T)$ is based on the statement: $G(T)$ is infinitely divisible [3]. To show this we follow the next reasoning. In microcirculation due to the absence of inertia pressure gradient and velocity are instantly connected $V = k \cdot \mathrm{grad} P$ [4]. Thus variations of pressure produce new velocities and also the variations of time to pass microcirculation. Now, if we divided each path within microcirculation into two about equal parts, then the $G(T)$ would become the convolution of two mutually independent distributions. Let us denote them as $G_{1/2}(T)$. We can continue this procedure thus $G(T)$ can be presented as convolution of any number of distributions. Thus $G(T)$ is infinitely divisible distribution and for simplicity Gamma distribution with density.

$$f_{\alpha,\nu}(T) = \frac{\alpha^\nu T^{\nu-1} \exp(-\alpha T)}{\Gamma(\nu)} \tag{5}$$

3 Equations of Math Model

Equations (1) through (5) are the "bricks" from which a model of the passage of diffusible or intravascular indicators is made.

3.1 The Passage of a Diffusing Indicator

Since time between two jumps out of vascular space has density of exponential distribution, Eq. (1), the n jumps during time s (time spent by particle within intravascular space) has a Poisson distribution $p_n = \exp(-\delta \cdot s) \cdot (\delta \cdot s)^n/n!$ [3]. If a particle is not consumable, then appearance in tissue follows by returning into vasculature. Thus $D(r, s)$ (with $f_\gamma^{0*}(r - s) = 1$; if $r = s$ and zero if $r > s$, with r as time to be in microcirculation, intra- and extra- vascular space) is:

$$D(r, s) = \sum_{n=0}^{\infty} p_n \cdot f_\gamma^{n*}(r - s)$$

$$= \exp(-\delta \cdot s) f_\gamma^{0*}(r - s) + \exp(-\delta \cdot s) \sum_{n=1}^{\infty} \frac{(\delta \cdot s)^n}{n!} f_\gamma^{n*}(r - s) \qquad (6)$$

Laplace transform of (6) with $\varphi(\lambda) = \lambda \dfrac{\gamma + \delta + \lambda}{\gamma + \lambda}$

$$d(\lambda, s) = \int \exp(-\lambda r) D(r, s) dr$$

$$= \exp\left(-s\lambda \left(\frac{\gamma + \delta + \lambda}{\gamma + \lambda}\right)\right) = \exp\left(-s\varphi(\lambda)\right) \qquad (7)$$

The (7) is a conditional Laplace transform. Now we need to perform randomization of s in $d(\lambda, s)$. The randomization of the expression (7) by distribution of s leads to the Laplace transform of the $V(s)$ with the replacement of λ in $\exp(-\lambda \cdot s)$ by $\varphi(\lambda)$:

$$d(\lambda) = \int \exp\left(-s\varphi(\lambda)\right) V(s) ds \qquad (8)$$

Thus our next step is to find the distribution for the s, $V(s)$.

3.2   An Intravascular Indicator: The Search for $V(s, T)$, with T as Time Spent by Particle Within Open Microvessels

The passage of an intravascular indicator is the composition of two processes (a) the change of the state of any microvessel, meaning that some closed microvessels become open and vice versa, and (b) a variation of the time T to pass through open microvessels. We start with T fixed.

Since time between two stops follows exponential distribution the probability of n stops is given by Poisson distribution, $p_n = \exp(-\beta T)(\beta T)^n/n!$. Every stop follows by resuming of flow. Thus we get a compound Poisson distribution for the transit time of an intravascular indicator with the density $V(s, T)$, and $f_\mu^{0*}(s - T) = 1$ if $s = T$ and 0, if $s > T$.

$$V(s, T) = \sum_{n=0}^{\infty} p_n \cdot f_\mu^{n*}(s - T)$$

$$= \exp(-\beta T) \sum_{n=0}^{\infty} \frac{(\beta T)^n}{n!} f_\mu^{n*}(s - T) \tag{9}$$

Laplace transform of (9) with $\phi(\lambda) = \lambda \left(\dfrac{\beta + \mu + \lambda}{\mu + \lambda} \right)$ is:

$$v(\lambda, T) = \int \exp(-\lambda s) V(s, T) ds$$

$$= \exp\left(-\lambda T \left(\frac{\mu + \beta + \lambda}{\mu + \lambda} \right) \right) \tag{10}$$

Thus we have conditional Laplace transform for intravascular indicator. Now final step: to obtain unconditional distributions for diffusing and intravascular indicators.

3.3   Unconditional Distributions to Pass Through Microcirculation

The randomization of T in $v(\lambda, T)$, Eq. (10), leads to Laplace transform for $f_{\alpha,\nu}(T)$, only parameter λ is replaced by $\phi(\lambda)$:

$$v(\lambda) = \int \exp\left(-T\phi(\lambda) \right) f_{a,\nu}(T) dT = a^\nu \left(a + \phi(\lambda) \right)^{-\nu} \tag{11}$$

The unconditional Laplace transform for $d(\lambda)$ is obtained in two steps:

(a) From $d(\lambda, s)$ as conditional Laplace transform is obtained $d(\lambda, T)$, with fixed T:

$$d(\lambda, T) = \int \exp\bigl(-s \cdot \phi(\lambda)\bigr) V(s, T) ds = \exp\bigl(-T \cdot \phi(\varphi(\lambda))\bigr) \qquad (12)$$

(b) From $d(\lambda, T)$, by randomizing T with $f_{\alpha,\nu}(T)$ is obtained $d(\lambda)$:

$$d(\lambda) = \int \exp\bigl(-T \cdot \phi(\varphi(\lambda))\bigr) f_{a,\nu}(T) dT = a^{\nu}\bigl(a + \phi(\varphi(\lambda))\bigr) \qquad (13)$$

It is possible to transform expressions (11) and (13) into corresponding distribution functions. However for all practical purposes it is better to analyze Laplace transformation itself.

3.4 Goresky Transform

Formally Goresky transform is performed in two steps:

(a) Obtaining of the coefficient a of the transform:

$$a(M_V \quad T) = M_D - T; \quad a = \frac{M_D - T}{M_V - T} \qquad (14)$$

where M_V is the mean time to pass investigated tissue by intravascular indicator, M_D is the mean time to pass by diffusing indicator, and T is the common delay;

(b) The distribution of diffusing indicator, $D(t)$, changes to $D_{GT}(t)$:

$$D_{GT} = aD\left(T + \frac{(t-T)}{a}\right), \ t > T; \ D_{GT} = 0, \ t = T, \ \text{or } t < T \qquad (15)$$

Thus we obtain the Goresky coefficient, a, and the new shape, $D_{GT}(t)$, for diffusing indicator.

Goresky Phenomenon If two distributions, $F(t)$ and $G(t)$ are coincide, then all their moments $M_k = (-1)^k f^{(k)}(0)$, where $f(\lambda)$ is the Laplace transform of $F(t)$, are equal, $M_k(F) = M_k(G)$, for each k. Practical coincidence can be reached by equalities of the first two moments, or, what is the same, the equality of the means and dispersions. In our case we have two dilution curves, from intravascular indicator, $V(t)$ and, after Goresky transform, $D_{GT}(t)$, the dilution curve obtained from diffusing indicator, $D(t)$. Goresky phenomenon takes place if applying Goresky coefficient, $D_D^2 = a^2 D_V^2$, in other words dispersions of $V(t)$ and $D_{GT}(t)$ are equal [3].

Experiments with Goresky Transform In Figs. 2 and 3 there are experiments on PC with math model of intravascular and diffusing indicators, respectively.

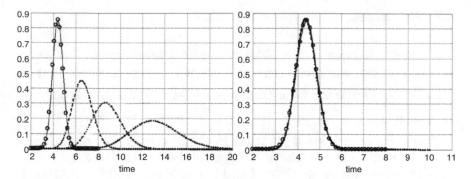

Fig. 2 Math model curves following Goresky phenomenon

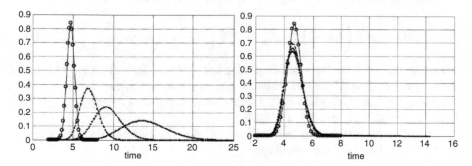

Fig. 3 Math model curves following Goresky phenomenon

The distribution to pass intravascular space is characterized by delay 2 s, and by binomial distribution given on $N = 40$ points between 2 and 8 s, thus with step $h = 0.15$ s, and $p = 0.4$ thus $p_i = \frac{40!}{i!(40-i)!} p^i (1 - p)^{40-i}$. The mean transit time $M_V = Nph + T = 2.4 + 2 = 4.4$; and dispersion $D_V^2 = Np(1 - p)(h)^2 = 0.216$.

The distributions of diffusing indicators additionally characterized by relation of extravascular/intravascular distribution, this is δ/γ. Thus we have relation between two means to pass microcirculation (diffusing and intravascular): $M_D = M_V(1 + \delta/\gamma)$. For experiment are chosen three types of diffusing indicator, with δ/γ equal 0.5, 1.0 and 2.0 (by other words with small, medium and expanded extravascular space). Goresky transform leads to the next three, corresponding Goresky coefficients (12): 1.8; 2.7, and 4.4.

In common case dispersions are not connected by Goresky coefficient $a^2 D_V^2 \neq D_D^2$. However if we choose γ to fulfill equality between dispersion: $D_D^2 = a^2 D_V^2$; then Goresky transform leads to the dilutions given in Fig. 2, right.

Figure 3 presents the same intravascular dilution but diffusing indicators are different. The probability to return into intravascular space is two times down (from presented in Fig. 1), thus dispersions to pass microcirculation are increased and application of Goresky transform does not lead to the coincidence dilution curves, Fig. 3, right.

3.5   Permeability by Goresky Transform

Our diffusing indicator has a distribution with Laplace transform $d(\lambda) = g\big(\phi(\varphi(\lambda))\big)$ (13), and intravascular indicator has Laplace transform $v(\lambda) = g\big(\phi(\lambda)\big)$ (11). Due to stochastic description of the diffusion the function $\varphi(\lambda)$ looks as: $\varphi(\lambda) = \lambda \dfrac{\gamma + \delta + \lambda}{\gamma + \delta}$.

With such a presentation of $\varphi(\lambda)$ Goresky transform leads to the determination of the characteristic of permeability of endothelial barrier, these are δ and γ. Also will be found out that the specificity of $g(\lambda)$ and $\phi(\lambda)$ has no role (but $\phi(\lambda)$ should be infinitely divisible).

Indeed, the mean and dispersion of diffusible indicator are

$$M_D = M_V \left(1 + \frac{\delta}{\gamma}\right); \quad D_D^2 = D_V^2 \left(1 + \frac{\delta}{\gamma}\right)^2 + M_V \frac{2\delta}{\gamma^2}; \qquad (16)$$

where M_V and D_V are mean and dispersion for intravascular indicator. The (14) is follow from next equations established connection between M and D^2 of any distribution, $f(t)$, and derivatives of its Laplace transform $F(\lambda) = \int \exp(-\lambda t) f(t) dt$. Thus $M = F'(\lambda)|_{\lambda=0}$; and $D^2 = F''(\lambda)|_{\lambda=0} - M^2$.

Now, if we put relations between two means, given by (14) into Eq. (12), we get, for Goresky coefficient, next equality:

$$a = \frac{M_D - T}{M_V - T} = \frac{M_V \left(1 + \frac{\delta}{\gamma}\right) - T}{M_V - T}$$

Thus the knowledge of Goresky coefficient leads to the obtaining of δ/γ:

$$\frac{\delta}{\gamma} = (a - 1) \frac{(M_V - T)}{M_V}; \qquad (17)$$

The use of the second relation in (14), the knowledge of dispersions, leads to the obtaining of γ:

$$\gamma = \frac{2\delta \cdot M_V}{\gamma \cdot D_V^2} \left(\frac{D_D^2}{D_V^2} - \left(1 + \frac{\delta}{\gamma}\right)^2\right)^{-1} \qquad (18)$$

4   Discussion

The comprehensive descriptions of stochastic models are given in [5, 6]. Application of stochastic approaches is based on the approximation of real dilution curves, meaning that parameters of chosen distributions become parameters of recorded curves. Such formal approach has problem with physiological interpretation of model's parameters.

In given manuscript it is shown that assumption of stochasticity leads to the uniqueness of math equations of model for blood flow.

Since four basic events: (1) to be in extravascular space, (2) to be in intravascular space, (3) a microvessel is closed, and (4) a microvessel is open, due to Markovian property, follow exponential distributions, we have a very effective application of Laplace transform. The combinations of these processes become compound Poisson distributions thus the combinations have Laplace transform as $\exp(-t \cdot f(\lambda))$. The randomization of t in $\exp(-t \cdot f(\lambda))$ by any distribution becomes Laplace transform of this distribution also, only λ is replaced by $f(\lambda)$.

5 Conclusion

From the assumption of stochasticity follows uniqueness of distributions that formed the passage of indicator through microcirculation. Thus exponential distributions and their generalization, gamma-distribution, become the motivations for introducing the permeability of endothelium.

References

1. C.A. Goresky, A linear method for determining liver sinusoidal and extravascular volumes. Am. J. Physiol. **204**, 626–640 (1963)
2. K. Zierler, Indicator dilution methods for measuring blood flow, volume, and other properties of biological systems: a brief history and memoir. Ann. Biomed. Eng. **28**(8), 836–848 (2000)
3. W. Feller, *An Introduction to Probability Theory and Its Applications. Volume II* (Wiley, New York, 1966)
4. Y.C. Fung, Stochastic flow in capillary blood vessels. Microvasc. Res. **5**(1), 34–48 (1973)
5. W.S. Kendal, A stochastic model for the self-similar heterogeneity of regional organ blood flow. Proc. Natl. Acad. Sci. U. S. A. **98**(3), 837–841 (2001)
6. J.H. Matis, H.D. Tolley, On the stochastic modeling of tracer kinetics. Fed. Proc. **39**(1), 104–109 (1980)

Influence of External Factors on Inter-City Influenza Spread in Russia: A Modeling Approach

V. N. Leonenko and Yu. K. Novoselova

1 Introduction

Influenza and acute respiratory infections (ARI) are the most frequent infections in the world. Epidemic outbreaks of ARI and influenza cause 3–5 million of severe cases of illness and 250–500 thousand deaths yearly [20]. For the sake of control of epidemic situation surveillance networks are used by healthcare organs, like ILINet in the USA, Sentinelles in France and the Russian surveillance system deployed at Research Institute of Influenza. Due to big delays of data acquisition, only partial coverage of the countries' territories by surveillance posts [5], under-reporting (for instance, due to reduced show-up during holidays) the incidence data delivered by these networks is of limited use for planning containment measures. Thus, the healthcare organs start to employ the influenza epidemic prediction frameworks based on the statistical methods and mathematical modelling.

In the late 1960s, multiple mathematical models of influenza-like illness outbreaks were created. While some of them relied on influenza modeling in local urban areas [17], the others took into account the fact that influenza epidemics are caused by flu virus strains circulating over the globe with migration flows. The simplest form of reproducing this mechanism mathematically is to rely on coupled SEIR epidemic models, i.e. to consider consequent flu epidemic outbreaks within a group of interconnected urban territories (for instance, a country) caused by inter-city movement of infected individuals. One of the first models of that type was implemented by Baroyan and Rvachev [3]. Initially designed to reproduce the spread of flu among the cities of Soviet Union, it was later applied to model flu outbreaks in France [4], Cuba [1] and worldwide pandemic flu circulation [14].

V. N. Leonenko (✉) · Yu. K. Novoselova
ITMO University, Saint Petersburg, Russian Federation
e-mail: international@mail.ifmo.ru

© Springer International Publishing AG, part of Springer Nature 2018
R. P. Mondaini (ed.), *Trends in Biomathematics: Modeling, Optimization and Computational Problems*, https://doi.org/10.1007/978-3-319-91092-5_26

375

The Baroyan-Rvachev approach demonstrated that the combination of Kermack-McKendrick SEIR model and a linear model of inter-city migration flows could satisfactory match the true incidence data and provide accurate forecasts of the outbreaks starts and peaks for Soviet cities. However, since early 1980s the Soviet modeling complex for flu forecasting, created in Research Institute of Influenza [13] and based on Baroyan-Rvachev model, showed the signs of growing incoherence with the seasonal epidemic outbreak patterns observed in Soviet cities [7]. The reason for that, according to one of the versions, is in the growing levels of collective immunity to flu due to increasing speed of its circulation around the globe. The dynamics of growth of collective immunity could be dependent from different factors, including the structure of contact networks within an urban area, that's why the original assumption that fraction of non-immune individuals is the same for all Soviet cities and depends only on currently circulating virus strain, seems to be less applicable [7].

Due to the circumstances described above the simplified models do not capture in detail the dynamics of within-country flu epidemic process. The experiments with the SEIR models calibration on the available Russian ARI incidence data [9, 11] demonstrated that using standard models one can obtain the prediction of prospected epidemic peak height with satisfactory accuracy, but there is no chance to assess in advance the epidemic peak day. To enhance the prediction accuracy it's vital to obtain more data, particularly the one on migration flows between the cities.

This paper represents the first stage of the broader investigation aimed at inter-city modeling of flu propagation with regard to heterogeneity of Russian cities caused by different herd immunity levels and other local factors. The objective of the current research is to assess the accuracy of the coupled models approach in contemporary conditions using the incidence data provided by Research Institute of Influenza. Particularly we want to establish a baseline accuracy for the sake of comparison with future corrected metapopulation models and to assess the levels of prediction biases caused by uncertainties in migration flow data. The novelty of the task is proved by the fact that the modeling of influenza dynamics in Russia using the modern incidence data was never performed. Nevertheless, the region-specific peculiarities in influenza dynamics [18] limit the usage of general models and demand for the accounting of distinct features of epidemic situation in each country. Due to that the creation of prediction models of seasonal ARI dynamics and influenza epidemics for the case of Russia is an important task.

2   Outbreak Incidence Data

The original dataset provided by the Research Institute of Influenza [13] contains weekly cumulative incidence for all the ARI types (including flu) in 41 Russian cities from 2000 to 2015. Before the model fitting, we have to refine the incidence data by restoring the missed values and fixing the under-reporting. We also need to extract flu incidence from the cumulative ARI incidence data. Corresponding algorithms are described in detail in [10], here we introduce briefly the sequence of operations.

- Under-reporting correction. Since infected people avoid visiting healthcare facilities during holidays, the corresponding weekly prevalence is lower than the actual number of newly infected. This under-reporting bias can be corrected by means of cubic interpolation [3] using the incidence registered in the adjacent weeks. The sporadic gaps in incidence data are filled in the same fashion.
- Bringing the incidence data to daily format. The daily incidence is found with the help of cubic interpolation of weekly incidence. We assume that $n_{inf}^{Thu} = n_{inf}^{W}/7$, where n_{inf}^{W} is the weekly incidence taken from the database and n_{inf}^{Thu} is the daily incidence for Thursday of the corresponding week.
- Extracting data on influenza outbreak from the cumulative seasonal ARI data with the help of a separate epidemic curve allocation algorithm. At first, the algorithm finds higher non-flu ARI incidence level a_2, which corresponds to the average number of newly infected in non-epidemic period. ARI epidemic curves, which are detected as flu outbreaks, should have their peaks well above the higher ARI level. They should also comply with the time period during which the ARI prevalence exceeds the non-epidemic ARI threshold assessed in the Flu Research Institute. The beginning and ending of the extracted curve is chosen to match the level a_2. The first incidence point of the curve is considered to be the first day of the epidemic outbreak.

3 Model Structure

3.1 Local Model

The local submodel used in the Baroyan-Rvachev prediction framework was represented by the system of difference equations, with the time step equal to 1 day. Let x_t be the fraction of susceptibles in the population, y_t be the number of newly infected individuals at the moment t and $\overline{y_t}$—the cumulative number of *infectious* persons (i.e. those who can transmit flu) by the time t. Then, the equation system may be written in the following manner [6]:

$$\overline{y_t} = \sum_{\tau=0}^{T} y_{t-\tau} g_\tau, \tag{1}$$

$$y_{t+1} = \frac{\beta}{\rho} x_t \overline{y_t},$$

$$x_{t+1} = x_t - y_{t+1},$$

$$x_0 = \alpha\rho. \tag{2}$$

The piecewise constant function g_τ gives a fraction of infectious individuals in the group of individuals infected τ days before the current moment t. The function reflects the change of individual infectiousness over time from the moment of acquiring influenza. It is assumed that there exists some moment $\bar{t}$: $\forall t \geq \bar{t}$ $g_\tau = 0$, which reflects the moment of recovery.

3.2 Global Model

Let us consider an influenza epidemic in n cities. We assume that the outbreak is started in one of those cities and transmits from one city to another with the migration flows. In this case the local model (1)–(2) is transformed by adding the description of this process, giving us the following set of equations:

$$\overline{y_j}(t) = \sum_{\tau=0}^{T} y_j(t, \tau) g(\tau), \tag{3}$$

$$y_j(t+1, 0) = \frac{\beta_j}{\rho_j} x_j(t) \overline{y_j}(t),$$

$$y_j(t+1, \tau) = y_i(t, \tau-1) + \sum_{t=1}^{n} \sigma_{ij} \frac{y_i(t, \tau-1)}{\rho_l},$$

$$x_j(t+1) = x_j(t) - y_j(t+1, 0),$$

$$x_j(0) = \alpha_j \rho_j,$$

$$j = 1, 2, \ldots, n; \tag{4}$$

Here j is the city index and σ_{ij} corresponds to the migration flow from the city i to the city j per time unit.

4 Predicting Outbreaks

4.1 Local Model Fitting

Let $Z^{(dat)}$ be the set of incidence data points loaded from the input file and corresponding to one particular outbreak. Assume that the number of points is t_1, which equals the observed duration of the outbreak. The algorithm varies the values of model parameters to achieve the model output, which minimizes the distance between the modeled and real incidence points [9]:

$$F\left(Z^{(mod)}, Z^{(dat)}\right) = \sum_{i=0}^{t_1} \left(z_i^{(mod)} - z_i^{(dat)}\right)^2, \tag{5}$$

Here $z_i^{(dat)}$ and $z_i^{(mod)}$ are the absolute incidence numbers for the i-th day taken from the input dataset and derived from the model correspondingly. The limited-memory BFGS optimization method is used to find the best fit [12]. Since the existence of several local minima is possible, the algorithm has to be launched several times with different initial values of input variables. The best fit is chosen as a minimum among the distances achieved from all the algorithm runs. To characterize the goodness of fit we utilize the coefficient of determination $R^2 \in (0, 1]$. This coefficient shows the fraction of the response variable variation that is explained by a model [19].

Before optimizing the model parameters, we need to match accurately the model timeline ($t = 0, 1, \ldots$) to the timeline of the epidemic outbreak incidence dataset. For that, we assume that the peak moment of the modeled epidemic curve coincides with the epidemic peak day from the dataset.

The parameter description of the fitting algorithm which corresponds to Baroyan-Rvachev model is given in Table 1. In addition to the parameters taken from the model (1)–(2), a curve positioning parameter Δ_p is introduced. In the ideal case, the best fit of the model curve to data should give the model curve peak occurring at the same day as the real peak seen from the incidence data. In that case, Δ_p is fixed and equals zero, so there is no need to vary it. However, sometimes the best fit is achieved when the modeled and the real peak moments differ by several days due to discrepancies between the real outbreak process and its theoretical model, that is why we made this parameter variable.

It has been proven [3] that without the loss of fit quality we can vary the sole auxiliary value $s = \alpha\beta$ instead of variation α and β separately. This fact results from the biologically plausible assumption that the virulence of the current circulating influenza strain and the immunity level to this strain in the population are interconnected [6]. Another idea of the algorithm is that $s = s(k)$, where k

Table 1 Parameters of the fitting algorithm for the local model

Definition	Description	Value	Unit
Model parameters			
α	Initial ratio of susceptible individuals in the population	Estimated	–
β	Intensity of infection	Estimated	–
I_0	Initial ratio of infected in the population	Estimated	–
T	Duration of infection	Fixed	Day
g_τ	A fraction of infectious individuals among those who were infected τ days before the current moment	Fixed	–
ρ	Population size	Fixed	Persons
Curve positioning parameters			
Δ_p	Absolute horizontal bias of the modeled incidence curve peak position compared to the data	Estimated	Day

is a semi-empirical parameter which approximates the initial fraction of infectious individuals in the first stage of the outbreak [3]. Finally we come to the idea that it is k that should be varied, and s is consequently calculated as a function of k (see the formula below).

The calibration algorithm structure is similar for the cases was first described in [9] and conforms to the following sequence of operations.

- For each fixed combination of values $\{k, I_0\}$ generated by BFGS optimization procedure, and for every Δ_p:

  1. Derive the value of s from the current value of k using the following formula:

     $$s = \frac{k^{T+1}}{\sum_{\tau=0}^{T} k^{T-\tau} g_\tau} \tag{6}$$

  2. Set the preliminary model parameter values, α' and β', to make them conform to the equation $\alpha'\beta' = s$, for instance:

     $$\alpha' = 1, \beta' = s$$

  3. Find the preliminary numerical solution of the model (1)–(2) with the parameter values $\alpha = \alpha'$, $\beta = \beta'$ and the initial conditions $y_0 = I(0)$, $y_{-1} = \cdots = y_{-T} = 0$.

  4. Derive the preliminary number of newly infected each day from the model output: $z_i^{(mod)'} := \overline{y_i}$, $i \in \overline{0, t_1 - 1}$.

  5. Derive the baseline level for the modeled outbreak start z_{base} from the value for higher ARI incidence level a_2 and subtract it from the data incidence points:

     $$z_i^{(dat)} := z_i^{(dat)} - a_2, i \in \overline{0, t_1 - 1}$$

  6. According to the algorithm, we need to match in time the model peak with the incidence data peak. For that purpose we find the shift δ_{adj}:

     $$\delta_{adj} = t_{peak}^{(mod)} - \left(t_{peak}^{(dat)} + \Delta_p\right), \tag{7}$$

where $t_{peak}^{(dat)}$ and $t_{peak}^{(mod)}$ are peak days for the data incidence and the modeled incidence correspondingly, Δ_p is the difference between the modeled peak and the data peak after the shift. After performing a shift, we are to compare the distance between the following datasets:

$$Z^{(dat)} = \left\{z_0^{(dat)}, z_1^{(dat)}, \ldots, z_{N-1}^{(dat)}\right\},$$

$$Z^{(mod)'} = \left\{z^{(mod)'}(\delta_{adj}), z^{(mod)'}(\delta_{adj} + 1), \ldots, z^{(mod)'}(\delta_{adj} + N - 1)\right\},$$

where N is the total number of incidence points.

7. Assigning optimal values to α, β. It was mathematically justified in [3] that for $\alpha, \beta : \alpha\beta = s$

$$\min_{\alpha,\beta} F(Z^{(mod)}(\alpha, \beta, \Delta_p), Z^{(dat)}) = F(Z^{(mod)}(\tilde{\alpha}, \tilde{\beta}, \Delta_p), Z^{(dat)}),$$

$$\max Z^{(mod)}(\tilde{\alpha}, \tilde{\beta}, \Delta_p) = \max Z^{(dat)} + \Delta_p;$$

where $\tilde{\alpha} = \alpha' a$, $\tilde{\beta} = \frac{\beta'}{a}$, a is the correction coefficient calculated according to the formula:

$$a = \frac{\sum_{i=0}^{t_1} z_i^{(dat)}}{\sum_{i=0}^{t_1} \left(z_i^{(mod)'}\right)^2}.$$

To avoid launching the simulation for the second time, now with the values $\alpha = \tilde{\alpha}, \beta = \tilde{\beta}$, one may obtain the new model incidence values $z_i^{(mod)}$ by multiplying the corresponding preliminary values by a [6], that is:

$$z_i^{(mod)} = a z_i^{(mod)'}, z_i^{(mod)} \in Z^{(mod)}(\tilde{\alpha}, \tilde{\beta}, \Delta_p), z_i^{(mod)'} \in Z^{(mod)'}.$$

In that manner we find the optimal parameter values and the corresponding model curve $Z^{(mod)}$.

It is worth mentioning that α has the sense of fraction, $\alpha \in [0; 1]$. In the case if $\tilde{\alpha} > 1$, we artificially set it to 1.

8. Calculate the value of the fit function $F(Z^{(mod)}, Z^{(dat)})$ according to the formula (5).

The BFGS algorithm finds the least distance F_{Δ_p} in the described manner for every value of Δ_p. We define $\Delta_p^{(min)}$: $F(\Delta_p^{(min)}, \dots) = \min F(\Delta_p, \dots)$, and the parameter set $\{\alpha, \beta, I_0\}$, corresponding to $\Delta_p^{(min)}$. These values are the final result of our optimization procedure.

The described fitting algorithm for the local model features some differences from the original one, firstly introduced in [6], namely:

- The curve positioning parameter Δ_p was introduced (the similar parameter was mentioned in [6], but it was not explicitly included into the fitting procedure).
- The iteration over the values of variable k with a fixed step was replaced by BFGS optimization algorithm.
- The value of I_0 was changed from fixed to varied. This change was made due to the ambiguity of flu epidemic outbreak start detection in seasonal incidence data [8]. This allows the algorithm to fit the model to the early outbreak stages more accurately.

The modifications described enhanced both the accuracy and the performance of the algorithm.

4.2 Transport Flow Matrices

One of the issues we faced during the work on the model is the absence of reliable transport flow data appropriate to be used for assessing σ_{ij}. The official statistics of Russian Federal Migration Service, which is the first source of necessary data that comes to mind, includes only the incoming number of people for a fixed city, but not their city of departure. Also it is subjected to under-reporting and includes only the statistics on people who moved to the city for a long stay, which makes it inapplicable for the modeling purposes.

The movements of people between the cities can be traced more accurately by calculating the number of sold airplane, train and bus tickets in different directions, like it was made by Baroyan and Rvachev for the sake of influenza modeling in USSR [3]. Due to the fact that now the data on sold tickets could be harder to get than in Soviet times, as it belongs to commercial companies rather than to state officials, the rougher way to assess the migration flows could be employed based on the capacities of planes, buses and trains performing regular transportation between the cities.

As the assessment biases are inevitable for any chosen way to assess the migration flows, in the current paper we have decided to compare the prediction accuracy of received with the help of calibrated global model with three different transport matrices. The following generation methods were used.

The first two matrices were generated using the airline flights statistics between the cities under consideration. The flight schedule was formed via an online service [21]. The numbers of passengers were derived from the maximum number of seats for the plane types used for particular flights, taking into account the partial load (for the sake of simplicity and due to the absence of accurate data the fraction of occupied seats was set constant for all the flights). The formed matrices differ by the time periods of flight schedules that were used as algorithm input:

- First matrix: daily flight schedule for October 20, 2016;
- Second matrix: weekly flight schedule for the period from February 27th to March 5th, 2017. The daily flows were assessed simply by dividing the weekly flows by 7.

The third matrix was generated using the so-called "gravitational model", according to which the migration between two areas is directly proportional to the product of the quantities of their inhabitants and inversely proportional to the squared distance between them [2]. The corresponding formula has the following form:

$$V_{ij} = \gamma \frac{\rho_i \rho_j}{d_{ij}^2},$$

where $V_{ij} = \sigma_{ij} + \sigma_{ji}$. In our case we assume that the transport flow matrix is symmetric, i.e. $\sigma_{ij} = \sigma_{ji}$, so $\sigma_{ij} = \frac{V_{ij}}{2}$.

The parameter γ is the correction coefficient of the model, it was assessed with the help of the open long-term data on the airline passengers flow between Moscow and Saint Petersburg [16] and kept constant for all the pairs of cities.

Strictly speaking, to model accurately the migration flow, we need to use dynamically formed transport matrices based on the schedule for every particular day of the modeled epidemic spread instead of applying constant transport matrices. Nevertheless, it is not worth additional efforts in case the order of bias caused by this simplification is less than the order of biases caused by inaccurate flow data assessment, or if the sensitivity of the model output on the transport flow data is low. Below we will demonstrate that this is apparently the case for our simulation.

4.3   Global Model Calibration

The application of the global model for the purpose of influenza outbreak prediction in Russian cities is possible under the following assumptions [3]:

- The all-Russian influenza epidemic starts from the outbreak in one of the Russian cities caused by an infectious person who came from abroad.
- Epidemic outbreaks in another Russian cities are caused by the migration of the infectious individuals within the country.
- The epidemic process in the country is mostly driven by a single dominant virus strain, and this strain is the same for all the cities.

Let the first influenza outbreak of the current season be registered in the city j, $j \in \overline{1, n}$ (for our case $n = 41$). In this case the prediction procedure can be performed according to the following sequence of actions:

- Wait until the outbreak reaches its peak and collect the corresponding raw incidence data.
- Correct the data with the algorithms described in Sect. 2.
- Calibrate the local model (see Sect. 4.1) and find the model parameters α, β.
- Assuming that $\forall i \in \overline{1, n} \ \alpha_i \equiv \alpha$, $\beta_i \equiv \beta$, and taking the values of σ_{ij}, $i, j \in \overline{1, n}$, from the transport flow matrix, launch the global model and find the quantities of infected in every city over time.
- Gather the statistics on the days of outbreak peaks and their height.

The preliminary predictions could also be made before the moment of the outbreak peak in the first city using the incomplete incidence data from that city [6], but in that case the accuracy of parameter assessment can drop dramatically [9], thus making implausible the predictions for another cities.

5   Numerical Experiment

We used the interpolated daily ARI incidence data for 41 Russian cities from 2000 to 2015 to perform the retrospective forecast of epidemic peak days and the maximum number of infected (peak heights) in every fixed city in a given epidemic season where the epidemic did take place (those that were not affected by flu in a fixed season were excluded). To assess the accuracy of prediction results, we have used the criteria employed in 1970s for the Soviet flu outbreak prediction framework [6]:

- 'Square'. The prediction is thought to be accurate if $dt \in -8 \ldots 8$ and $dh \in$ (0.5; 2.0).
- 'Vertical stripe'. Accurate prediction should have $dt \in -7 \ldots 7$.
- 'Horizontal stripe'. Accurate prediction should have $dh \in$ (0.7; 1.5).

As it was stated in Sect. 4.2, to find the impact of transport flow assessment accuracy on the accuracy of predictions we have prepared three different transport matrices in the experiment. Also, to reduce the effect of poor model calibration in some intricate cases of the first outbreaks, we have decided to launch simulations with the local model parameters estimated by calibrating it to incidence data of the first, the second and the third cities affected by the epidemics during the season. Thus, in sum nine simulation runs were performed for each epidemic season.

The example of prediction for eight different cities in a fixed epidemic season is given in Fig. 1. One can see that for some cities (Barnaul, Perm, Chelyabinsk) the peak day is predicted better than the height, for some others (Norilsk, Nizhny Novgorod, Kazan) it is the other way round, and for a few cities (Kursk, Irkutsk) the model was unable to give satisfactory predictions neither for a peak day nor for its height.

The experiment brought us to the following results.

- The influence of the choice of transport matrix on the accuracy of peak predictions could be considered negligible. In most of the cases the percentage of accurate predictions is the same for all three matrices, save for several epidemic seasons when slight difference was demonstrated (see Table 2). Although this result is somewhat unexpected, it complies to the earlier experiments with the model on USSR data performed by Baroyan and Rvachev [3].
- As one can see from Table 3, the accuracy of the predictions varies dramatically depending on the city chosen for model calibration. At the same time, these discrepancies are not related directly to poor model fitting (in most of the epidemic seasons the value of R^2 was higher than 0.9 and more or less equal for the different choice of city for local model calibration). The possible explanation of this effect is that the first outbreak within the country registered during the season can be not the one that caused a full-fledged epidemic, so the global epidemic parameters should be derived from the outbreak in another city. Also it's worth mentioning that on the scale of the week which the initial incidence

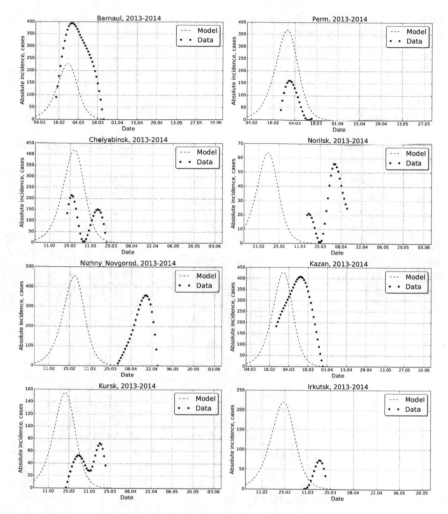

Fig. 1 Model trajectories compared to real data, 2013–2014, gravitation model, calibration to the first outbreak city data (in this particular case—Magadan)

data has it is sometimes impossible to understand what of the cities infected during the week 1 is really "the first city".

- The achieved peak prediction accuracy for any of the criteria is sufficiently lower than those achieved in Soviet times. For instance, the average compliance to vertical stripe criterion for the predictions made in the 1970s is said to be 87.4% [6], whereas in our case it varies from 3.85% to 72.73%. Also the prediction accuracy shown for Moscow, Saint Petersburg and Novosibirsk is lower than the one for the same cities demonstrated by another approaches which do not use explicitly the transport flow data [9, 11]. Some examples of peak prediction biases are given in (Fig. 2).

Table 2 Overall compliance of peak predictions to accuracy criteria for different transport matrices and different cities chosen for calibration, 2012–2013 epidemic season

Transport matrix generation	Vertical stripe	Horizontal stripe	Square
Daily schedule	14.71 2.94 11.76	5.88 50.0 44.12	2.94 2.94 11.76
Weekly schedule	14.71 2.94 11.76	5.88 47.06 44.12	2.94 2.94 11.76
Gravitational model	0.0 2.94 11.76	41.18 47.06 44.12	0.0 2.94 11.76

Table 3 Overall compliance of peak predictions to accuracy criteria, transport matrix No. 2 (averaged weekly schedule)

Year	Vertical stripe	Horizontal stripe	Square
2000	3.85 0.0 **3.85**	7.69 11.54 **30.77**	3.85 3.85 **3.85**
2001	**23.08** 11.54 3.85	15.38 0.0 **73.08**	**7.69** 7.69 3.85
2002	**72.73** 6.06 9.09	**54.55** 54.55 54.55	**60.61** 3.03 12.12
2003	0.0 **33.33** 33.33	**45.45** 27.27 24.24	3.03 27.27 **33.33**
2004	3.33 10.0 **43.33**	**43.33** 23.33 36.67	3.33 6.67 **26.67**
2005	7.41 **14.81** 3.7	44.44 18.52 **48.15**	7.41 **7.41** 3.7
2006	21.43 **21.43** 3.57	3.57 **39.29** 28.57	7.14 **17.86** 3.57
2007	3.23 **38.71** 3.23	**38.71** 32.26 35.48	6.45 **22.58** 3.23
2008	23.53 **41.18** 14.71	**41.18** 35.29 20.59	11.76 **23.53** 2.94
2009	**5.41** 2.7 2.7	**45.95** 13.51 27.03	**5.41** 2.7 2.7
2010	**10.81** 2.7 2.7	2.7 **5.41** 0.0	**2.7** 0.0 0.0
2011	**45.45** 18.18 9.09	36.36 **54.55** 0.0	**36.36** 18.18 9.09
2012	**14.71** 2.94 11.76	5.88 **47.06** 44.12	2.94 2.94 11.76
2013	35.29 **35.29** 17.65	41.18 **52.94** 23.53	29.41 **29.41** 17.65
2014	3.23 **3.23** 3.23	35.48 **51.61** 3.23	3.23 **3.23** 3.23
2015	**18.42** 10.53 5.26	34.21 31.58 **55.26**	7.89 **10.53** 5.26

The highest accuracy is marked with bold

6 Discussion and Future Works

As the experiment demonstrated, the accuracy of peaks predictions made by homogeneous coupled SEIR models on the contemporary incidence data is much lower than it used to be in Soviet times and does not allow to use this approach for the disease forecasting. The reason for low accuracy does not seem to be connected with the accuracy of assessing the migration flows, as the change of transport matrix generation doesn't change it significantly. The hypothesis that still not proved or disproved is that the main cause is the difference between the values of α and β for the different cities. We are planning to conduct several separate investigations to shed light on that matter, using additional data on virus spread from Russian Institute of Influenza.

In any case, calculating α, β separately for each city will inevitably cause issues with the early disease prediction, that is why we want to try to implement another

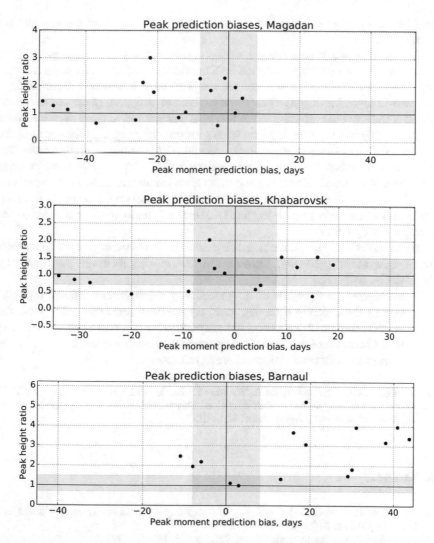

Fig. 2 Scatterplots reflecting the peak prediction biases for the 15 cities affected by flu during the epidemic season of 2013–2014, gravitation model. The model parameters were found by calibrating to incidence data from Magadan, Khabarovsk and Barnaul correspondingly

model updates before proceeding with that idea. Particularly one of the changes in flu transmission over the years that is mentioned in healthcare reports and may possibly account for decreased efficiency of homogeneous coupled models is the growing role of age structure of the population [7]. This fact was also mentioned in the modeling studies conducted using the data of French Sentinelles flu surveillance network [15]. Enhancing the model prediction accuracy by incorporating age structure seems like an interesting challenge, although the odds of success are hard to be estimated.

In case of α and β remaining the same constants for all the cities yet another issue arises, which was demonstrated both by old studies [3] and by our numerical experiment described in the previous section. This issue is dramatic dependence of the model output on the city chosen for calibration. Apparently the choice of the city should be made according to the real epidemic course, and the question arises how to distinguish separate local outbreaks caused either by the resurgence of the old seasonal virus strain or some severe non-flu ARI infection from the circulation of current epidemic strain. For this purpose the data of laboratory virus studies may serve well. We plan additional investigations aimed at understanding the role of first city chosen for calibration on the resulting model parameter values and establishing the indicators that may help understand what city was really the first affected by the incoming flu epidemic.

By means of conducting the research described above we hope to consequently improve the model prediction performance, thus making its usage worthwhile by the healthcare officials.

The enhanced performance of the model may require more detailed transport data, such as dynamically changing transport matrices connected with particular days, or the account of other means of transport beside planes. In case of necessity we plan to obtain these data by collaborating with another scientific groups in our facility who deal with urban informatics and transport flow models.

Acknowledgements The authors thank Vladislav Karbovskii and Vladislav Shmatkov (ITMO University) for providing the transport flow matrices. This research is financially supported by The Russian Science Foundation (Agreement #14-21-00137).

References

1. A. Aguirre, E. Gonzalez, The feasibility of forecasting influenza epidemics in Cuba. Mem. Inst. Oswaldo Cruz **87**(3), 429–432 (1992)
2. J.E. Anderson, The gravity model. Annu. Rev. Econ. **3**, 133–160 (2011)
3. O.V. Baroyan, L.A. Genchikov, L.A. Rvachev, V.A. Shashkov, An attempt at large-scale influenza epidemic modelling by means of a computer. Bull. Int. Epid. Assoc. **18**, 22–31 (1969)
4. A. Flahault et al., Modelling the 1985 influenza epidemic in France. Stat. Med. **7**(11), 1147–1155 (1988)
5. J. Greenspan, S. Valkova, Documenting the missed opportunity period for influenza vaccination in office-based settings. Online J. Public Health Inform. **7**(1), e26 (2015)
6. Yu.G. Ivannikov, A.T. Ismagulov, Epidemiologiya grippa (the epidemiology of influenza). Almaty, Kazakhstan, 1983 (in Russian)
7. Yu.G. Ivannikov, P.I. Ogarkov, An experience of mathematical computing forecasting of the influenza epidemics for big territory. Zhurnal Infectologii **4**(3), 101–106 (2012) (in Russian)
8. V.N. Leonenko, S.V. Ivanov, Fitting the SEIR model of seasonal influenza outbreak to the incidence data for Russian cities. Russ. J. Numer. Anal. Math. Model. **31**, 267–279 (2016)
9. V.N. Leonenko, S.V. Ivanov, Influenza peaks prediction in Russian cities: comparing the accuracy of two SEIR models. Math. Biosci. Eng. **15**(1), 209–232 (2018). https://doi.org/10.3934/mbe.2018009

10. V.N. Leonenko, S.V. Ivanov, Yu.K. Novoselova, A computational approach to investigate patterns of acute respiratory illness dynamics in the regions with distinct seasonal climate transitions. Proc. Comput. Sci. **80**, 2402–2412 (2016)
11. V.N. Leonenko, Yu.K. Novoselova, K.M. Ong, Influenza outbreaks forecasting in Russian cities: is Baroyan-Rvachev approach still applicable? Proc. Comput. Sci. **101**, 282–291 (2016)
12. D. Liu, J. Nocedal, On the limited memory BFGS method for large-scale optimization. Math. Program. **45**, 503–528 (1989)
13. Research Institute of Influenza website. http://influenza.spb.ru/en/
14. L.A. Rvachev, I.M. Longini, A mathematical model for the global spread of influenza. Math. Biosci. **75**(1), 3–22 (1985)
15. C. Segolene, K. Pakdaman, P.-Y. Boëlle, Commuter mobility and the spread of infectious diseases: application to influenza in France. PloS one **9**(1), e83002 (2014)
16. Slishkom populyarnyy marshrut Moskva-Sankt-Peterburg (in Russian) // ATO.RU. http://www.ato.ru/content/slishkom-populyarnyy-marshrut-moskva-sankt-peterburg
17. C.C. Spicer, C.J. Lawrence, Epidemic influenza in greater London. J. Hyg. **93**(01), 105–112 (1984)
18. J. Tamerius, M.I. Nelson, S.Z. Zhou, C. Viboud, M.A. Miller, W.J. Alonso, Global influenza seasonality: reconciling patterns across temperate and tropical regions. Environ. Health Perspect. **119**(4), 439 (2011)
19. S.P. Van Noort, R. Aguas, S. Ballesteros, M.G.M. Gomes, The role of weather on the relation between influenza and influenza-like illness. J. Theor. Biol. **298**, 131–137 (2012)
20. WHO, Influenza (seasonal). Fact sheet No. 211, March 2014. http://www.who.int/mediacentre/factsheets/fs211/en/
21. Yandex.Raspisaniya. http://rasp.yandex.ru

Two Views on the Protein Folding Puzzle

**Alexei V. Finkelstein, Oxana V. Galzitskaya, Sergiy O. Garbuzynskiy,
Azat J. Badretdin, Dmitry N. Ivankov, and Natalya S. Bogatyreva**

1 Introduction

1.1 Overview of the Protein Folding Problem

The ability of proteins to fold spontaneously puzzled protein science for a long time.
It is well known that a protein chain (actually, the chain of a globular protein) can
spontaneously fold into its unique native 3D structure [1, 2]. In doing so, the protein
chain has to find its native (and seemingly the most stable) fold among zillions of
others within only minutes or seconds given for its folding.

Indeed, the number of alternatives is vast [3, 4]: it is at least 2^{100} but rather
may be 3^{100} or 10^{100} (or even 100^{100}) for a 100-residue chain, because at least 2
("right" and "wrong"), but more likely 3 (α, β, "coil") or 10 [5] (or even (10_for_φ)
$\times$ (10_for_Ψ) = 100 [3, 4]) conformations are possible for each residue (Fig. 1).

Since the chain cannot pass from one conformation to another faster than within
a picosecond (the time of a thermal vibration), the exhaustive search would take at

A. V. Finkelstein (✉) · O. V. Galzitskaya · S. O. Garbuzynskiy
Institute of Protein Research, Russian Academy of Sciences, Pushchino, Moscow Region,
Russian Federation
e-mail: afinkel@vega.protres.ru

A. J. Badretdin
National Center for Biotechnology Information, National Library of Medicine, National Institutes
of Health, Bethesda, MD, USA

D. N. Ivankov · N. S. Bogatyreva
Institute of Protein Research, Russian Academy of Sciences, Pushchino, Moscow Region,
Russian Federation

Bioinformatics and Genomics Programme, Centre for Genomic Regulation (CRG),
The Barcelona Institute of Science and Technology, Barcelona, Spain

Universitat Pompeu Fabra (UPF), Barcelona, Spain

© Springer International Publishing AG, part of Springer Nature 2018   391
R. P. Mondaini (ed.), *Trends in Biomathematics: Modeling, Optimization and
Computational Problems*, https://doi.org/10.1007/978-3-319-91092-5_27

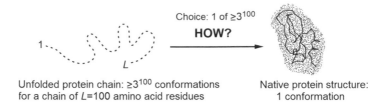

Fig. 1 The Levinthal's choice problem

least $\sim 2^{100}$ ps (or 3^{100} or 10^{100} or even 100^{100} ps), that is, $\sim 10^{10}$ (or 10^{25} or even 10^{80} or 10^{180}) years. And it looks like the sampling has to be really exhaustive, because the protein can "feel" that it has come to the stable structure only when it hits it precisely, while even a 1 Å deviation can strongly increase the chain energy in the closely packed globule.

Then, how does the protein choose its native structure among zillions of possible others, asked Levinthal [3, 4] (who first noticed this paradox), and answered: It seems that the protein folding follows some specific pathway, and the native fold is simply the end of this pathway, no matter if it is the most stable chain fold or not. In other words, Levithal suggested that the native protein structure is determined by kinetics rather than stability and corresponds to the easily accessible local free-energy minimum rather than the global one.

However, computer experiments with lattice models of protein chains strongly suggest that the chains fold to their stable structure, i.e., that the "native protein structure" is the lowest-energy one, and protein folding is under thermodynamic rather than kinetic control [6, 7].

Nevertheless, most of the suggested hypotheses on protein folding are based on the "kinetic control assumption."

Ahead of Levinthal, Phillips [8] proposed that the protein folding nucleus is formed near the N-end of the nascent protein chain, and the remaining chain wraps around it. However, successful in vitro folding of many single-domain proteins and protein domains does not begin from the N-end [9, 10].

Wetlaufer [11] hypothesized formation of the folding nucleus by adjacent residues of the protein chain. However, in vitro experiments show that this is far not always so [12].

Ptitsyn [13] proposed a model of hierarchical folding, i.e., a stepwise involvement of different interactions and formation of different folding intermediate states. This hypothesis has some important advantages and drawbacks; below, it will be considered in detail, together with some interesting implications [14, 15] that follow from the Ptitsyn's model.

Alongside with approaches based on various hypotheses of "kinetic choice" of the protein native structure, some models and theories are based on the idea of the "stability choice" of this structure.

In particular, various "folding funnel" models [16–19] have become popular for illustrating and describing fast folding processes. These models, which have their own important advantages and drawbacks, will be considered below in detail as well.

On the top of that, the free-energy barrier separating the folded (native) and unfolded states of protein chains has been investigated, and the estimated rate of overcoming of this barrier [20, 21] turned out to be in a good concordance with experimental results (see below).

The difficulty of the "kinetics vs. stability" problem, underlined by Levinthal, is that it hardly can be solved by direct experiment. Indeed, suppose that a protein has some structure that is more stable than the native one. How can we find it if the protein does not do so itself? Shall we wait for $\sim 10^{10}$ (or even $\sim 10^{180}$) years?

On the other hand, the question as to whether the protein structure is controlled by kinetics or stability arises again and again when one has to solve practical problems of protein physics and engineering. For example, in predicting a protein's structure from its sequence, what should we look for? The most stable or the most rapidly folding structure? In designing a protein de novo, should we maximize stability of the desired fold, or create a rapid pathway to this fold?

However, is there a real contradiction between "the most stable" and the "rapidly folding" structure? Maybe, the stable structure automatically forms a focus for the "rapid" folding pathways, and therefore it is automatically capable of fast folding?

1.2   Overview of the Basic Thermodynamic Facts Related to Protein Folding

Before considering the kinetic aspects of protein folding, let us recall some basic experimental facts concerning protein thermodynamics (as usual, we will consider single-domain proteins only, i.e., chains of ~ 100 residues). These facts will help us to understand what chains and what folding conditions we have to consider. The facts are as follows:

1. The denatured state of proteins, at least that of small proteins treated with a strong denaturant, is often the unfolded random coil [22].
2. Protein unfolding is reversible [2]; moreover, the denatured and native states of a protein can be in a kinetic equilibrium [23]; and there is an "all-or-none" transition between them [5]. The latter means that only two states of the protein molecule, native and denatured, are present (close to the midpoint of the folding–unfolding equilibrium) in a visible quantity, while all others, "semi-native" or misfolded, are virtually absent.

   (Notes: (1) the "all-or-none" transition makes the protein function reliable: like a light bulb, the protein either works or not; (2) the physical theory shows that such a transition requires the amino acid sequence that provides a large "energy gap" between the most stable structure and the bulk of misfolded ones [6, 24–27].)
3. Even under normal physiological conditions, the native (i.e., the lowest-energy) state of a protein is only more stable than its unfolded (i.e., the highest-entropy) state by a few kilocalories per mole [5] (and these two states have equal stability at mid-transition, naturally).

(Notes: For the below theoretical analysis, it is essential that (1) as is customary in the literature on this subject, the term "entropy" as applied to protein folding means only conformational entropy of the chain without solvent entropy; (2) accordingly, the term "energy" actually implies "free energy of interactions" (often called the "mean force potential"), so that hydrophobic and other solvent-mediated forces, with all their solvent entropy [22], come within "energy." This terminology is commonly used to concentrate on the main problem of sampling the protein chain conformations.)

The abovementioned "all-or-none" transition means that the native (N) and denatured (U) states are separated by a high free-energy barrier. It is the height of this barrier that limits the kinetics of this transition, and just this height is to be estimated to solve the Levinthal's paradox.

1.3 Is the Levinthal's Paradox a Paradox Indeed?

However, to begin with, it is not out of place considering whether the "Levinthal's paradox" is a paradox indeed. Bryngelson and Wolynes [28] mentioned that this "paradox" is based on the absolutely flat (and therefore unrealistic) "golf course" model of the protein potential energy surface (Fig. 2a), and somewhat later Leopold et al. [16], following the line of Gō and Abe [29], considered more realistic (tilted and biased to the protein's native structure) energy surfaces and introduced the "folding funnels" (Fig. 2b), which seemingly eliminate the "paradox" at all.

Its not as simple as that, though...

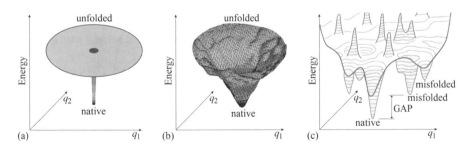

Fig. 2 Schematic illustration of basic models of the energy landscape of protein chains. (**a**) The "golf course" (Levinthal's) model of the protein potential energy landscape. (**b**) The "funnel" model of the protein potential energy landscape. The funnel is centered in the lowest-energy ("native") structure. (**c**) In more detail: the bumpy potential energy landscape of a protein chain. A wide (of many $k_B T_{melt}$, where k_B is Boltzmann's constant and T_{melt} is protein melting temperature) energy gap between the global and other energy minima is necessary to provide the "all-or-none" type of decay of the stable protein structure. Only two coordinates (q_1 and q_2) can be shown in the figures, while the protein chain conformation is determined by hundreds of coordinates

The problem of huge sampling does exist even for realistic energy surfaces. It has been mathematically proven that, despite the folding funnels and all that, finding the lowest free-energy conformation of a protein chain is the so-called NP-hard problem [30, 31], which, loosely speaking, requires an exponentially large time to be solved (by a folding chain or by a man).

Anyhow, various "folding funnel" models became popular for explaining and illustrating protein folding [17, 32, 33]. In the funnel, the lowest-energy structure (formed, thus, by a set of most powerful interactions) is the center surrounded by higher-energy structures containing only a part of these interactions. The "energy funnels" are not perfectly smooth due to some "frustrations," i.e., contradictions between optimal interactions for different links of a heteropolymer forming the protein globule, but a stable protein structure is distinguished by minimal frustrations (that is, most of its elements have enhanced stability) [28, 34–36]. Anyhow, the "energy funnel" directs movement toward the lowest-energy structure, which seems to help the protein chains to avoid the "Levinthal's" sampling of all conformations.

However, it can be shown that the energy funnels per se do not solve the Levinthal's paradox. Strict analysis [37] of the straightforwardly presented funnel models [19, 38] shows that close to the midpoint of the folding–unfolding equilibrium they cannot *simultaneously* explain both the major features observed in protein folding: (1) its nonastronomical time, and (2) the "all-or-none" transition, i.e., coexistence of native and unfolded protein molecules during the folding process.

By the way, the stepwise mechanism of protein folding [13], taken per se, also cannot [39] *simultaneously* explain these two major features observed in protein folding. Rather, it states that the folding must be fast *if* each subsequent folding intermediate is much more stable than the preceding one (and thus, if the native fold is much more stable than the unfolded state of the chain).

Thus, neither stepwise nor simple funnel mechanisms solve the Levinthal's problem, although they give a hint as to what accelerates protein folding.

The basic solution of the paradox is provided by very special nucleation funnels [20, 21]: those, considering the separation of the unfolded and native phases within the folding chain (now called the "capillarity theory" [40]).

It will be described in the next part of this review.

2 Physical Estimate of the Height of Free-Energy Barrier Between the Folded and Unfolded States: View at the Barrier from the Side of the Folded State

To solve the "Levinthal's paradox" and to show that the most stable chain fold can be found within a reasonable time, we could, to a first approximation, consider only the rate of the "all-or-none" transition between the coil and the most stable structure. And we may consider this transition only for the crucial case when the most stable fold is as stable as (or only a little more stable than) the coil, with all other states of the chain being unstable, i.e., close to the "all-or-none" transition midpoint. Here, the analysis can be made in the simplest form, without accounting for accumulating intermediates. True, the maximum folding rate is achieved when the native fold is considerably more stable than the coil [23, 41], and then observable intermediates often arise; but let us consider not the fastest but the simplest case... (We have to note that this special attention to the mid-transition conditions differs our approach [20, 21, 42] from those that prevailed from 1960s to the middle of 1990s.)

Since the "all-or-none" transition requires a large energy gap between the most stable structure and misfolded ones [6, 24–27] (Fig. 2c), we will assume that the considered amino acid sequence provides such a gap. Our aim is to estimate the rate of the "all-or-none" transition and to prove (if possible) that the most stable structure of a normal size domain (~100 residues) can fold within minutes or seconds or even faster.

To prove that the most stable chain structure is capable of rapid folding, it is sufficient to prove that *at least one* rapid folding pathway (i.e., passing the low-free-energy barrier) leads to this structure. Additional pathways can only accelerate the folding since the rates of parallel reactions are additive. And we can avoid considering folding of other, non-native structures. They have high energy because of the "energy gap," and, near the point of the "all-or-none" transition between the most stable globule and the unfolded chain, they are unstable even taken together, and therefore, they cannot serve as "folding traps" that absorb folding chains. (One can imagine water leaking from a full pool to an empty one through cracks in the wall between them: when the cracks cannot absorb all the water, each additional crack accelerates filling of the empty pool.)

To be rapid, the pathway must consist of not too many steps, and most importantly, it must not require overcoming of a too high-free-energy barrier.

An L-residue chain can, in principle, attain its lowest-energy fold in L steps, each adding one fixed residue to the growing structure (Fig. 3). *If* the free energy went downhill along the entire pathway, a 100-residue chain would fold in ~100–1000 ns, since the growth of a structure (e.g., an α-helix) by one residue is known to take a few nanoseconds [43].

Protein folding takes minutes or seconds or even milliseconds rather than a fraction of a microsecond because of the free-energy barrier: most of the folding time is spent on climbing up this barrier and falling back, rather than on moving along the folding pathway.

The key role in this process is played by the transition state, i.e., the least stable ("barrier") state on the reaction pathway. According to the conventional transition state theory [44–46], the time of the multistep process of overcoming the barrier is estimated as

$$\text{TIME} \sim \tau \times \exp(+\Delta F^{\#}/RT), \qquad (1)$$

where τ is the time of one elementary step, and $\Delta F^{\#}$ is the height of the free-energy barrier.

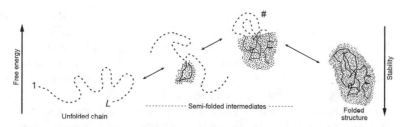

Fig. 3 A scheme [21] of a sequential folding pathway of some globular structure (if passed in the opposite direction, it is the sequential unfolding pathway of this structure). At each step of sequential folding, one residue leaves the coil and takes its final position in the structure. The free energy of intermediates is elevated due to the interface of folded and unfolded phases. The # sign indicates the most unstable ("transition") state. The folded part (shaded) of semi-folded intermediates which constitute the optimal ("low-free-energy") pathway must be compact (having a small boundary between the folded and unfolded phases). The bold lines show the backbone fixed in the already folded part; fixed side chains are not shown for the sake of simplicity (the volume that they occupy is shaded). The broken line shows the yet unfolded chain

As for $\Delta F^{\#}$, this is our main question: how high is the free- energy barrier $F^{\#}$ on the pathway leading to the lowest-energy structure? Formation of this structure decreases both the chain entropy (because of an increase in the chain's ordering) and its energy (due to formation of contacts stabilizing the lowest-energy fold). The former increases and the latter decreases free energy of the chain.

If fold-stabilizing contacts start to arise only when the chain comes very close to its final structure (i.e., if the chain has to lose almost all its entropy *before* the energy starts to decrease), the initial free-energy increase would form a very high-free-energy barrier (proportional to the *total* chain entropy loss). The Levinthal's paradox claiming that the lowest-energy fold cannot be found within any reasonable time since this involves exhaustive sampling of all chain conformations originates exactly from this picture (loss of the entire entropy *before* the energy gain).

However, this paradox can be avoided if there is a folding pathway where the entropy decrease is immediately or nearly immediately compensated for by the energy decrease [29].

Let us consider a *sequential wetlaufer1* folding pathway (Fig. 3). More specifically, we will consider a process at each step of which one residue leaves the coil and takes its final position in the lowest-energy 3D structure. True, this pathway may look a bit artificial, but actually the outlined pathway is exactly the pathway that one expects to see watching the movie on unfolding, but in the opposite direction.

According to the well-known in physics *detailed balance* law [47], the direct and reverse reactions follow the same pathway and have equal rates when both the end states have equal stability. (This law follows from the second law of thermodynamics. It is proved by contradiction: if, in thermodynamic-equilibrium ambient conditions, the pathway A → 1 → B is faster than A → 2 → B for the A → B reaction, while the pathway A ← 2 ← B is faster than A ← 1 ← B for the reciprocal A ← B reaction under the same conditions, one obtains a *permanent* flow $A \underset{2}{\overset{1}{\rightleftharpoons}} B$, which contradicts to the second law of thermodynamics.)

Thus, one can use the detailed balance law to find the transition state for folding by finding the optimal transition state for *un*folding! An advantage of analysis of the unfolding pathway is that it is much easier: for any final globular structure, one can easily figure out its sequential unfolding passing through the least unstable semi-unfolded states, i.e., those where the compact globular phase is separated from the unfolded one (Fig. 3) by minimal interfaces [20, 21, 48, 49].

(In this connection, it is not out of place mentioning that, odd enough, protein unfolding, in contrast to folding, has been never treated as a "puzzle," although it is well known for a long time that these two states, unfolded and folded, can be in kinetic equilibrium! Despite all that, nobody asked a question complementary to Levithal's one, that is, how the protein gains a huge energy required for unfolding...This shows that it is easier to imagine how to unfold any protein structure than how to fold it.)

Thus, let us consider the energy change ΔE, the entropy change ΔS, and the resultant free-energy change $\Delta F = \Delta E - T \Delta S$ along the *sequential* (Fig. 3) folding pathway (reconstructed from the way of sequential *un*folding).

When a piece of the final globule grows sequentially, the interactions that stabilize its final fold are restored sequentially as well. If the folded piece remains compact, as in Figs. 3 and 4a, the number of restored interactions grows (and their total energy decreases, see Fig. 4c) approximately in proportion to the number n of residues that have taken their final positions.

Approximately in proportion—but with one significant deviation: At the beginning of folding, the energy decrease is a little slower, since the contact of a newly joined residue with the surface of a small globule is, on average, smaller than its contact with the surface of a large globule. This results in a nonlinear *surface* term (the surface being proportional to $\approx n^{2/3}$) in the energy ΔE of the growing globule.

Thus, the maximal deviation from the linear energy decrease is proportional to $L^{2/3}$, while the total energy decrease is proportional to the total number L of residues. The deviation is still greater, see dotted line in Fig. 4c, if the folded parts do not form a compact piece, as in Fig. 4b.

The entropy decrease is also *approximately* proportional to the number n of residues that have taken their final positions (Fig. 4d).

At the beginning of folding, though, the entropy decrease can be a little faster owing to disordered but closed loops protruding from the growing globule (Figs. 3 and 5). The maximal number of such loops is proportional to the interface between the folded and unfolded phases, and the free energy of a loop is known [50, 51] to have a very slow, logarithmic dependence on its length. This again results in a nonlinear *surface* term in the entropy ΔS of the growing globule. The overall entropy decrease is proportional to L again, and the maximal deviation from the linear entropy decrease again is proportional to $L^{2/3}$ (actually, it is proportional to $\sim L^{2/3} \times \ln(L^{1/3})$ at the most, but the multiplier $\ln(L^{1/3})$ is insignificant, about 1–2 when L is 10–1000) [20]; see also the later rigorous mathematical papers [52, 53].

Both linear and surface constituents of ΔS and ΔE enter the free energy $\Delta F = \Delta E - T \Delta S$ of the growing (or unfolding) globule. However, when the final globule is in thermodynamic equilibrium with the coil, the large linear terms *annihilate* each

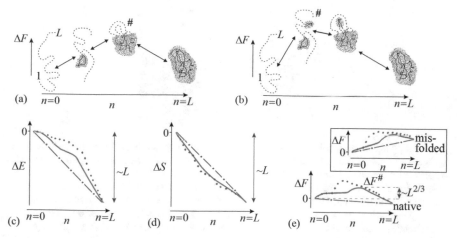

Fig. 4 Schematic illustration of sequential folding/unfolding with compact (**a**) and non-compact (**b**) semi-folded intermediates and the change of energy (**c**), entropy (**d**), and free energy (**e**) along these sequential folding/unfolding pathways close to the point of thermodynamic equilibrium between the coil ($n = 0$) and the final structure ($n = L$: all the L chain residues are folded). The full energy and entropy changes, $\Delta E(L)$ and $\Delta S(L)$, are approximately proportional to L. The bar-dotted lines show the linear (proportional to the number of already folded residues n) parts of $\Delta E(n)$ and $\Delta S(n)$. The nonlinear parts of $\Delta E(n)$ and $\Delta S(n)$ result mainly from the surface of the folded part of the molecule (solid lines: for a pathway with compact intermediate structures; dotted lines: for that with non-compact intermediates). The maximal deviations of the $\Delta E(n)$ and $\Delta S(n)$ values from linear dependences are proportional to only $L^{2/3}$. As a result, $\Delta F(n) = \Delta E(n) - T\Delta S(n)$ also deviates from the linear dependence (bar-dotted line) by a value of only $\sim L^{2/3}$ for compact intermediate structures (while for non-compact intermediates, the deviations are greater). Thus, at the equilibrium point (where $\Delta F(0) = \Delta F(L)$), the maximal on this pathway free-energy excess $\Delta F^{\#}$ ("the barrier") over the bar-dotted free-energy baseline is also proportional to only $L^{2/3}$ for compact intermediate structures. The change $\Delta F(n)$ on the pathway to other structures looks similar (see inset in panel (**e**)), but these pathways can be neglected, because all these structures are unstable with $\Delta F(n = 0) < \Delta F(L)$ in the presence of the energy gap and the "all-or-none" transition between the unfolded and the most stable globular state of the chain. Adapted from [20, 21]

other in the difference $\Delta E - T\Delta S$ (since $\Delta F = 0$ both in the coil (i.e., at $n = 0$) and in the final globule (at $n = L$)), and only the surface terms remain: $\Delta F(n)$ would be *zero* all along the pathway in the absence of surface terms.

Thus, the free-energy barrier (Figs. 4e and 6) on a sequential folding pathway with compact semi-folded structures depends only on relatively small globule surface effects, and its height is proportional *not to L* (as Levinthal's estimate implies), but to $L^{2/3}$ only.

In the most simplified form (for details, see [20, 21, 42, 49]), free energy of the barrier is estimated as follows.

The fastest folding pathway is that having the lowest free-energy barrier; the barrier, on a given pathway, corresponds to the intermediate with the highest free energy, that is, the maximal for this pathway interface between the folded and unfolded phases; this interface contains about $L^{2/3}$ residues.

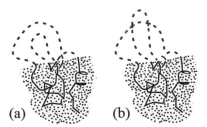

Fig. 5 A compact semi-folded intermediate with protruding unfolded loops. Its growth corresponds to a shift of the boundary between the folded (globular) and unfolded parts. Successful folding requires correct knotting (**a**) of loops: the structure with incorrect knotting (**b**) cannot change directly to the correct final structure: first it has to unfold and achieve the correct knotting. However, since a chain of ∼100 residues can only form one or two knots [42], the search for correct knotting can only slow down the folding twofold or at most fourfold; thus, the search for correct chain knotting does not limit the folding rate of normal size protein chains. Adapted from [42]

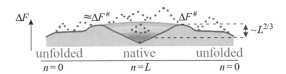

Fig. 6 This purely illustrative drawing shows how entropy converts the energy funnel (illustrated in Fig. 2b) into a "volcano-shaped" (as it is called now [54]) *free-energy* folding landscape with free-energy barriers (Fig. 4e) on each pathway leading from an unfolded conformation to the native fold. Any pathway from the unfolded state to the native one first goes uphill, and only then, from the barrier (i.e., crater edge), descends into the "free-energy funnel." The smooth free-energy landscape corresponds to compact semi-folded intermediate structures (shown in Fig. 4a), the rocks (denoted by dotted lines) present a landscape including non-compact semi-folded intermediate structures (shown in Fig. 4b). More accurate but less beautiful scheme of a free-energy landscape is shown in Fig. 2 in [48]

The energy constituent $\Delta E^{\#}$ of the barrier free energy results from interactions lost by the interface residues; it is about

$$L^{2/3} \cdot \frac{1}{4}\varepsilon,\tag{2a}$$

where $\varepsilon \approx 1.3\,\text{kcal/mol} \approx 2k_B T_{mel}$ is the average heat of protein melting per residue [5] (this is the first empirical parameter used by the theory), and $\approx \frac{1}{4}$ is the fraction of interactions lost by an interface residue. Thus,

$$\frac{\Delta E^{\#}}{k_B T_{melt}} \approx 0.5\, L^{2/3}.\tag{2b}$$

The entropy constituent $\Delta S^{\#}$ of the barrier free energy is caused by entropy loss in closed loops protruding from the globular into the unfolded phase (see Fig. 5).

The upper limit of $\Delta S^{\#}$ is zero (when the interface contains no such loops). The lower limit of $\Delta S^{\#}$ is about

$$(\Delta S^{\#})_{\text{lower}} = \frac{1}{6}L^{2/3}\left[-\frac{5}{2}k_B \ln(3\,L^{1/3})\right], \qquad (3a)$$

where $\frac{1}{6}L^{2/3}$ corresponds to the maximal number of closed loops protruding from the optimal (minimally covered by loops) globule/coil interface (actually, this is the average number for one globule cross section (Fig. 5), since the interface residue can have 6 directions—4 along the surface, 1 inside, and only 1 outside; and the folding-involved interface must be covered by a minimal, never exceeding the average, number of loops). $3L^{1/3} \equiv \left(\frac{L}{2}\right)/\left(\frac{1}{6}L^{2/3}\right)$ is the average number of residues in such a loop (equal to the number of unfolded residues divided by the number of loops), and $-\frac{5}{2}k_B \ln(3L^{1/3})$ is entropy lost by such a closed loop (the interior parts of which do not penetrate inside the globule; this changes the conventional Flory's coefficient, 3/2 to 5/2 [20, 21]). Having $L \sim 100$ (actually, this approximation is good for the whole range of $L = 10$–1000), we obtain

$$(\Delta S^{\#})_{\text{lower}} \approx -k_B L^{2/3} \qquad (3b)$$

As a result, the time of both folding and unfolding of the most stable chain structure grows with the number of chain residues L *not* "according to Levinthal" (i.e., *not* as 2^L, or 10^L, or any exponent of L), but, in mid-transition conditions, as

$$\text{TIME} \sim \tau \times \exp\left[(1 \pm 0.5)L^{2/3}\right] \qquad (4)$$

where $\tau \approx 10\,\text{ns}$ [43] (this is the second and the last empirical parameter used in the theory).

The folding time depends on the size and the shape (see above) of the folding protein's native structure.

The physical reason for this "non-Levinthal" estimate is that (1) during folding, the entropy decrease is almost immediately and almost completely compensated for by an energy decrease along the sequential folding pathway (and, likewise, the energy increase is almost immediately and almost completely compensated for by an entropy increase along the same sequential *un*folding pathway), and (2) the free energy results only from surface effects which are relatively weak.

The observed protein folding times span (Fig. 7) 11 orders of magnitude (which is akin to the difference between the life span of a mosquito and the age of the universe). The range of folding times at mid-transition (where $\Delta F = 0$) is from $10\,\text{ns} \times \exp(0.5L^{2/3})$ to $10\,\text{ns} \times \exp(1.5L^{2/3})$, in accordance with the estimate obtained. Under more physiological conditions ("in water", where $\Delta F < 0$), $L^{2/3}$ is replaced by $L^{2/3} + 0.4\Delta F/RT$ (see Sect. 4), but in all other respects the range remains the same.

It is noteworthy that the outlined sequential folding pathways do not require any rearrangement of the dense globular part (which could take a lot of time): all rearrangements occur in the coil.

Anyhow, the obtained Eq. (4) illustrated in Fig. 7 shows that a chain of $L \lesssim$ 80–90 residues will find its most stable fold within minutes (or faster) even under "nonbiological" mid-transition conditions, where folding is known [23, 41] to be the slowest. Native structures of such relatively small proteins are under thermodynamic control: they are the most stable among all structures of these chains.

Native structures of larger proteins (of $\approx$90–400 residues) are, in addition, under a "structural control," in a sense that too entangled folds of their long chains cannot be achieved within days or weeks even if they are thermodynamically stable; and indeed, greatly entangled folds of long protein chains have been never observed [49]: they seem to be excluded from the repertoire of existing protein structures. This also explains why larger proteins should be far from spherical or consist (according to the "divide and rule" principle) of separately folding domains: otherwise, chains of more than 400 residues would fold too slowly. This is a "structural control" again. Its effect, in some sense, resembles that of Levinthal's "kinetic control," though at another level and only for large proteins. The above estimates ($\approx$80– 90 and $\approx$400 residues) are somewhat elevated when the native fold free energy ΔF is lower than that of the unfolded chain (see below), but essentially they remain the same [49].

One thing is left to be said:

The "quasi-Levinthal" search over intermediates with different chain knotting (Fig. 5) can, in principle, be a "Levinthal-like" rate-limiting factor, since knotting cannot be changed without a decay of the globular part. However, since the computer experiments show that one knot involves about a hundred residues, the search for correct knotting can only be rate-limiting for extremely long chains (see [42] and references therein) which cannot fold within a reasonable time (according to Eq. (4)) in any case.

3   Estimating Dependence of the Sampling Volume on Protein Size: View at the Barrier from the Side of Unfolded State

The above given estimate of the folding time is based on consideration of protein _unfolding_ rather than _folding_. We have considered _unfolding_ because it is easier to outline a good _un_folding pathway (and time, see above) of any structure than a good folding pathway leading to the lowest-energy fold, while the free-energy barrier at both pathways is the same.

In other words, we considered the free-energy barrier between the unfolded and folded states (Figs. 5 and 6) with the focus on its _un_folding side (connected with energy increase on the pathway from the volcano throat to the crater edge) and did not consider its folding side (connected with entropy loss on the pathway from the unfolded state to the crater edge). Since the rates of direct and reverse reactions

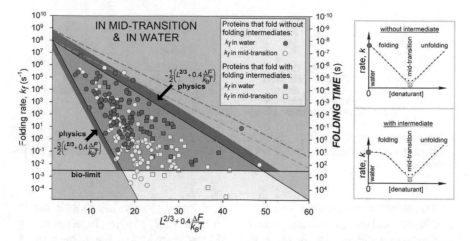

Fig. 7 Main panel: experimentally measured in vitro folding rate constants in water (under approximately "biological" conditions) and at mid-transition for 107 single-domain proteins (or separate domains) without SS bonds and covalently bound ligands (though the rates for proteins with and without SS bonds are principally the same [55]). Triangle: the region allowed by physics; its gray part (with the dark belt) corresponds to biologically reasonable folding times (≤10 min); the larger folding times (i.e., the smaller folding rates) are observed (for some proteins) only under mid-transition, i.e., nonbiological conditions. The light-gray dashed line limits the area allowed only for oblate (1:2) and oblong (2:1) globules at mid-transition; the dark-gray dashed line means the same for "biologically normal" conditions. L is the number of amino acid residues in the protein chain under study. ΔF is the free-energy difference between the native and unfolded states of the chain. Adapted from [49]. Supplementary panels: Typical forms of "chevron plots" for the folding/unfolding kinetics of proteins that fold without and with folding intermediates (after [41])

are equal under mid-transition conditions (as follows from the physical "detailed balance" principle), here the "*un*folding" and "folding" sides of the barrier are of equal heights, and therefore, examination of only one ("*un*folding") side is sufficient to estimate the barrier height.

However, a complete analysis of folding urges us to look at the barrier from its folding (connected with entropy loss) side, which is most interesting for the biological audience, and obtain the second view on the protein folding puzzle.

To analyze folding, we have to analyze sampling of conformations of the protein chain.

The total volume of the protein conformation space estimated at the level of amino acid residues by Levinthal [4] is huge indeed: as many as from 3^{100} to 100^{100} conformations for a 100-residue chain.

However, should the chain sample all these conformations in search for its most stable fold? No: the conformation space is covered by local energy minima, each surrounded by a local energy funnel (Fig. 2b) providing fast downhill decent to this local minimum.

Actually, the folding protein chain has to sample not all its possible conformations, but only various ways of packing the chain in the compact protein globule.

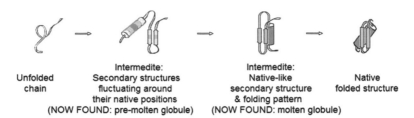

Intermedite:
Unfolded   Secondary structures   Intermedite:
chain   fluctuating around   Native-like   Native
                      their native positions   secondary structure   folded structure
                                                         & folding pattern
                      (NOW FOUND: pre-molten globule)   (NOW FOUND: molten globule)

Fig. 8 A scheme of Ptitsyn's [13] hypothetical mechanism of stepwise protein folding. Cylinders: α-helices; arrows: β-strands. Both predicted in 1973 intermediates have been observed in 1980s–1990s [62]

Therefore, to estimate the actual volume of sampling, one has to estimate the number of local energy minima (and also the time taken by jumping from one energy minimum to another). In some sense, this is similar to the idea to enumerate possible "topomers" that a protein chain can form [56, 57], but our aim is not to calculate the protein folding rate, but to estimate its lower limit only (which is very different from the somewhat contradictive [58] theory of the native-like topomer search by simulation).

An overview of protein structures shows that interactions occurring in the chains are mainly connected with secondary structures [13, 59–61]. Thus, a question arises as to how large the total number of energy minima is, if considered at the level of formation and assembly of secondary structures into a globule, that is, at the level considered by Ptitsyn [13] in his model (Fig. 8) of stepwise protein folding.

It turns out that the number of conformations at the level of secondary structures is by many orders of magnitude smaller than that of conformations of amino acid residues of the chain [14]: the latter, according to Levinthal's estimate, scales up as something like 100^L or 10^L or 3^L with the number L of residues in the chain, while the former scales up not faster than $\sim L^N$ with the number of residues L and the number N of the secondary structure elements. N is much less than L, and this is the main reason for the drastic decrease of the conformation space.

The estimate L^N was obtained as follows (see Fig. 9).

The number of architectures (i.e., types of dense stacks of secondary structures) is small (cf. [59, 60, 63]), usually ~ 10 or less for a given set of secondary structures (Fig. 9a), since the architectures are packings of a few secondary structure layers (each containing several secondary structures), and therefore combinatorics of the layers is very small as compared to combinatorics of much more numerous secondary structure elements (see Fig. 9b–e).

The maximal number of packings, i.e., all combinations of positions of N elements in the given protein architecture, cannot exceed $N!$ (Fig. 9b).

The maximal number of topologies, i.e., all combinations of directions of these elements cannot exceed 2^N (Fig. 9c).

Transverse shifts and tilts of an element within each dense packing are prohibited (Fig. 9d).

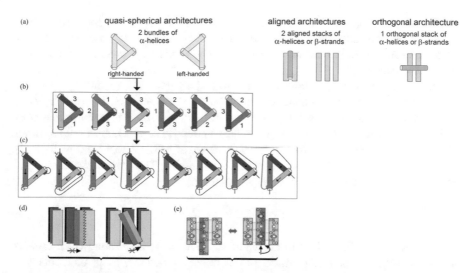

Fig. 9 A scheme of estimate of the conformation space volume at the level of secondary structure assembly. Adapted from Supplement to [14]. (**a**) Architectures of $N = 3$ structural elements. (**b**) Packings: $N! = N \times (N-1) \times \ldots \times 2 \times 1$. (**c**) Topologies: 2^N. (**d**) Transverse shifts and tilts: prohibited. (**e**) Coupled shift and rotation

Longitudinal shifts and turns about the axes of secondary structure elements within a dense packing are coupled (this is shown in Fig. 9e using a β-sheet as the best illustrative example, but this is also true for α-helices—remember "knobs in the holes" close packings by Crick [64]); as a result, each α or β element can have about L/N (that is, about the mean chain length per element) possible shift/turns in the globule formed by N secondary structures of the L-residue chain.

All this limits the number of energy minima in the conformational space to $\sim 10 \times (L/N)^N \times 2^N \times N!$ conformations; this (using Stirling's approximation $N! \sim (N/e)^N$) gives

$$\text{NUMBER of energy minima to be sampled} \sim L^N \qquad (5)$$

in the main term (if $L \gg N \gg 1$) [14].

This number can be somewhat reduced by symmetry of the globule; also, no α-helix can take the place of a β-strand without rearrangement of other elements, and vice versa, because the β-strand needs a partner to form hydrogen bonds, while the α-helix avoids such a partnership. Further, short or crossing loops between secondary structures can prevent these from taking arbitrary positions and directions in the globule, etc. [65]. However, this reduction is not important to us, because our aim now is to estimate the upper limit of the number of conformations.

Here, a question may arise as to how the chain knows where to form a secondary structure and what secondary structure is to be formed there. The answer seems to be as follows. Most of secondary structures are determined by local amino acid sequences [13, 66]. Anyhow, the choice of "to be or not to be" for a secondary structure element adds only 1 state to the number L/N of its possible shift/turn states (already taken into account), and conversion only duplicates it, which is not significant (see [67]).

In a compact globule of not too small size, the length of a secondary structure element should be proportional to the globule's diameter, i.e., to $\sim L^{1/3}$. More specifically, the globule's volume is about $150\,\text{Å}^3 \times L$ (and thus its diameter is $\approx 5\,\text{Å} \times L^{1/3}$), while the shift per residue is about $1.5\,\text{Å}$ in a helix and $3\,\text{Å}$ in an extended strand [61]. Therefore, a helix consists of $\approx 3L^{1/3}$ residues, while a β-strand, as well as a loop, comprises $\approx 1.5L^{1/3}$ residues. Thus, the mean number of residues in "secondary structure + loop" element is

$$L/N \approx 4.5L^{1/3} - 3L^{1/3}, \tag{6}$$

(which, at $L \sim 100$, is close to the value of $L/N = 15 \pm 5$ found from protein statistics [54]), and the mean number of "secondary structure + loop" elements is

$$N \approx \frac{L^{2/3}}{4.5} - \frac{L^{2/3}}{3}. \tag{7}$$

Thus, the value L^N (the sampling volume) is within the range

$$\sim L^{\frac{L^{2/3}}{4.5}} \equiv \exp\left(\left[\frac{\ln(L)}{4.5}\right] \times L^{2/3}\right) -\!\!\!-\sim L^{\frac{L^{2/3}}{3}} \equiv \exp\left(\left[\frac{\ln(L)}{3}\right] \times L^{2/3}\right) \tag{8}$$

Analogous scaling was obtained [52, 53] from mathematical consideration of complexity of the choice problem. Also, one can see that, since $\ln(L)/4.5 \approx 1$ and $\ln(L)/3 \approx 1.5$ for $L \sim 100$, the estimate given by Eq. (8) is, eventually, more or less close to the upper limit outlined by Eq. (4).

Taking, from experiments on folding of the smallest proteins [68, 69], a few microseconds as a rough estimate of the time necessary to sample one conformation and the value $L/N = 15\pm5$ from protein statistics, we see that the time theoretically needed to sample the whole conformation space at the level of secondary structure formation and assembly closely approaches (Fig. 10) the upper limit of experimental folding times observed for small ($L \lesssim 80$–90) residue proteins. It is also close to the upper limit of the folding time estimate given by Eq. (4), earlier obtained from consideration of unfolding and illustrated in Fig. 7; note that folding of these small proteins is, as we have concluded, under complete thermodynamic control.

The above consideration does not mean, of course, that a folding protein samples the *entire* conformation space at the level of secondary structure formation and packings (though a chain of 80–90 residues or less can do this within minutes (or faster), as Fig. 10 shows for the most slowly folding proteins of such size). It means only that the native fold-leading "energy funnel," working at the level of secondary structures, has to accelerate (for some, rapidly folding proteins) the folding process by several orders of magnitude (as Fig. 10 shows for the majority of proteins), rather

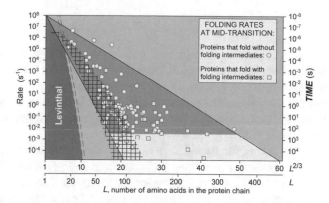

Fig. 10 Sampling rate and folding rate. Folding rates (circles and squares) are shown for proteins experimentally studied at mid-transition (i.e., at equal stability of their folded and unfolded states); the dark- gray/light-gray triangle shows the predicted (from consideration of *unfolding!*) range of these rates (cf. Fig. 7). The netted shading shows a theoretical estimate of the minimal rate of exhaustive sampling, at *folding*, of all possible packings of protein secondary elements (helices and strands). The maximal "Levinthal-like" sampling rate (10^{12} s^{-1}/3^L, allowing for 3 possible states: α, β, and coil) is shown by the double dashed line; the lines for "Levinthal-like" sampling rates with 10 or 100 possible states of a residue would have been much below (in the dark-gray zone). Adapted from [70]

than for all proteins and by many tens or hundreds of orders, which would have been the case *if* the funnel were to start working from the level of amino acid residues (cf. with the theory of searching for topomers [56, 57]). Figure 10 shows that the "funnel-due" acceleration is pronounced for chains of >100 residues, but even then the main work is done by secondary structures.

Bird's-eye view of the obtained estimates (4)–(8) of the number of chain conformations (or rather, of all kinds of chain packing in a compact globule), which have to be enumerated when searching for the most stable protein structure is as follows. This number scales, in the main term, in proportion to the globule's surface, i.e., to the number of surface residues or—nearly the same—to the number of the secondary structures N, which are both proportional to $L^{2/3}$. The physical reason is that in a dense globule all independent degrees of freedom are connected only with its surface, because the globule's density prohibits independent rearrangements of residues in its interior [24, 27], just like the secondary structure prohibits independent movements of residue backbones inside it. From this point of view, the used secondary structure elements are not necessary for estimating the scaling law (estimates by Fu and Wang [52] and Steinhofel et al. [53], as well as our estimates based on unfolding pathways [20, 21, 48, 49], did not use secondary structures), though these structures do form the protein core, and they are useful for refinement of the principal law.

4 Discussion and Conclusion

We have viewed the pathways through the "volcano-shaped" (illustrated in Fig. 5) folding landscape both from outside, i.e., from the "volcano" foot, and from inside, that is, from its crater. In this way, we investigated the free-energy barrier separating the folded and unfolded states of a protein chain from its both sides. We have passed it there and back again and obtained two views on the protein folding puzzle; these two views solve the Levinthal's paradox.

The barrier side facing the folded state is easier for investigation because it is easier to outline a reasonable *un*folding pathway from any given fold than a good folding pathway to a fold that is still unknown for the chain. The view from inside of the folding funnel gave us an estimate of the range of unfolding times, and then we used the detailed balance principle to find the folding time.

The view from outside of the folding funnel gave us only the upper limit of the folding (or rather, sampling) time.

It is worth mentioning that the unfolding-based estimate gives both the upper and lower estimates of the folding time, while the folding-based estimate gives its upper limit only.

The same scheme can be applicable to formation of the native protein structure not only from the coil (which we used in this study for simplicity) but also from the molten globule or from another intermediate. However, for these scenarios, all the estimates would be much more cumbersome due to more complicated nature of the denatured state of the protein, while these processes do not demonstrate (in experiment, see Fig. 7) any drastic advantage in the folding rate. Therefore, we now will not go beyond the simplest case of the coil-to-native globule transition.

It is not out of place mentioning that something similar to the Levinthal's problem must exist in crystallization (which resembles protein folding, because atoms of a few sorts have to acquire a particular conformation among plentiful others in "yet unknown" for them crystal; though, to our best knowledge, it did not attract there as much attention as in the protein science (cf. [71, 72]).

A few more things remain to be said:

1. Our estimate of the number of the secondary structure ensembles (i.e., the energy minima to be sampled) is independent (see Eqs. (5)–(8)) on stability of these ensembles. The influence of the native state stability (ΔF) on the folding *time* is considered below.

2. Our basic estimate of the folding *time*, Eq. (4), referred, up to now, to $\Delta F = 0$, i.e., to the point of equilibrium between the unfolded and native states—but here the observed folding time is at a maximum and can exceed by orders of magnitude the folding time under native conditions [41].

   How will the folding time change when the native state becomes somewhat more stable than the coil (that is, $\Delta F < 0$)? In accordance with experiment (see [41]), the theoretical analysis [21, 61, 73, 74] shows that as long as $-\Delta F$ is small, about a few $k_B T$, so that no stable intermediates arise, the folding time decreases with increasing stability, and, theoretically, it can be estimated [49] as

$$\text{TIME} \sim \tau \times \exp\left[(1 \pm 0.5) \times (L^{2/3} + 0.4 \times \Delta F/RT)\right] ; \qquad (9)$$

the multiplier 0.4 corresponds to the approximate theoretical estimate of the average fraction of a chain involved in the folding nucleus, so that $0.4 \times \Delta F$ is the approximate change of the nucleus free energy. (The overview of other details of folding nuclei is out of the scope of this paper; one can find them in [41, 48, 61, 73, 74]). Equation (9) gives a unified approximate estimate of folding rates occurring under various conditions (see Fig. 6).

For the case of a very high native fold stability ($-\Delta F \gg k_B T$), another but similar to Eq. (4) scaling law ($\ln(\text{TIME}) \sim L^{1/2}$) was obtained [75]. Then, protein folding is the fastest, because it essentially goes "downhill" in energy all the way; but the "downhill slope" has (due to protein heterogeneity) random bumps, whose energy is proportional to $L^{1/2}$. However, numerical experiments with lattice protein chains have shown [27, 76] that, at the temperature providing the fastest folding, the folding time grows with the chain length as $\ln(\text{TIME}) \sim A \times \ln(L)$, where the coefficient A equals to 6 for chains with "random" sequences and 4 for sequences selected to fold most rapidly (i.e., for chains having a large energy gap between the most stable fold and other ones). This emphasizes once again the dependence of the folding rate on experimental conditions and on the difference in stability between the lowest-energy fold and its competitors [26, 40].

3. Here, it is worth mentioning that some, quite rare proteins are "metamorphic" [77]: they are observed in two or more distinct folds. Of interest for us are those very few in number (e.g., serpin) that first obtain some "native," that is, working structure, work in the cell or a test tube for an hour or so, and then acquire another, non-working but more stable structure [78]. Significantly, this transition is not connected with a change in the protein's environment (aggregation, as in amyloids, or formation of some complexes). Thus, the chain of such a protein has two stable folds: one of them folds faster, the other is more stable. It seems, though, that such "metamorphic" (or "polymorphous") proteins are and must be very rare: theoretical estimates [61, 74] show that the amino acid sequence coding for one stable chain fold (i.e., whose energy is separated by a wide gap from energies of others) is a kind of wonder by itself, but the sequence coding for two stable folds is a squared wonder. . .

4. Equations (4), (9) estimate the range of possible folding rates rather than folding rates of an individual protein, which, even for proteins of the same size, may differ (Fig. 6) from one another by orders of magnitude. The influence of a particular protein chain fold shape upon the folding rate can be estimated using a phenomenological "contact order" parameter (CO%) [79]. CO% is equal to the average distance along the chain between residues that are in contact in the native protein fold, divided by the chain length (see also [33, 80]). A high CO% value reflects the presence of many long closed loops in the protein fold, while a high

value of (1 ± 0.5) factor in Eqs. (4), (9) reflects their presence at the surface of the semi-folded globule (Fig. 5). Therefore, CO% is more or less proportional to this factor (1 ± 0.5) [81]. CO% by itself is a good predictor of folding rates of proteins equal in size, but it fails to compare folding rates of small proteins with those of large ones, because CO% decreases approximately in proportion to $L^{-1/3}$ with increasing protein size L [49, 81, 82] (which reflects a low entangling of chains forming large domains)—while the folding rate decreases, on the average, with increasing protein size (Fig. 6)).

Therefore, a really good predictor of protein folding rates is AbsCO $=$ CO% $\times$ L, which scales as $L^{2/3}$ [81] and combines the effect of protein fold shape [79, 82] with the main effect of protein size.

5. The attempts to use machine learning and information provided by protein sequences to raise the quality of predictions over the level achieved with AbsCO (or ln(AbsCO) [83]) were not quite successful up to now [84].
6. Coming back to the Levinthal's paradox, we can conclude that it is solved for protein chains of less than 100 amino acid residues (provided that sequences of these chains ensure a significant stability to only one of their folds); this is because (1) these chains can overcome free-energy barrier at the pathway to their most stable folds, independently of their complexity (Fig. 7), and (2) they are able to sample all their folds at the level of secondary structure formation and assembly (Fig. 10) and find the most stable one. As to the chains of larger proteins, they can sample only relatively simple (not too entangled) folds, and it remains a question whether some another fold can be more stable than the native one (which is indeed observed for some "exceptional" proteins like serpin, having a 400-residue chain).
7. All told above is also applicable to in vivo folding, because NMR studies of ^{15}N, ^{13}C-labeled nascent chains of small protein state that "polypeptides [at ribosomes] remain unstructured during elongation but fold into a compact, native-like structure when the entire sequence is available" [85, 86]; thus, there is no principal difference between in vivo and in vitro protein folding.

Acknowledgements We are grateful to O.B. Ptitsyn, A.M. Gutin, and E.I. Shakhnovich for seminal discussions at the initial stages of our work.

The first part of this work has been partially supported by the Howard Hughes Medical Institute Awards and the Russian Academy of Sciences Program "Molecular and Cell Biology" (Grant Nos. 01200957492, 01201358029); the second part has been partially supported by the Russian Science Foundation Grant No. 14-24-00157.

References

1. C.B. Anfinsen, E. Haber, M. Sela, F.H. White Jr., Proc. Natl. Acad. Sci. U. S. A. **47**, 1309 (1961)
2. C.B. Anfinsen, Science **181**, 223 (1973)
3. C. Levinthal, J. Chim. Phys. Chim. Biol. **65**, 44 (1968)

4. C. Levinthal, in *Mössbauer Spectroscopy in Biological Systems: Proceedings of a Meeting Held at Allerton House, Monticello*, ed. by P. Debrunner, J.C.M. Tsibris, E. Munck (University of Illinois Press, Urbana-Champaign, 1969), p. 22
5. P.L. Privalov, Adv. Protein Chem. **33**, 167 (1979)
6. A. Šali, E. Shakhnovich, M. Karplus, J. Mol. Biol. **235**, 1614 (1994)
7. V.I. Abkevich, A.M. Gutin, E.I. Shakhnovich, Biochemistry **33**, 10026 (1994)
8. D.C. Phillips, Sci. Am. **215**, 78 (1966)
9. D.P. Goldenberg, T.E. Creighton, J. Mol. Biol. **165**, 407 (1983)
10. V.P. Grantcharova, D.S. Riddle, J.V. Santiago, D. Baker, Nat. Struct. Biol. **5**, 714 (1998)
11. D.B. Wetlaufer, Proc. Natl. Acad. Sci. U. S. A. **70**, 697 (1973)
12. K.F. Fulton, E.R.G. Main, V. Dagett, S.E. Jackson, J. Mol. Biol. **291**, 445 (1999)
13. O.B. Ptitsyn, Dokl. Akad. Nauk SSSR (Moscow, in Russian) **210**, 1213 (1973)
14. A.V. Finkelstein, S.O. Garbuzynskiy, ChemPhysChem **16**, 3373 (2015)
15. A.V. Finkelstein, A.J. Badretdin, O.V. Galzitskaya, D.N. Ivankov, N.S. Bogatyreva, S.O. Garbuzynskiy, Phys. Life Rev. **20** (2017). https://doi.org/10.1016/j.plrev.2017.01.025 [Epub ahead of print]
16. P.E. Leopold, M. Montal, J.N. Onuchic, Proc. Natl. Acad. Sci. U. S. A. **89**, 8721 (1992)
17. P.G. Wolynes, J.N. Onuchic, D. Thirumalai, Science **267**, 1619 (1995)
18. K.A. Dill, H.S. Chan, Nat. Struct. Biol. **4**, 10 (1997)
19. D.J. Bicout, A. Szabo, Protein Sci. **9**, 452 (2000)
20. A.V. Finkelstein, A.Ya. Badretdinov, Mol. Biol. (Moscow, Eng. Trans.) **31**, 391 (1997)
21. A.V. Finkelstein, A.Ya. Badretdinov, Fold. Des. **2**, 115 (1997)
22. C. Tanford, Adv. Protein Chem. **23**, 121 (1968)
23. T.E. Creighton, Prog. Biophys. Mol. Biol. **33**, 231 (1978)
24. E.I. Shakhnovich, A.M. Gutin, Nature **346**, 773 (1990)
25. A.M. Gutin, E.I. Shakhnovich, J. Chem. Phys. **98**, 8174 (1993)
26. O.V. Galzitskaya, A.V. Finkelstein, Protein Eng. **8**, 883 (1995)
27. E.I. Shakhnovich, Chem. Rev. **106**, 1559 (2006)
28. J.D. Bryngelson, P.G. Wolynes, J. Phys. Chem. **93**, 6902 (1989)
29. N. Gō, H. Abe, Biopolymers **20**, 991 (1981)
30. J.T. Ngo, J. Marks, Protein Eng. **5**, 313 (1992)
31. R. Unger, J. Moult, Bull. Math. Biol. **55**, 1183 (1993)
32. M. Karplus, Fold. Des. **2**(Suppl. 1), S69 (1997). http://www.sciencedirect.com/science/journal/13590278
33. B. Nölting, *Protein Folding Kinetics: Biophysical Methods*, Chaps. 10, 11, 12 (Springer, Berlin, 2010)
34. J.D. Bryngelson, P.G. Wolynes, Proc. Natl. Acad. Sci. U. S. A. **84**, 7524 (1987)
35. J.D. Bryngelson, J.N. Onuchic, N.D. Socci, P.G. Wolynes, Proteins **21**, 167 (1995)
36. A.V. Finkelstein, A.Ya. Badretdinov, A.M. Gutin, Proteins **23**, 142 (1995)
37. N.S. Bogatyreva, A.V. Finkelstein, Protein Eng. **14**, 521 (2001)
38. R. Zwanzig, A. Szabo, B. Bagchi, Proc. Natl. Acad. Sci. U. S. A. **89**, 20 (1992)
39. A.V. Finkelstein, J. Biomol. Struct. Dyn. **20**, 311 (2002)
40. P.G. Wolynes, Proc. Natl. Acad. Sci. U. S. A. **94**, 6170 (1997)
41. A. Fersht, *Structure and Mechanism in Protein Science: A Guide to Enzyme Catalysis and Protein Folding*, Chaps. 2, 15, 18, 19 (W.H. Freeman & Co, New York, 1999)
42. A.V. Finkelstein, A.Ya. Badretdinov, Fold. Des. **3**, 67 (1998)
43. R. Zana, Biopolymers **14**, 2425 (1975)
44. H. Eyring, J. Chem. Phys. **3**, 107 (1935)
45. L. Pauling, *General Chemistry*, Chap. 16 (W.H. Freeman & Co., New York, 1970)
46. N.M. Emanuel, D.G. Knorre, *The Course in Chemical Kinetics*, 4th Russian edn., Chaps. III (§2), V (§§2, 5) (Vysshaja Shkola, Moscow, 1984)
47. L.D. Landau, E.M. Lifshitz, *Statistical Physics*. A Course of Theoretical Physics, vol. 5, 3rd edn. (Elsevier, Amsterdam, 1980) §§7, 8, 150
48. O.V. Galzitskaya, A.V. Finkelstein, Proc. Natl. Acad. Sci. U. S. A. **96**, 11299 (1999)

49. S.O. Garbuzynskiy, D.N. Ivankov, N.S. Bogatyreva, A.V. Finkelstein, Proc. Natl. Acad. Sci. U. S. A. **110**, 147 (2013)
50. H. Jacobson, W. Stockmayer, J. Chem. Phys. **18**, 1600 (1950)
51. P.J. Flory, *Statistical Mechanics of Chain Molecules*, Chap. 3 (Interscience Publishers, New York, 1969)
52. B. Fu, W. Wang, Lect. Notes Comput. Sci. **3142**, 630 (2004)
53. K. Steinhofel, A. Skaliotis, A.A. Albrecht, Lect. Notes Comput. Sci. **4175**, 252 (2006)
54. G.C. Rollins, K.A. Dill, General mechanism of two-state protein folding kinetics. J. Am. Chem. Soc. **136**, 11420 (2014)
55. O.V. Galzitskaya, D.N. Ivankov, A.V. Finkelstein, FEBS Lett. **489**, 113 (2001)
56. D.A. Debe, M.J. Carlson, W.A. Goddard 3rd., Proc. Natl. Acad. Sci. U. S. A. **96**, 2596 (1999)
57. D.E. Makarov, K.W. Plaxco, Protein Sci. **12**, 17 (2003)
58. S. Wallin, H.S. Chan, Protein Sci. **14**, 1643 (2005)
59. M. Levitt, C. Chothia, Nature **261**, 552 (1976)
60. C. Chothia, A.V. Finkelstein, Ann. Rew. Biochem. **59**, 1007 (1990)
61. A.V. Finkelstein, O.B. Ptitsyn, *Protein Physics. A Course of Lectures*, Chaps. 7, 10, 13, 17–21 (Academic, An Imprint of Elsevier Science, Amsterdam, 2002)
62. O.B. Ptitsyn, Adv. Protein Chem. **47**, 83 (1995)
63. A.G. Murzin, A.V. Finkelstein, J. Mol. Biol. **204**, 749 (1988)
64. F.H.C. Crick, Acta Crystallogr. **6**, 689 (1953)
65. O.B. Ptitsyn, A.V. Finkelstein, Quart. Rev. Biophys. **13**, 339 (1980)
66. D.T. Jones, J. Mol. Biol. **292**, 195 (1999). Current version of the program: http://bioinf.cs.ucl.ac.uk/psipred/
67. A.V. Finkelstein, S.A. Garbuzynskiy, Biofizika (in Russian) **61**, 5 (2016)
68. V. Muñoz, P.A. Thompson, J. Hofrichter, W.A. Eaton, Nature **390**, 196 (1997)
69. S. Mukherjee, P. Chowdhury, M.R. Bunagan, F. Gai, J. Phys. Chem. B **112**, 9146 (2008)
70. A.V. Finkelstein. http://atlasofscience.org/are users-satisfied-with-single-sign-on-technologies-in-er/
71. A.R. Ubbelohde, *Melting and Crystal Structure*, Chaps. 2, 5, 6, 10–12, 14, 16 (Clarendon Press, Oxford, 1965)
72. V.V. Slezov, *Kinetics of First-Order Phase Transitions*, Chaps. 3–5, 8 (Wiley-VCH, Weiheim, 2009)
73. A.V. Finkelstein, O.V. Galzitskaya, Phys. Life Rev. **1**, 23 (2004)
74. A.V. Finkelstein, O.B. Ptitsyn, *Protein Physics. A Course of Lectures*, 2nd edn., Chaps. 7, 10, 13, 18, 19–21 (Academic, An Imprint of Elsevier Science, Amsterdam, 2016)
75. D. Thirumalai, J. Phys. I. (Orsay, Fr.) **5**, 1457 (1995)
76. A.M. Gutin, V.I. Abkevich, E.I. Shakhnovich, Phys. Rev. Lett. **77**, 5433 (1996)
77. A.G. Murzin, Science **320**, 1725 (2008)
78. Y. Tsutsui, R.D. Cruz, P.L. Wintrode, Proc. Natl. Acad. Sci. U. S. A. **109**, 4467 (2012)
79. K.W. Plaxco, K.T. Simons, D. Baker, J. Mol. Biol. **277**, 985 (1998)
80. B. Nölting, W. Schälike, P. Hampel, F. Grundig, S. Gantert, N. Sips, W. Bandlow, P.X. Qi, J. Theor. Biol. **223**, 299 (2003)
81. D.N. Ivankov, S.O. Garbuzynskiy, E. Alm, K.W. Plaxco, D. Baker, A.V. Finkelstein, Protein Sci. **12**, 2057 (2003)
82. D.N. Ivankov, N.S. Bogatyreva, M.Yu. Lobanov, O.V. Galzitskaya, PLoS One **4**, e6476 (2009)
83. A.V. Finkelstein, N.S. Bogatyreva, S.O. Garbuzynskiy, FEBS Lett. **587**, 1884 (2013)
84. M. Corrales, P. Cuscó, D.R. Usmanova, H.C. Chen, N.S. Bogatyreva, G.J. Filion, D.N. Ivankov, PLoS One **10**, e0143166 (2015)
85. C. Eichmann, S. Preissler, R. Riek, E. Deuerling, Proc.Natl. Acad. Sci. U. S. A. **107**, 9111 (2010)
86. Y. Han, A. David, B. Liu, J.G. Magadan, J.R. Bennink, J.W. Yewdell, S.-B. Qian, Proc. Natl. Acad. Sci. U. S. A. **109**, 12467 (2012)

Index

Printed in the United States
By Bookmasters